Advances in
MARINE BIOLOGY

VOLUME 42

MOLLUSCAN RADIATION – LESSER-KNOWN BRANCHES

Advances in MARINE BIOLOGY

MOLLUSCAN RADIATION – LESSER-KNOWN BRANCHES

Edited by

A. J. SOUTHWARD

Marine Biological Association, The Laboratory, Citadel Hill, Plymouth, PL1 2PB, UK

P. A. TYLER

School of Ocean and Earth Science, University of Southampton, Southampton Oceanography Centre, European Way, Southampton, SO14 3ZH, UK

C. M. YOUNG

Marine Science Division, Harbor Branch Oceanographic Institution, 5600 US Highway 1N, Fort Pierce, Florida 34946, USA

and

L. A. FUIMAN

Marine Science Institute, University of Texas at Austin, 750 Channel View Drive, Port Aransas, Texas 78373, USA

ACADEMIC PRESS

An Elsevier Science Imprint

Amsterdam – Boston – London – New York – Oxford – Paris – San Diego
San Francisco – Singapore – Sydney – Tokyo

This book is printed on acid-free paper.

Academic Press
An Elsevier Science Imprint
Harcourt Place, 32 Jamestown Road, London NW1 7BY, UK
http://www.academicpress.com

Academic Press
An Elsevier Science Imprint
525 B Street, Suite 1900, San Diego, California 92101-4495, USA
http://www.academicpress.com

ISBN 0-12-026142-1

A catalogue record for this book is available from the British Library

Typeset by Keyset Composition, Colchester, Essex, UK
Printed and bound in Great Britain by MPG, Bodmin, Cornwall, UK

02 03 04 05 06 07 MP 9 8 7 6 5 4 3 2 1

CONTRIBUTORS TO VOLUME 42

M. G. Harasewych, *Department of Systematic Biology, National Museum of Natural History, Smithsonian Institution, Washington, DC 20560-0118, USA*

P. M. Mikkelsen, *Division of Invertebrate Zoology, American Museum of Natural History, Central Park West at 79th Street, New York, NY 10024-5192, USA*

P. D. Reynolds, *Biology Department, Hamilton College, 198 College Hill Road, Clinton, NY 13323, USA*

J. D. Zardus, *Pacific Biomedical Research Center, Kewalo Marine Laboratory, University of Hawaii, 41 Ahui St, Honolulu, HI 96813, USA*

CONTENTS

Protobranch Bivalves

John D. Zardus

Shelled Opisthobranchs

Paula M. Mikkelsen

The Scaphopoda

Patrick D. Reynolds

Pleurotomarioidean Gastropods

M. G. Harasewych

The color plate section appears between pages 242 and 243.

Series Contents for Last Ten Years*

VOLUME 28, 1992.

Heath, M. R. Field investigations of the early life stages of marine fish. pp. 1–174.

James, M. A., Ansell, A. Q. D., Collins, M. J., Curry, G. B., Peck, L. S. and Rhodes, M. C. Biology of living brachiopods. pp. 175–387.

Trueman, E. R. and Brown, A. C. The burrowing habit of marine gastropods. pp. 389–431.

VOLUME 29, 1993.

Kiørboe, T. Turbulence, phytoplankton cell size, and the structure of pelagic food webs. pp. 1–72.

Kuparinen, K. and Kuosa, H. Autotrophic and heterotrophic picoplankton in the Baltic Sea. pp. 73–128.

Subramoniam, T. Spermatophores and sperm transfer in marine crustaceans. pp. 129–214.

Horwood, J. The Bristol Channel sole (*Solea solea* (L.)): a fisheries case study. pp. 215–367.

VOLUME 30, 1994.

Vincx, M., Bett, B. J., Dinet, A., Ferrero, T., Gooday, A. J., Lambshead, P. J. D., Pfannküche, O., Soltweddel, T. and Vanreusel, A. Meiobenthos of the deep Northeast Atlantic. pp. 1–88.

Brown, A. C. and Odendaal, F. J. The biology of oniscid Isopoda of the genus *Tylos*. pp. 89–153.

Ritz, D. A. Social aggregation in pelagic invertebrates. pp. 155–216.

Ferron, A. and Legget, W. C. An appraisal of condition measures for marine fish larvae. pp. 217–303.

Rogers, A. D. The biology of seamounts. pp. 305–350.

VOLUME 31, 1997.

Gardner, J. P. A. Hybridization in the sea. pp. 1–78.

Egloff, D. A., Fofonoff, P. W. and Onbé, T. Reproductive behaviour of marine cladocerans. pp. 79–167.

Dower, J. F., Miller, T. J. and Leggett, W. C. The role of microscale turbulence in the feeding ecology of larval fish. pp. 169–220.

Brown, B. E. Adaptations of reef corals to physical environmental stress. pp. 221–299.

Richardson, K. Harmful or exceptional phytoplankton blooms in the marine ecosystem. pp. 301–385.

*The full list of contents for volumes 1–37 can be found in volume 38.

VOLUME 32, 1997.

Vinogradov, M. E. Some problems of vertical distribution of meso- and macroplankton in the ocean. pp. 1–92.

Gebruk, A. K., Galkin, S. V., Vereshchaka, A. J., Moskalev, L. I. and Southward, A. J. Ecology and biogeography of the hydrothermal vent fauna of the Mid-Atlantic Ridge. pp. 93–144.

Parin, N. V., Mironov, A. N. and Nesis, K. N. Biology of the Nazca and Sala y Gomez submarine ridges, an outpost of the Indo-West Pacific fauna in the eastern Pacific Ocean: composition and distribution of the fauna, its communities and history. pp. 145–242.

Nesis, K. N. Goniatid squids in the subarctic North Pacific: ecology, biogeography, niche diversity and role in the ecosystem. pp. 243–324.

Vinogradova, N. G. Zoogeography of the abyssal and hadal zones. pp. 325–387.

Zezina, O. N. Biogeography of the bathyal zone. pp. 389–426.

Sokolova, M. N. Trophic structure of abyssal macrobenthos. pp. 427–525.

Semina, H. J. An outline of the geographical distribution of oceanic phytoplankton. pp. 527–563.

VOLUME 33, 1998.

Mauchline, J. The biology of calanoid copepods. pp. 1–660.

VOLUME 34, 1998.

Davies, M. S. and Hawkins, S. J. Mucus from marine molluscs. pp. 1–71.

Joyeux, J. C. and Ward, A. B. Constraints on coastal lagoon fisheries. pp. 73–199.

Jennings, S. and Kaiser, M. J. The effects of fishing on marine ecosystems. pp. 201–352.

Tunnicliffe, V., McArthur, A. G. and McHugh, D. A. Biogeographical perspective of the deep-sea hydrothermal vent fauna. pp. 353–442.

VOLUME 35, 1999.

Creasey, S. S. and Rogers, A. D. Population genetics of bathyal and abyssal organisms. pp. 1–151.

Brey, T. Growth performance and mortality in aquatic macrobenthic invertebrates. pp. 153–223.

VOLUME 36, 1999.

Shulman, G. E. and Love, R. M. The biochemical ecology of marine fishes. pp. 1–325.

VOLUME 37, 1999.

His, E., Beiras, R. and Seaman, M. N. L. The assessment of marine pollution – bioassays with bivalve embryos and larvae. pp. 1–178.

Bailey, K. M., Quinn, T. J., Bentzen, P. and Grant, W. S. Population structure and dynamics of walleye pollock, *Theragra chalcogramma*. pp. 179–255.

VOLUME 38, 2000.

Blaxter, J. H. S. The enhancement of marine fish stocks. pp. 1–54.

Bergström, B. I. The biology of *Pandalus*. pp. 55–245.

VOLUME 39, 2001.

Peterson, C. H. The "Exxon Valdez" oil spill in Alaska: acute indirect and chronic effects on the ecosystem. pp. 1–103.

Johnson, W. S., Stevens, M. and Watling, L. Reproduction and development of marine peracaridans. pp. 105–260.

Rodhouse, P. G., Elvidge, C. D. and Trathan, P. N. Remote sensing of the global light-fishing fleet: an analysis of interactions with oceanography, other fisheries and predators. pp. 261–303.

VOLUME 40, 2001.

Hemmingsen, W. and MacKenzie, K. The parasite fauna of the Atlantic cod, *Gadus morhua* L. pp. 1–80.

Kathiresan, K. and Bingham, B. L. Biology of mangroves and mangrove ecosystems. pp. 81–251.

Zaccone, G., Kapoor, B. G., Fasulo, S. and Ainis, L. Structural, histochemical and functional aspects of the epidermis of fishes. pp. 253–348.

VOLUME 41, 2001.

Whitfield, M. Interactions between Phytoplankton and Trace Metals in the Ocean. pp. 1–128.

Hamel, J.-F., Conand, C., Pawson, D. L. and Mercier, A. The Sea Cucumber *Holothuria scabra* (Holothuroidea: Echinodermata): Its Biology and Exploitation as Beche-de-Mer. pp. 129–223.

PREFACE

Because seashells have been popular with collectors for thousands of years, the phylum Mollusca has always been one of the best-known taxa of marine animals. Amateur zeal notwithstanding, there still remain several groups distributed among the various classes of mollusks that are poorly known because of their size, inconspicuousness or inaccessibility. The present volume assembles reviews on a selection of these lesser known groups: the mostly diminutive protobranch bivalves, with greatest abundance in deeper waters; the slug-like shelled opisthobranchs, lesser known relatives of the colorful nudibranchs; the highly specialized and evolutionarily advanced tusk shells; and the beautiful, priceless, yet frustratingly hard-to-collect slit shells. Except for their status as "lesser-known" taxa within a single phylum, the groups selected for coverage in this volume have little in common. Each attains its peak abundance and diversity at a different depth horizon: shelled opisthobranchs in the shallows, scaphopods on continental shelves, slit shells down continental slopes and protobranchs in the abyss. The lifestyles, habitats and modes of existence are equally diverse. All four articles in this volume provide wonderful examples of the mechanisms contrived by evolution to meet the day-to-day demands of biotic existence.

Descriptions of shell form, which dominated the molluskan literature of the 19th century, gradually gave way to studies of physiology, form and function during the early half of the 20th century. Numerous excellent scientists were involved in this movement, but perhaps none was more prolific than the late C. M. Yonge, one of the founding editors of *Advances in Marine Biology*. Beginning in the 1920s his studies of feeding, digestion, mantle cavities, locomotion, burrowing and reproduction conducted on a wide variety of bivalves and prosobranch gastropods, all of them quite poorly known at the time, provided a springboard for modern malacology as represented by the reviews in the present volume. Although all of the authors in this volume discuss shell form and function, they focus their attention more sharply on the functioning of the soft parts, placing the current state of knowledge in a modern ecological and evolutionary context. The papers contain the latest details needed by professional biologists, but should also be of tremendous interest to the army of dedicated amateurs hungry to know the state of the art and eager to push the frontiers. It is hoped that the readership of this volume will be as diverse as the animals described herein. This volume is dedicated to the memory of Sir Maurice Yonge.

Craig M. Young

Protobranch Bivalves

John D. Zardus*

Department of Biology, University of Massachusetts, Boston, MA 02125, USA
email: zardus@hawaii.edu

*Currently at: Pacific Biomedical Research Center, Kewalo Marine Laboratory, University of Hawaii, 41 Ahui St, Honolulu, Hawaii 96813, USA. email: zardus@hawaii.edu.

ADVANCES IN MARINE BIOLOGY VOL. 42
ISBN 0-12-026142-1

The subclass Protobranchia comprises more than 600 species of bivalves that occur throughout the world ocean. Mostly deposit feeders in soft sediments, they are abundant in the deep sea. Apomorphies that unite them as a group include gill structure, hinge conformation, shell microstructure, larval development, foot morphology, respiratory pigments, trophic mode and digestion. They are relatively small and highly conserved in form, originating in the Cambrian era. They may represent an ancestral, derived or paraphyletic group of the Bivalvia. The protobranchs include two orders, the Nuculoida and Solemyoida, which previously were classified separately in the subclasses Paleotaxodonta and Cryptodonta, respectively. They are of ecological interest and have a unique functional morphology. They feed mostly under the surface of the sediment with highly modified labial palps, but the degree to which they are selective in diet remains difficult to determine. They are important bioturbators in many soft-sediment assemblages; their feeding and locomotion affects sediment structure and community development. Solemyoids are unusual in inhabiting reducing environments and hydrocarbon seeps and in deriving their nutrition from endosymbiotic chemosynthetic bacteria. A variety of species of protobranchs are found in oceanic trenches, near hydrothermal vents, and in submarine caves. Protobranchs produce a lecithotrophic larval stage, the pericalymma, making their development unique among bivalves. The pericalymma remains in the plankton for a short time and presumably has low dispersal ability. Recruitment may be intermittent. Growth is rapid in post-larvae but decreases with age, though rates may not necessarily be slow, especially in continental shelf species. Life spans are commonly 1 to 2 decades, but deep-sea representatives may grow more slowly and live longer. Bottom fish, seastars and gastropods are their major predators and a few parasites and commensals have been documented. The predominance of protobranchs in deep-sea sediments may be a result of deep-sea origin or displacement from shallow waters by lamellibranchs. Their ability to deposit-feed, digest food extracellularly, and develop by means of lecithotrophic larvae make them particularly well adapted to cold and oligotrophic habitats.

1. INTRODUCTION

1.1. Characteristics of the subclass

The simplicity in form and feature of protobranch bivalves belies the complexity of their function and phylogeny. The subclass encompasses upward of 600 species of relatively small, deposit-feeding bivalves occurring in oceans worldwide and represents an ancient branch of the molluscan tree. Morphologically distinct from the remainder of the bivalves or Autobranchia (Waller, 1998), the Protobranchia are united by gill structure, hinge conformation, shell microstructure, larval development, foot morphology, respiratory pigments and digestion.

Important members of marine soft-sediment assemblages since their appearance in early to middle Cambrian times, protobranchs are particularly diverse in the deep sea (Sanders *et al.*, 1965; Sanders and Allen, 1973; Allen, 1983). Though possibly the oldest clade of the Bivalvia, they have changed little over time and have radiated less than any other subclass (Allen, 1983) with some early Ordovician examples being very similar to present day forms (Morris and Fortey, 1976). Though similar morphotypes occur over vast areas of the sea bottom, great genetic variation has been found in some representative protobranch taxa at both small and large geographic scales (Chase *et al.*, 1998; Etter *et al.*, 1999; Zardus *et al.*, in preparation), suggesting their radiation may include cryptic species that are phenotypically similar but genetically divergent.

The subclass derives its name from the elementary or spare design of the putatively primitive ctenidium of its members (Pelseneer, 1889). Each demibranch is composed of a row of gill "filaments" or leaflets that are plate-like and unattached to their adjacent neighbors, unlike other bivalves in which long thread-like gill filaments connect to form a sheet or lamella that may fold or recurve upon itself. Gill structure in bivalves is intimately related to feeding and the majority of genera, including infaunal burrowers, are suspension feeders. Only tellinaceans and protobranchs deposit feed (Stanley, 1970). The protobranch gill is used almost solely as a respiratory organ; for feeding they employ uniquely modified labial palps to acquire and sort large quantities of sediment (Yonge, 1939; Stasek, 1961). Though small, protobranchs often occur in high numbers and their feeding activities can exert an important influence on the sediment characteristics and ecology of the benthos (Rhoads and Young, 1970).

One of the more unusual features of protobranchs is their larval development. The veliger is the typical bivalve larval form; but protobranchs possess a lecithotrophic pericalymma larva (see Zardus and Martel, 2002). The embryo develops within a thin covering of test cells that is cast off at settlement following a brief larval period. Protobranch post-larvae are relatively small, as are adults, which may be an adaptation to environments of low productivity such as the deep sea. The smallest known representatives, however, are two shallow water species from the Caribbean: *Condylonucula maya* Moore that matures at a length of 0.5 mm and *C. cynthiae* Moore that matures at 0.6 mm (Moore, 1977). A number of shallow and deep water species, approximately 1.0 mm in length, are also known (Gofas and Salas, 1996). These tiny animals possibly achieve small body size by decreasing cell size (Sanders and Allen, 1973).

1.2. Classification

Higher order bivalve systematics has long been in a state of flux, but most classification schemes based on morphological features have recognized protobranchs as a natural group, except for divided opinion on the placement of the solemyaceans, which lack hinge teeth (Pelseneer, 1889; Atkins, 1937a, b; Cox, 1959; Yonge, 1959; Cox, 1960; Newell, 1969; Taylor *et al.*, 1969; Taylor, 1973; Owen, 1978; Runnegar and Pojeta, 1985; Cope, 1996; Morton, 1996). The solemyaceans have variously been placed in the subclass Cryptodonta (Newell, 1969) or Lipodonta (Cope, 1996, 1997) and the remaining protobranchs placed in the Paleotaxodonta. Solemyaceans are now well regarded as members of the Protobranchia belonging to the order Solemyoida (Liljedahl, 1984; Allen and Hannah, 1986; Maxwell, 1988; Waller, 1998). The taxodont Manzanellidae (formerly Nucinellidae) are also placed in this order along with a few recent species in the genera *Huxleyia* and *Nucinella* (Vokes, 1956; Allen and Sanders, 1969; Maxwell, 1988; Pojeta 1988), which previously had been assigned to the Pteriomorphia (Newell, 1969).

Protobranch taxonomy has been greatly aided by recent investigations of the deep-sea forms of the Atlantic Ocean by John Allen and co-workers (Allen and Sanders, 1973; Sanders and Allen, 1973, 1977; Allen and Sanders, 1982; Sanders and Allen, 1985; Allen and Hannah, 1989; Rhind and Allen, 1992; Allen *et al.*, 1995; Allen and Sanders, 1996a), including a comprehensive zoogeographic summary (Allen and Sanders, 1996b). The classification followed here is based on that of Allen and Hannah (1986) as revised by Maxwell (1988) (but see additional comments by Ockelmann and Warén, 1998), wherein subclass Protobranchia is divided into the orders Solemyoida and Nuculoida. These in turn are each divided into two superfamilies: Solemyacea and Manzanellacea (order Solemyoida) and Nuculacea and Nuculanacea (order Nuculoida) with 13 recognized families occurring in the subclass (Table 1). Scarlato and Starobogatov (1985) present an alternate classification in which bivalve subclasses are not recognized and many protobranch families are added. The following sections review the morphology, natural history, ecology, zoogeography and evolution of this interesting group.

2. ANATOMY AND MORPHOLOGY

2.1. The shell

Variation in the protobranch shell is "never extravagant" (Allen, 1985) and is highly conserved, particularly within genera (Cowan, 1968; Allen

Table 1 Classification of subclass Protobranchia (after Maxwell, 1988). Families are in bold type.

Taxon	Name	Authority
Order	Solemyoida	Dall, 1889
Suborder	Solemyina	Dall, 1889
Superfamily	Solemyacea	Gray, 1840
Family	**Solemyidae**	Gray, 1840
Family	**Acharacidae**	Scarlato and Starobogatov, 1979
Suborder	Nucinellina	Scarlato and Starobogatov, 1971
Superfamily	Manzanellacea	Chronic, 1952
Family	**Manzanellidae**	Chronic, 1952
Order	Nuculoida	Dall, 1889
Superfamily	Nuculacea	Gray, 1824
Family	**Nuculidae**	Gray, 1824
Subfamily	Nuculinae	Gray, 1824
Subfamily	Nuculominae	Maxwell, 1988
Family	**Pristiglomidae**	Sanders and Allen, 1973
Superfamily	Nuculanacea	H. and A. Adams, 1858
Family	**Nuculanidae**	H. and A. Adams, 1858
Subfamily	Nuculaninae	H. and A. Adams, 1858
Subfamily	Ledellinae	Allen, 1978
Subfamily	Spinulinae	Allen and Sanders, 1982
Family	**Sareptidae**	A. Adams, 1860
Subfamily	Sareptinae	A. Adams, 1860
Subfamily	Yoldiinae	Habe, 1977
Subfamily	Yoldiellinae	Allen, 1978
Family	**Tindariidae**	Verrill and Bush, 1897
Family	**Neilonellidae**	Allen, 1978
Family	**Malletiidae**	H. and A. Adams, 1858
Family	**Phaseolidae**	Scarlato and Starobogatov, 1971
Family	**Lametilidae**	Allen and Sanders, 1973
Family	**Siliculidae**	Allen and Sanders, 1973

and Hannah, 1986). This may be due to a slow rate of evolution, especially in the deep sea (Allen, 1983), or to strict constraints imposed by lifestyle and habitat and the development of cryptic species that differ genetically but not morphologically (see Etter *et al.*, 1999). Non-allometric growth is also common among protobranchs, which adds to the difficulty of differentiating species from morphological variants among specimens of varying size (Allen *et al.*, 1995). Four subspecies of *Ledella pustulosa* (Jeffreys), for example, are quite similar in appearance yet vary in bathymetric range and differ morphometrically on a fine but significant scale (Fuiman *et al.*, 1999). Protobranch shells lack elaborate external ornamentation and most species are smooth or only slightly sculptured with faint concentric growth lines and radial ribs (Figure 1). The nuculids have been divided by Schenck (1934) into three taxonomic groups based on shell sculpture: those with

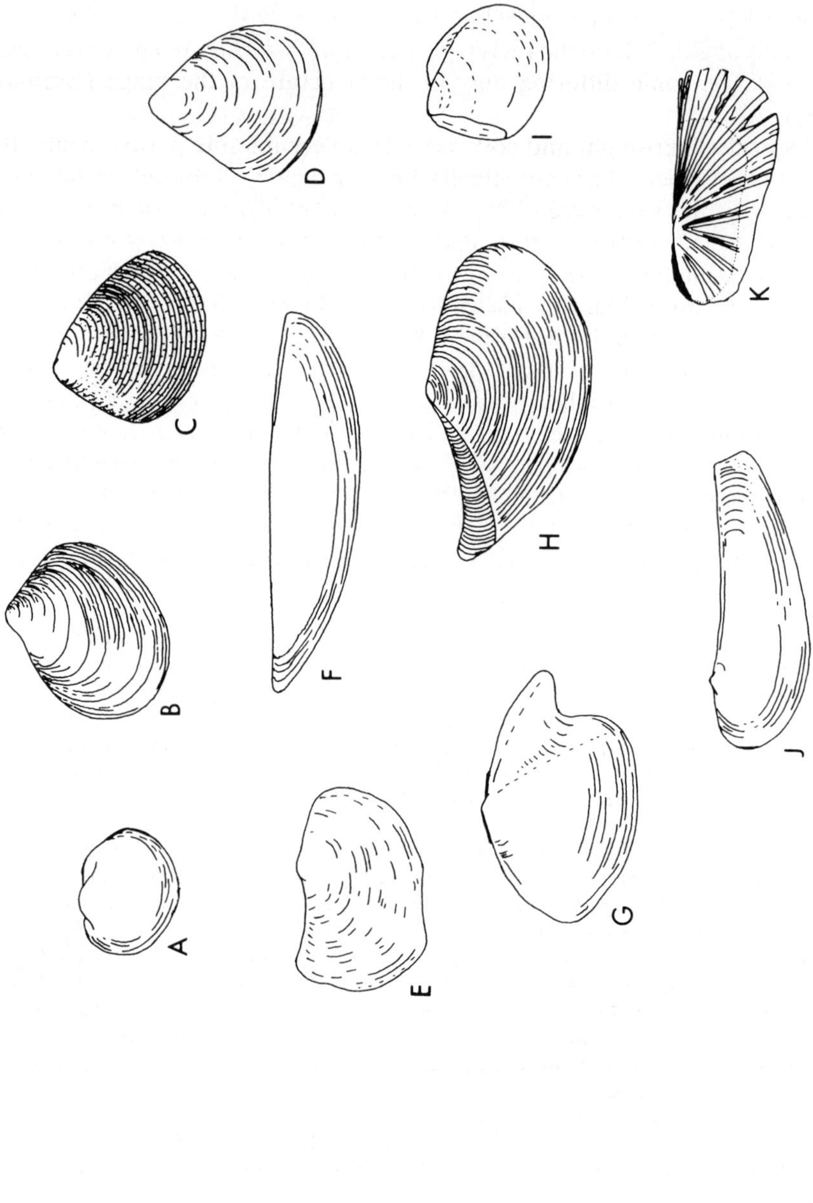
A
B
C
D
E
F
G
H
I
J
K

a crenulate inner ventral margin, those with a smooth inner ventral margin and those possessing pronounced divaricate ribbing (see Rhind and Allen, 1992 for a listing of genera in these groups). Separation of *Nucula* species can be made on the basis of shell morphometrics and position of the umbo (Allen, 1954). Some sareptids (formerly Yoldiidae) exhibit concentric growth lines superimposed with secondary ridges that cross obliquely at an acute angle. Solemyoids have perhaps the most highly conserved shell morphology, some differing only in the structure of the hinge (Bernard, 1980).

Typically lightweight and covered with a very durable periostracum, the protobranch shell characteristically bears a heavily toothed or taxodont hinge, with the exception of the edentulous Solemyidae and Acharacidae (Beedham and Owen, 1965). Protobranchs rely on both posterior and anterior adductor muscles to close the shell valves but members of the solemyoid family Manzanellidae (formerly Nucinellidae) are unusual in being monomyarian. In this family, the posterior adductor is missing rather than the anterior, as is usual among monomyarian lamellibranchs (Allen and Sanders, 1969). Lighter shells tend to occur in species that inhabit fine, loose sediments and heavier shells are found in those from coarser sediment. In the majority of species, the shell form tends toward lateral compression and elongation except among the Nuculoida which are often rotund and relatively heavy shelled. Elongation and compression is particularly pronounced in the sareptid genus *Adrana*, whereas rostrate

Figure 1 Radiation in shell form among protobranch bivalves. (A) *Microgloma*, family Pristiglomidae (redrawn with permission of the Conchological Society of Great Britain and Ireland, after Allen and Hannah, 1986). (B) *Neilonella hadalis*, family Neilonellidae (redrawn with permission after Knudsen, 1970). (C) *Nucula exigua*, family Nuculidae (drawn with permission of the Paleontological Research Institution, from photograph in Olsson, 1961). (D) *Brevinucula*, family Nuculidae (redrawn with permission of the Conchological Society of Great Britain and Ireland, after Allen and Hannah, 1986). (E) *Megayoldia*, family Sareptidae (redrawn with permission of the Conchological Society of Great Britain and Ireland, after Allen and Hannah, 1986). (F) *Adrana superba*, family Nuculanidae (drawn with permission of the Paleontological Research Institution, from photograph in Olsson, 1961). (G) *Bathyspinula*, family Spinulidae (redrawn with permission of the Conchological Society of Great Britain and Ireland, after Allen and Hannah, 1986). (H) *Nuculana*, family Nuculanidae (redrawn with permission of the Conchological Society of Great Britain and Ireland, after Allen and Hannah, 1986). (I) *Nucinella serrei*, family Manzanellidae (redrawn with permission of Malacologia, after Allen and Sanders, 1969). (J) *Lamellileda*, family Siliculidae (redrawn with permission of the Conchological Society of Great Britain and Ireland, after Allen and Hannah, 1986). (K). *Acharax*, family Acharacidae (redrawn with permission of the Conchological Society of Great Britain and Ireland, after Allen and Hannah, 1986).

forms are well developed among the Ledellinae. Umbo position generally relates to feeding mode; the Nuculanacea, which employ posteriorly positioned siphons for feeding, have a more anteriorly positioned umbo whereas nuculaceans, which have anterior labial palps, and solemyoids, which have a large anterior foot, bear posteriorly shifted umbones (Starobogatov, 1992).

Periostracum color varies among individuals and is not a reliable character separating species. In many, there is a background cast of yellow that is obscured or tinted by other colors which overly it (Allen, 1954). The surface ranges from glossy to matte, the latter frequently collecting deposits of manganese and other metals while those that are glossy remain clean (Allen, 1954; Christensen and Kanneworff, 1985).

Solemyids and acharacids possess a remarkable periostracum that is water repellent and extends beyond the calcareous margin of the shell in radiating bands of thick and thin pleats, allowing it to flex beyond the shell to its point of insertion in the mantle margin, covering the otherwise exposed mantle margins. This is especially pronounced in *Acharax tokunagai* (Yokoyama) (see Okutani, 1968). The periostracum in these two families is usually a rich brown color (Reid, 1980; Cavanaugh, 1983; Conway *et al.*, 1992), but *Solemya occidentalis* Deshayes, a species inhabiting light-colored calcareous sediment, possesses a periostracum the color of the sediment, presumably as camouflage (Krueger *et al.*, 1996a).

Shell formation and structure among protobranchs is relatively uniform and similar to that of other bivalves. The shells of all protobranchs are composed of calcium carbonate entirely in the aragonite form but differ in being composed of one or two layers, even within genera (Taylor *et al.*, 1969; Gilkinson *et al.*, 1986). The Nuculacea have outer prismatic and inner and middle nacreous layers, the Nuculanacea have inner and outer homogeneous layers and the solemyoids have unique outer prismatic and inner homogenous layers (Taylor *et al.*, 1969). In most solemyid species, the inner layer is reduced to a small area around the posterior adductor muscle (Beedham and Owen, 1965). Carter (1990) and Carter *et al.* (1990) have given a detailed analysis of shell microstructure for some fossil and many recent protobranchs.

The ligament of the protobranch shell is similar to that of the majority of bivalves and mirrors formation of the shell with a superficial periostracal layer, an outer layer secreted by the outer fold of the mantle margin, and an inner layer secreted by the general mantle epithelium (Owen, 1956). In nuculoids the ligament is composed of an outer tensional region and an inner compressional region or resilium that may have been lacking in the ancestral form (Carter, 1990). It grows in an anterior direction rather than a posterior direction, as is the case in most other bivalves (Trueman, 1952).

In solemyids and acharacids a portion of the ligament, the outer layer, is extended for a great distance along the hinge (Owen, 1956). The ligament of the Solemyoida also differs in possessing a medial interior region and a posterior internal region (Carter, 1990). In the monomyarian Manzanellidae, the ligament is shifted posteriorly to compensate for the anterior placement of the adductor muscle (Allen and Sanders, 1969).

The protobranch ligament is not particularly strong, no doubt a consequence of developing in conjunction with a very strong taxodont hinge mechanism, which often does not allow the valves to separate without breaking teeth (Allen, 1985). Heavier shelled forms increase valve stability by adding more teeth. The actinodont hinge form is thought to be the primitive condition, which gave rise to a taxodont hinge (Allen and Sanders, 1973). Evidence for this transition is supported by *Nuculoidea pernambucensis* (Smith), a living protobranch with dentition intermediate between taxodont and actinodont forms (Rhind and Allen, 1992). This condition parallels the hinge structure of the fossil nuculoid *Tironucula jugata* Morris & Fortey from the Lower Ordovician (Morris and Fortey, 1976).

2.2. Soft tissues

2.2.1. *Gills*

Organization of the soft tissues of protobranchs is similar to that of other bivalves, with a few characteristic differences in the organs of feeding, respiration and locomotion (Figure 2). The protobranch gills are paired structures lying at an oblique angle on either side of the body in the posterior region of the mantle cavity. In most species, they meet medially below the body along their anterior two thirds. The gills function primarily as respiratory organs rather than as suspension-feeding devices and their structure is distinctly different from those of other bivalve groups (Ridewood, 1903; Atkins, 1937a; Yonge, 1939). Each ctenidium consists of two demibranchs, which are composed of stacks of broad, plate-like filaments, attached together along a suspensory membrane. Gills of the symbiont-harboring solemyids are thick, fleshy and expansive and may occupy up to half of the mantle cavity. In the Nuculanacea, gill plates occur in alternate arrangement on either side of the gill axis (Ridewood, 1903; Yonge, 1939; Sanders and Allen, 1973) but occur in opposite arrangement in the Nuculacea and Solemyoida (Ridewood, 1903; Allen and Sanders, 1969; Sanders and Allen, 1973; Allen and Hannah, 1986), which may give the advantage of accommodating more gill plates (Yonge, 1939). The number of gill plates varies with species and tends to decrease with

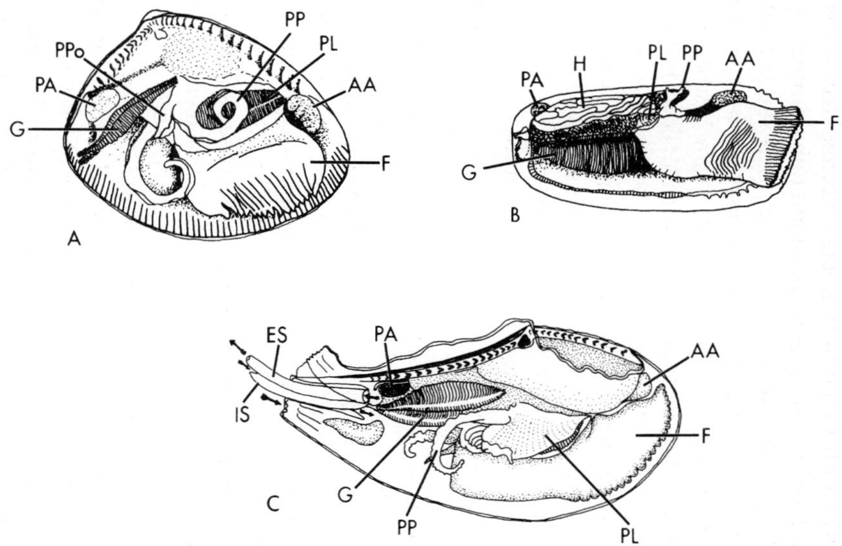

Figure 2 Gross anatomy of representatives from the three major protobranch superfamilies. (A) *Nucula nucleus*, superfamily Nuculacea, family Nuculidae (redrawn with permission from Cambridge University Press, after Hirasaka, 1927). (B) *Solemya reidi*, superfamily Solemyacea, family Solemyidae, with right ctenidium removed (redrawn with permission from the National Research Council of Canada, after Reid, 1980). (C) *Nuculana minuta*, superfamily Nuculanacea, family Nuculanidae (redrawn with permission from the Company of Biologists Ltd., after Atkins, 1936). AA, anterior adductor; ES, exhalant siphon; F, foot; G, gill; H, hypobranchial gland; IS, inhalant siphon; PA, posterior adductor; PL, right palp lamella; PP, right palp proboscis; PPo, palp pouch.

depth (Allen, 1971). The fewest plates are found in the minute, abyssal species *Pristigloma alba* Sanders & Allen which has only two pairs of platelets per gill (Sanders and Allen, 1973).

Ciliation of the gills is remarkably complex despite their relatively small surface area (Figure 3). According to Atkins (1936), this complex ciliation may represent the retention of a primitive suspension-feeding condition. Short frontal cilia cover the outer surface of each plate margin and longer latero-frontal cilia border the edges. Patches of lateral cilia occurring on the broad plate surfaces generate current flow between adjacent plates (see Morse and Zardus, 1997). Gill function in nuculanaceans may allow them to exploit suspension feeding to a greater degree than other groups. *Nuculana minuta* Müller, for instance, sorts particles on the gill with two tracts of adjacent frontal cilia that beat in opposite directions, possibly sorting food in one tract and removing debris in the other (Atkins, 1936). The finest particles are transported to the food groove but larger particles are collected at the inner gill margin and bound in a string of mucus that either falls to the mantle surface and is rejected, or is picked up by the palp proboscides and sorted on the labial palps (Atkins, 1936). In *Yoldia*, two of the gill plates are modified to dispose of material that is collected by the gills but not sent to the mouth (Drew, 1899a; Kellogg, 1915). Some nuculanaceans also direct a feeding current, along the axis of the gill, unlike other protobranchs. In conjunction with this axial current, the bands of latero-frontal cilia of each gill platelet pair is separated across the gill axis; however, in nuculids and solemyids the latero-frontal ciliary band is continuous across the gill axis (Atkins, 1936). The ultrastructure of protobranch cilia has been examined in several species and compares favorably with other metazoans, but exhibits slight intraorganism variation relative to location or function (Lundin and Schander, 2001).

Gill currents are stronger in *Nucula* than *Nuculana* species but the latter can also pump water to augment ciliary flow by rapidly contracting the gills dorsally, ejecting water out the exhalant siphon and drawing it in through the inhalant siphon (Drew, 1899a; Atkins, 1936). Ciliary junctions between the left and right gills are strong and the gills fit snugly within the mantle cavity to form an effective septum across the branchial chamber, dividing it into supra- and infrabranchial chambers (Atkins, 1936). In solemyoids, weak ciliary currents through the mantle chamber are intermittently increased by contractions of the foot.

In solemyoids, water flow traces a relatively straight path through the mantle cavity from its entrance anteriorly, around the foot, and into the infrabranchial chamber. From there it passes between the platelets of the gill into the suprabranchial chamber and exits posteriorly through an exhalant aperture or "siphon". Water flow in nuculaceans is also relatively simple but not as linear. Incoming water enters antero-ventrally along the

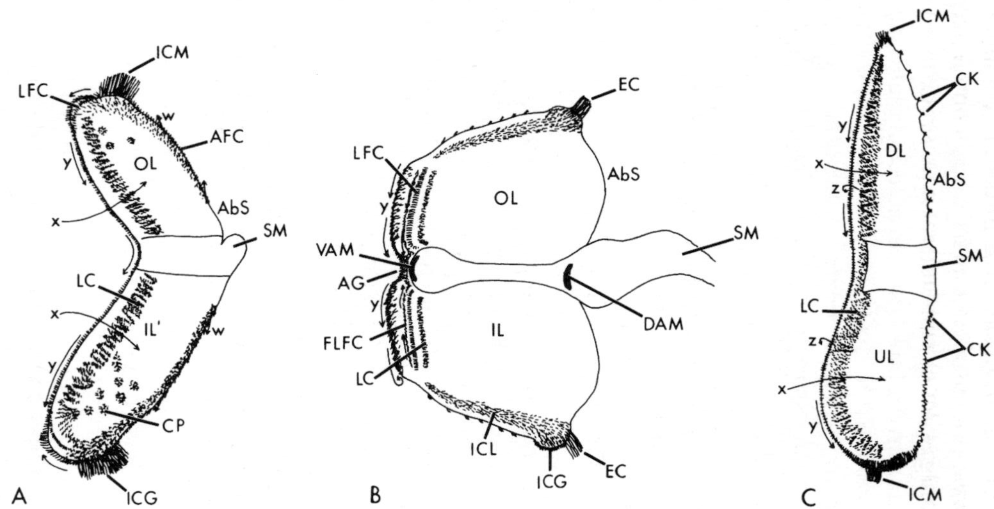

Figure 3 Middle gill leaflets from the right ctenidium of representatives from the three major protobranch superfamilies. (A) *Nucula*, superfamily Nuculacea, family Nuculidae (redrawn with permission from the Cambridge University Press, after Orton, 1912). (B) *Nuculana minuta,* superfamily Nuculanacea, family Nuculanidae (redrawn with permission from the Company of Biologists Ltd., after Atkins, 1936). (C) *Solemya togata*, superfamily Solemyacea, family Solemyidae (redrawn with permission from the Cambridge University Press, after Orton, 1913). AbS, ab-frontal surface; AFC, ab-frontal cilia; AG, axial food groove; CK, ciliated knob; CP, ciliary patch; DAM, dorsal longitudinal axial muscle; DL, dorsal leaflet; EC, elongate cilia across the dorsal edge of leaflets; FLFC, fine latero-frontal cilia and associated arrows indicating effective stroke direction; ICG, interlocking cilia that connect to the other gill; ICL, interlocking cilia that connect to the adjacent leaflet; ICM, interlocking cilia that connect to the mantle; IL, inner leaflet; LC, lateral cilia; LFC, latero-frontal cilia; OL, outer leaflet; SM, suspensory membrane; VAM, ventral longitudinal axial muscle; w, arrow indicating direction of effective stroke of ab-frontal cilia; x, arrow indicating direction of water flow across leaflets as driven by the effective stroke of the lateral cilia; y, arrow indicating direction of effective stroke of frontal cilia; z, arrow indicating direction of effective stroke of latero-frontal cilia acting as food strainers.

valve gape and around the foot into the infrabranchial chamber. Moving dorsally, it passes between the gill platelets into the suprabranchial chamber then is deflected posteriorly to exit at a postero-dorsal exhalant region between the valves. Paths of water flow are most complex in nuculanaceans, where water is drawn into the infrabranchial chamber posteriorly at a narrow inhalant region, often through a ventral siphon. The water passes dorsally between the gill plates and recurves posteriorly to exit through a dorsal exhalant region or siphon (when present).

2.2.2. *Labial palps*

The labial palps in all protobranchs (except the solemyoids) are highly modified and are one of the most complex structures known among deposit feeders for collecting and sorting sediment (Stasek, 1961). Though not connected to the palps, the gills cooperate with them by supplying them with stray particles (Orton, 1912; Morse, 1913; Atkins, 1936). The palps are paired structures, one on either side of the body. Each consists of a double-layered lamella subtending from the dorsal region of the body; on casual observation, these may be mistaken for the ctenidia. The inner surfaces of the palp lamellae are corrugated and have as many as 16 tracts of cilia for sorting particles (Stasek, 1961). Extending from the posterior end of the palp lamella is a grooved palp proboscid (Hirasaka, 1927), a tentacle-like appendage that is thrust out between the valves into the substratum and conveys sediment in via cilia and mucus for sorting (Drew, 1899a, b; Kellogg, 1915). Unique to the Nuculidae is a palp pouch, which bridges the transition between the palp proboscid and palp lamellae and assists in transfer of sediment received from the palp proboscid (Hirasaka, 1927; Stasek, 1961; Rhind and Allen, 1992). The palp proboscides are prehensile and very extensible. Across species there is a general trend of increased size and sorting capacity of the sorting organ with depth (Allen, 1971; Kuznetsov and Ohta, 1989).

2.2.3. *The mantle*

Modifications of the mantle in protobranchs tend to be less elaborate than in other bivalves. The mantle is fused ventrally in *Nuculana minuta* (Atkins, 1936) and partially so in solemyoids. In solemyoids, a short exhalant siphon (in some cases two) occurs posteriorly and bears fringing papillae and small tentacle-like extensions (Reid, 1980). Well-developed tubular siphons are found in some members of the Nuculanacea. These may be incomplete or split along the ventral edge and joined by ciliary junctions, as is the inhalant siphon of *Nuculana minuta* (Atkins, 1936), or entire and fused as in *Propeleda carpenteri* (Dall) (Allen and Sanders, 1996a). In *Spinula* (Allen and Sanders, 1982), *Ledella* (Allen and Hannah, 1989), *Yoldiella* (Allen *et al.*, 1995) and other genera, there is a single siphon with combined inhalant and exhalant function. Such variety in siphon morphology may have developed by a series of fusion events between separate siphons (Yonge, 1939) or by secondary loss of siphon partitions (Allen, 1985). Siphons do not occur among the Nuculacea and mantle margins are not fused ventrally in this group.

2.2.4. *The foot*

The foot of protobranchs, unlike those of other bivalves, is divided ventrally to form two wing-like halves that can extend laterally into the sediment to increase purchase in burrowing (Drew, 1900). The margin of the foot in most species is fringed with papillae that are muscular extensions of the sole and may be used to grasp and probe the sediment (Waller, 1998). As noted by Allen (1985) the foot has been important in the evolution of protobranchs and has facilitated their exploitation of soft sediment. Protobranchs burrow rapidly, anterior end first, by thrusting and pulling with their foot. When fully extended, the divided foot effectively anchors itself in the substratum, allowing the body to be pulled forward as foot muscles contract. Upon retracting its lateral extensions, the foot becomes wedge-shaped and is thrust into the sediment for another cycle of pulling (Driscoll, 1964). Bivalves such as nuculids, with a length to height ratio near 1, also use a rocking motion that assists in burrowing (Stanley, 1970). A conclusion by Morse (1913), that *Solemya* species burrow in the unusual attitude of posterior end first has not been upheld by other observers (Drew, 1900; Yonge, 1939; Owen, 1961; Reid, 1980). Solemyoids, unlike other protobranchs, are also able to swim short distances by simultaneously contracting the valves and retracting the foot to rapidly expel water through the posterior exhalant siphon to provide propulsion (Drew, 1900; Stanley, 1970).

2.2.5. *Digestive system*

Alimentary tract function and digestive efficiency are also key features of protobranch biology that have been critical to their success in soft-sediments, particularly in the oligotrophic deep sea (Allen, 1992). The stomach is similar among all protobranchs except the solemyoids (in which it is reduced or absent) and can be classed as Purchon's Type I (Purchon, 1956, 1987). Digestion is primarily extracellular, with absorption occurring at the site of digestion as opposed to the intracellular digestion with absorption found in the digestive diverticula of other bivalves (Owen, 1959; Allen, 1985). In the Nuculanidae, the digestive diverticula may be capable of some intracellular digestion, as indicated by lysosomal systems in the cells lining the digestive diverticula (Yonge 1939; Owen 1973).

Deposit-feeding protobranchs ingest sediment in high volume and pack it tightly into the stomach. The material passes over a sorting area in the stomach and is ground against the gastric shield, which secretes carbohydrases (Purchon, 1956; Halton and Owen, 1968). Instead of a crystalline style, protobranchs possess a viscid, cone-shaped mass mixed with mucus. Stiff bristle-like cilia aid in turning the cone and mixing the stomach contents (Owen, 1956; Purchon, 1956; Halton and Owen, 1968; Owen, 1973; Allen, 1985). Protobranchs also use extrinsic muscles to apply pressure on the gut to help triturate material (Allen, 1992). Periodically a sphincter at the intestine relaxes to allow processed sediment to exit the stomach. The efficiency of this system is greater than that found in lamellibranchs and one in which a firm crystalline style would be an impediment (Owen, 1956). Extracellular digestion allows large volumes of sediment to be processed through the gut rapidly, obviating the need to send material into blind tubules of the digestive gland for absorption. Extracellular digestion is also an efficient means of obtaining nutrients from microorganisms attached to sediment grains. One constraint of the system is the residence time required for particles to be processed (Owen, 1956; Lopez, 1980). As a consequence, the alimentary tracts of protobranchs are generally long and convoluted. This is especially true in deep-sea species, some of which also have a typhlosole (Allen, 1971), an involution of the gut wall that serves to increase digestive surface area. Processed sediment is egested as compacted fecal rods and, in nuculids at least, these are marked with longitudinal grooves corresponding to the profile of the intestinal bore and are characteristic for each species (Galliher, 1931; Moore, 1931b, c; Allen, 1953).

The configuration of hindgut coiling can also be diagnostic of species and families (Allen and Hannah, 1989; Allen, 1992). In the simplest case, the intestine makes a simple loop to the right side of the animal but in some species it is lengthened and forms multiple loops. In some species, it forms multiple tight coils on the right side of the body and in others multiple coils on both right and left sides of the body (Allen, 1973). The point at which the hindgut crosses over from right to left occurs either anteriorly between the esophagus and anterior adductor or dorsally, posterior to the stomach, depending on the species (Allen, 1992). An unusual situation is found in the genus *Prelametila* in which there are loops within loops (Allen and Sanders, 1973). These multiple coils and loops in deep-sea species provide expanded surface area for absorption of nutrients.

2.2.6. *Glands*

Protobranchs excrete nitrogenous waste as ammonia (Follum and Gray, 1987) and ultrafiltration is carried out in the heart–kidney complex via podocytes of the pericardial gland (Meyhöfer *et al.*, 1985; Morse, 1987; Morse and Meyhöfer, 1990). A number of other glands occur in protobranchs, particularly along the mantle margin. In nuculanaceans, they are positioned posteriorly and in nuculaceans, anteriorly, in conjunction with regions of inhalant flow (Sanders and Allen, 1973). The hypobranchial gland consists of a layer of swollen columnar cells covering the surface of the suprabranchial chamber and the outer surface of the suspensory membrane of the gills (Drew, 1901; Yonge, 1939; Rhind and Allen, 1992). It functions to keep the suprabranchial chamber clean and debris-free by secreting a substance that binds fine material passing between the gill filaments (Yonge, 1939). The composition of the secretion, as evidenced by differential uptake of histological stains, may vary among species (Heath, 1937), or within species on a temporal scale (Rhind and Allen, 1992). In *Nucula delphinodonta* Mighels & Adams, the hypobranchial gland has the unique function of producing a brood sac in the female, which attaches to the shell exterior and in which larvae are incubated (Drew, 1901).

A byssal gland is mentioned in the description of some protobranchs (Heath, 1937). Although possibly homologous to the byssal gland of lamellibranchs, the protobranch byssal gland produces no fibers, nor does it function in attachment. Located in the heel of the foot and often quite obvious it may secrete mucilage but its function remains unknown.

2.2.7. *Sense organs*

A number of sense organs have been described in protobranch species. Possible tactile or chemoreceptive structures include papillae fringing the foot in most species, large papillae of the ventral mantle in *Yoldia limatula* Say (Drew, 1899a) and enlarged papillae located along the entire mantle margin in *Solemya* (Morse, 1913). An anterior mantle sense organ (Allen, 1985), unique to nuculanaceans, is presumably chemosensory in function. It is composed of swollen epithelial cells with glandular structure, originating in the middle mantle fold (Allen and Sanders, 1973, 1982). Among the Nuculanidae, sensory mantle flaps develop from the inner folds of the mantle, near to where the proboscides emerge (Allen, 1985). Among most nuculanaceans there is also a solitary, retractable siphonal tentacle that arises near the union of the mantle with the siphon (Brooks, 1875; Pelseneer, 1888; Morse, 1919; Allen and Sanders, 1973, 1982). This siphonal tentacle may be homologous to a similar structure in *Macoma balthica* (Pelseneer, 1888); it is probably a specialized marginal tentacle with tactile function (Drew, 1899a). It has been observed to probe the bottom or slip between the unfused ventral edges of the inhalant siphon and extend out the siphonal opening.

Stenta's organ (Franc, 1960) is an organ of unknown function described from *Nuculana commutata* (Philippi) (Stenta, 1909), occurring as a ciliated depression of the inner mantle fold. A similar structure is found in *Malletia gigantea* Smith (Stoll, 1939) and other nuculanids (Yonge, 1939). This is not to be confused with Stempell's organ, also known as the cephalic sense organ or adoral sense organ (Stempell, 1898; Vlés, 1905; Hirasaka, 1927), which is a paired structure that lies just below the cephalic ganglion, near the junction of the palp lamellae and mouth, and is known to be present in a number of protobranchs (see Schaefer, 2000). Stenta's organ is an enclosed tubular structure located just dorsal to the anterior adductor muscle. It has been variously described as a relictual eye with degenerate cornea (Hirasaka, 1927) or an olfactory organ (Vlés, 1905), but most likely serves a proprioceptive function in detecting muscular contractions associated with inhalant feeding (Haszprunar, 1985).

The adoral sense organ of *Nucula nucleus* and *N. nitidosa* is made up of three types of bipolar receptor cells which appear similar to those found in mollusks and other invertebrates (Schaefer, 2000). The function of this organ is uncertain but most likely it is chemoreceptive (Schaefer, 2000). Paired statocysts with statoliths are known in a number of species (Rhind and Allen, 1992; Allen *et al.*, 1995).

In some nuculids, a unique sensory structure termed the post-anal organ appears during the larval period. It originates from internal mantle and visceral tissue not associated with the external larval test. It may provide a chemo- or mechanoreceptive function, possibly assisting in orientating or in selecting a suitable substratum during settlement (Zardus and Morse, 1998).

3. FEEDING

3.1. Deposit feeding

Protobranchs (with the exception of solemyids, described later) are primarily deposit feeders, although some are capable of suspension feeding to a limited degree. They frequently combine methods of inhalant deposit feeding with ciliary-mucoid particle capture to acquire sediment, and some collect material both above and beneath the sediment surface. Elaboration of the labial palps into palp proboscides and palp lamellae is clearly essential to protobranch feeding, though these two organs differ somewhat in function among taxa. In one study of various protobranchs, the palp proboscides of nuculids were never observed collecting food at the surface, whereas those of the sareptid *Nuculana minuta* were frequently seen doing so (Edwards, 1965). A more fundamental difference concerns the flow of water through the mantle cavity. By channeling the entry and exit of water posteriorly through siphons, nuculanaceans may better be able to separate feeding and respiratory currents. Lacking siphons and maintaining an anterior inhalant flow and a ventral exhalant flow through gaped valves, nuculaceans may have to cope with greater quantities of sediment-laden water moving across the gills (Stasek, 1961; Sanders and Allen, 1973).

Feeding has been studied in detail in only a few species of deposit-feeding protobranchs. The nuculid *Acila castrensis* (Hinds) burrows in a horizontal position with the umbones just at the sediment surface (Figure 4), and ingests subsurface sediment (Schenck, 1936). The palp proboscids extend out into the sediment, directly conveying material onto the palp lamellae by ciliary transport or resuspending sediment in subsurface cavities for inhalation. Ciliation of the palp lamellae and mantle create the requisite inhalant flow, and incoming sediment is intercepted primarily by the palp lamellae before reaching the ctenidia. Fine particles that collect on the ctenidia are passed to the palp lamellae or to the lateral oral grooves. Once gathered onto the palp lamellae, material is rigorously sorted by size over the folds of the lamellae and the finest particles are sent to the mouth for ingestion (Stasek, 1961).

In some species, such as those of the genus *Silicula*, there is interaction between gills and palps with coordinated respiratory and feeding activities,

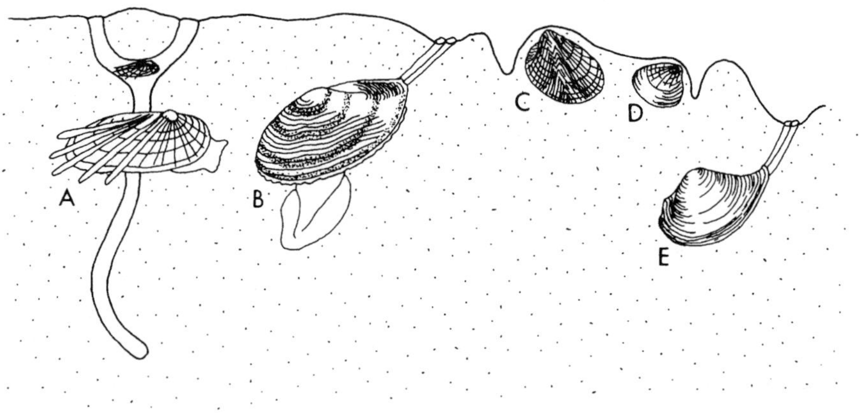

Figure 4 Life modes of several protobranchs (redrawn with permission from the Paleontological Society, after Hickman, 1984). (A) *Solemya.* (B) *Yoldia.* (C) *Acila.* (D) *Nucula.* (E) *Nuculana.*

whereas in lametilids, for example, these activities are completely separate (Allen and Sanders, 1973). *Yoldia ensifera* Dall is capable of raising and lowering its ctenidia in order to generate flow through the inhalant siphon for suspension feeding and respiration. This species also possesses palp proboscides for deposit feeding (Stasek, 1965). Drew (1899a, b) suggested a feeding pose for *Y. limatula* that has been commonly cited but is not entirely accurate. Rather than extending its valves somewhat above the sediment surface, as Drew suggested, the organism remains buried and feeds below the sediment in a semi-vertical attitude at depths of up to 8 cm (Figure 4). This orientation is similar to that of nuculanids, where only the siphons extend beyond the sediment surface (Stanley, 1970). *Yoldia limatula* uses its palp proboscides to excavate cavities with grooved walls, moving frequently to new locations as the cavity collapses (Rhoads and Young, 1970; Bender and Davis, 1984). This species may occasionally feed at the surface with its palp proboscides. This is accomplished by extending one proboscid up alongside the siphons and repeatedly laying it down on the sediment in a radial pattern, removing a narrow strip of sediment each time and creating a radially grooved, fan-shaped pit in the sediment (Bender and Davis, 1984). Only a small portion of sediment is actually ingested and most is rejected as pseudofeces. Larger particles are ejected at depth between the valves into surrounding sediment and smaller particles are expelled along with feces through the exhalant siphon (Yonge, 1939; Stasek, 1965; Bender and Davis, 1984; Davenport, 1988a). Large animals expel large amounts of sediment as feces and pseudofeces but smaller animals expel sediment more frequently. A compartmentalized view of feeding in *Yoldia limatula* suggests that the rate-limiting stage in feeding occurs with sediment sorting in the palp lamellae and that when one lamella is full, sediment may be shunted to the other (Levinton *et al.*, 1996). The Antarctic species *Y. eightsi* (Couthouy) behaves somewhat differently. It maintains an essentially horizontal attitude only 2–3 cm below the sediment surface with its palps on the surface but does not create fan-shaped pits. Instead, a large amount of sediment is engulfed between the valves in the mantle cavity where it is sorted by the palp proboscids. Fine material is passed to the mouth from the palp lamellae and coarser material is reversibly expelled through the inhalant siphon frequently and in large volume (Davenport, 1988a).

3.2. Diet

Given the impressive sorting capabilities of protobranchs, it is easily assumed they are highly selective in their diet. However, the gut contents of protobranchs often contain a representative sample of the surrounding sediment and include the inorganic remains of a variety of organisms including diatoms, dinoflagellates, spores, silicoflagellates, coccolithophores, tintinnids, radiolarians and foraminifera (Kellogg, 1915; Allen and Sanders, 1973; Sanders and Allen, 1977; Allen, 1978; Tyler *et al.*, 1992) along with unidentified scale fragments, fibrous material, hairs, cysts, ossicles and spicules (Tyler *et al.*, 1992). Their physical distribution in the field and in the laboratory usually does not correlate with the organic content of the sediment, suggesting that grain size may be the primary criterion for selection in some species (Wilson and Shelley, 1986). Particles approximately 125 μm in diameter and smaller have been shown to be favored by *Nucula turgida* Leckenby & Marshall (Davis and Wilson, 1985), whereas *Yoldia scissurata* Dall selects particles in the range of 2–20 μm in diameter. Specific gravity may also be a factor in particle selection, but conclusive evidence on this point is lacking (Self and Jumars, 1988).

On the other hand, evidence for selectivity, based on gut content analysis, has been found in some species, nuculids in particular (Driscoll and Brandon, 1973; Rachor, 1976; Cheng and Lopez, 1991). The skeletal remains of foraminifera at greater than ambient abundance have been observed in the guts of some species (Sanders and Allen, 1985). Members of the genus *Silicula* have also been found with guts containing almost exclusively diatom frustules of a similar size (Allen and Sanders, 1973). *Nucula proxima* Say appears to derive most of its nutrition from sedimentary detritus and seems able to selectively ingest organic rich sediment (Lopez *et al.*, 1989). However, the efficiency of energy extraction from sedimentary organic matter is low, from 5–20%, suggesting that much of this refractory material is unavailable to animals (Davis and Wilson, 1985). *Nucula annulata* Hampson has been found to selectively ingest organic and microbe-rich sediment, and exhibits high efficiency in absorbing bacteria (Lopez and Cheng, 1983). Though able to utilize bacterial sources of carbon more efficiently than from sedimentary organic matter, the latter is thought to contribute 65–98% of the bulk carbon demands (Lopez and Levinton, 1987; Cheng and Lopez, 1991). Relatively long gut-clearance times (approximately 15 h in *Nucula turgida* (Davis and Wilson, 1985)) may also aid in nutrient absorption. It has been shown that *Yoldia limatula* selectively grazes algal detritus and bacteria, and by enhancing degradation of algal lipids increases the availability of phytol to bacteria (Sun *et al.*, 1999). Microbes supply important proteins and vitamins but lack necessary fatty acids and sterols which are provided by diatoms, thus a mixed diet of microbes and detritus provides a balance of essential nutrients (Phillips, 1984). However, it is unknown whether protobranchs are involved in "gardening" of microbes (Hylleberg, 1975), through ventilating and enriching the sediment with fecal material then selectively removing microbial production. Microbial stripping, the digestion of microbes from ingested sediment with the organic detrital component eliminated for subsequent recolonization by microbes, probably does not occur (Lopez and Levinton, 1987).

In temperate climates, both microbial and detrital nutrients vary seasonally, and therefore diet may also co-vary. Deposit feeders may even be food-limited during certain times of the year (Lopez and Levinton, 1987). In the shallow water species *Nuculoma tenuis* (Montagu), phaeopigments in the tissue that correspond with seasonal pulses of phytoplankton, indicate a phytogenic food source (Christensen and Kanneworff, 1985). Energy absorption may also change seasonally in at least some species (Bayne and Newell, 1983). *Nucula proxima* has been found to absorb nutrients from bacterial sources most efficiently during warmer months, and this species does not feed when temperatures fall below 6°C (Cheng and Lopez, 1991). In the deep sea, feeding activity may vary with season but seasonal changes in the diets of protobranchs have not been observed (Tyler *et al.*, 1994).

3.3. Chemosynthetic nutrition

Nutrition is perhaps most remarkable among the solemyoid protobranchs. Early investigations determined that *Solemya velum* Say was a deposit feeder (Yonge, 1939); then later studies suggested it was a suspension feeder (Stanley, 1970). Not until it was discovered that some solemyoids completely lack an alimentary tract (Reid, 1980; Reid and Bernard, 1980; Kuznetsov and Shileiko, 1984), and that *S. velum* possesses chemolithoautotrophic symbionts (Cavanaugh, 1983), did it become clear that nutrition in solemyoids was supplied in whole or part by endosymbiotic bacteria (Reid, 1990; Krueger *et al.*, 1996b).

As far as it is known all solemyoids either have a reduced gut (Yonge, 1939; Owen, 1961; Reid and Brand, 1987; Kuznetsov *et al.*, 1991) or lack an alimentary tract altogether (Reid, 1980; Reid and Bernard, 1980; Kuznetsov and Shileiko, 1984; Conway *et al.*, 1992), and harbor chemosynthetic bacteria within their gills. Solemyoid gills are unusually large, accounting for as much as 30% of the body weight, greater than in any suspension-feeding bivalve. In contrast, the labial palps, which otherwise are highly developed in protobranchs, are very small (Kuznetsov *et al.*, 1991). *Solemya velum* is capable of suspension feeding to a nominal degree, or ingesting cyanobacteria as an additional source of nutrition but not particulate matter or single-celled green algae (Krueger *et al.*, 1992). *Solemya velum* obtains approximately 3% of its respiratory carbon budget from planktonic sources and possibly some necessary polyunsaturated fatty acids, which cannot be synthesized by most bacteria (Krueger *et al.*, 1992). Solemyoid species lacking a gut rely on symbiotic bacteria for much of their organic carbon needs and may also absorb dissolved organic molecules to some degree (Felbeck, 1983). Studies with *S. reidi* Bernard, *S. velum* and *S. occidentalis* indicate that organic products from the symbiotic sulfur-oxidizing bacteria can provide 90–98% of carbon and 100% of nitrogen requirements (Cavanaugh, 1983, 1985; Fisher and Childress, 1986; Anderson *et al.*, 1987; Cavanaugh *et al.*, 1988; Conway *et al.*, 1989; Conway and Capuzzo, 1990; Krueger *et al.*, 1996b). The symbionts may also synthesize lipids and essential amino acids (Conway and McDowell-Capuzzo, 1991, 1992). The free amino acid composition in solemyoids is unusually high in taurine (Conway *et al.*, 1992; Conway and McDowell-Capuzzo, 1992), more so in the foot than the gill. Taurine is not usually considered an essential amino acid but in solemyoids it may act as a sulfur sink in the cycling of this element through the host–symbiont system (Conway and McDowell-Capuzzo, 1992).

The bacterial symbionts occur intracellularly along the distal margin of the gill in special cells called bacteriocytes, which alternate with symbiont-free intercalary cells. Observed under the transmission electron microscope (TEM), the bacteria are located in vacuoles at the apices of the bacteriocytes (Cavanaugh, 1980; Felbeck, 1983; Cavanaugh *et al.*, 1988; Conway *et al.*, 1992). In *Solemya occidentalis*, the symbionts are not localized at the apex of each bacteriocyte but instead fill it, probably owing to this animal's small size (1–3 mm) (Krueger *et al.*, 1996b). Evidence for the role of these bacteria as chemoautotrophs comes from hybridization of the gene coding for the enzyme ribulose-1,5-bisphosphate carboxylase/oxygenase (RubisCO) with gill tissue only, and the presence of carboxysome-like inclusions (RubisCO storage sites) in the symbionts (Krueger *et al.*, 1996b). The RubisCO enzyme is known only from autotrophs that use the Calvin–Benson cycle (plants, green algae, cyanobacteria, most anaerobic photosynthetic bacteria and aerobic chemoautotrophs) (Tabita, 1988).

The bacterial symbionts are thiotrophic (Cavanaugh, 1985; Vetter, 1991), requiring their hosts to inhabit sulfide-rich, reducing sediments such as hydrocarbon seeps (Suess *et al.*, 1985; Embley *et al.*, 1990; Métivier and von Cosel, 1993), pulp mill effluents or deep sediments, where their burrows span the oxic/anoxic interface (Stanley, 1970). Sulfide in low concentrations is toxic to most organisms because it disrupts the terminal enzyme cytochrome C oxidase in the electron transport chain (Anderson *et al.*, 1987). However, chemolithoautotrophic bacteria are able to utilize sulfide as an energy source to generate ATP and fix CO_2, detoxifying the sulfide molecule in the process. The success of the symbiosis depends on spatially separating the supplies of sulfide and oxygen. This separation is achieved by the animal's orientation within its unique Y-shaped burrow (Figure 4) (Frey, 1967; Stanley, 1970). The upper two branches of the burrow are excavated first, then the downward extension is dug (Stanley, 1970). The burrow as a whole may reach a depth of 50 cm (Owen, 1961). In *Solemya reidi*, the burrow is mucus-lined (Reid, 1980) and the animal resides at the triple junction of the burrow, where it can extend its foot down into the lower branch. It alternates between uptake of oxygen and sulfide by ventilating the burrow to pump oxygenated water through the upper branches of the burrow or by keeping still to let sulfide-laden porewater well up into the lower branch (Reid, 1990). *Solemya occidentalis*, which is small and lives in coarse-grained calcareous sand where sulfide is probably patchy, migrates vertically in the sediment as needed (Krueger *et al.*, 1996b).

Another important aspect of the host–symbiont association lies in the ability of solemyoid blood to transport both oxygen and sulfide and inhibit the spontaneous oxidation of sulfide by oxygen before it reaches the symbiont. Several forms of hemoglobin occur in the blood cytoplasm, some that bind sulfide and others that bind oxygen (Doeller *et al.*, 1988; Kraus, 1995). The transport of sulfide to the symbiont involves a two-stage process where it is first oxidized by the animal in the mitochondria of the gill and foot to produce thiosulfate and some ATP (Powell and Somero, 1985, 1986). The thiosulfate diffuses into the hemolymph and is carried to the symbionts in the bacteriocytes for further oxidation, supplying energy for organic carbon fixation (Cavanaugh, 1983; Anderson *et al.*, 1987). The solemyoid system differs from other known chemosynthetic associations in that it does not require a separate molecular carrying system to directly deliver sulfide to the symbionts. Nutritive organic carbon in the form of lipids and amino acids produced by the symbionts is probably translocated to the host cells in the blood (Fisher and Childress, 1986; Lee *et al.*, 1999). It is also possible that the bacteria are actively "farmed" in the gills and lysed or digested to release their organic compounds (Fisher, 1990; Conway *et al.*, 1992).

The oxidation of sulfur by mitochondria is an aerobic process and the rate at which it occurs depends on the levels of oxygen and sulfur present, with maximum autotrophy occurring at 100 μM sulfide (Anderson *et al.*, 1987). Thiosulfate has been found to lower the oxygen affinity of hemocyanin in *S. reidi*. This may be a mechanism to increase off-loading of oxygen to respiring mitochondria under conditions of high sulfide (Sanders *et al.*, 1998). *Solemya reidi* is able to maintain aerobic or anaerobic metabolism depending on ambient levels of sulfide (Anderson *et al.*, 1990).

Chemosynthesis occurs in just five groups of bivalves and is believed to have been acquired independently by each (Lucinidae, Thyasiridae,

Vesicomyidae, Mytilidae and Solemyoida) (Reid and Brand, 1986; Fisher, 1990). Chemosynthesis may have already been the mode of nutrition in solemyoids when they arose in the middle Ordovician (460–480 mybp) (Pojeta, 1988). Chemolithoautotrophic bacteria are potentially symbiotic with other protobranchs but a likely candidate, a recently discovered hydrothermal-vent nuculanid with a large volume of red-pigmented blood (most likely containing hemoglobin), appears to lack symbionts (Allen, 1993). Molecular analysis of 16S rRNA in bacterial symbionts indicates that those within the Solemyidae are more closely related to each other than to those occurring in other bivalve hosts (Krueger *et al.*, 1996b), and that there is evolutionary congruence between symbiont and host (Eisen *et al.*, 1992; Distel *et al.*, 1994). A single species of bacteria occurs in each host, is specific to the host species (Krueger *et al.*, 1996b) and is in obligate association with its host (Cavanaugh *et al.*, 1988).

4. REPRODUCTION AND DEVELOPMENT

4.1. Fecundity

Sexes are separate in most protobranchs with the exceptions of the simultaneous hermaphroditic genera *Pristigloma* and *Microgloma* (Sanders and Allen, 1973; Ockelmann and Warén, 1998). Protandry or protogyny are not known. Sex ratios are approximately even except in *Tindaria callistiformis* Verrill & Bush where there are five males for every female (Sanders and Allen, 1977). Instantaneous fecundity is highest in shelf species, ranging from 1233–4120 eggs per individual, whereas in deep-sea species a range of 217–260 is common (Scheltema, 1972). *Pseudotindaria erebus* (Clarke), with as many as 800 oocytes per individual, is unusually fecund for a deep-sea species (Sanders and Allen, 1977). *Microgloma pusilla* (Jeffreys), also a deep-sea species, has the lowest fecundity known, with just one egg developing per individual at a time (Sanders and Allen, 1973; Ockelmann and Warén, 1998). The cosmopolitan deep-sea species *Malletia cuneata* Jeffreys harbors 30 oocytes per individual, but these are large (240 μm) compared to *Ledella pustulosa*, which contains 240 small (120 μm) oocytes (Tyler *et al.*, 1994). Lower fecundity in the deep sea may be offset by higher survival rates or longer periods of reproduction (Scheltema, 1972).

More than 50% of the energy budget may be allocated to gonadal production each year in mature individuals (Rachor, 1976; Davis and Wilson, 1985), which may mostly be a consequence of the high cost of lecithotrophic development. This particularly affects the females. In *Nucula turgida*, there is a biochemical difference between males and females during gonadal development, the males having a higher protein content and females having a higher lipid content (Davis and Wilson, 1983a, b). In *Nucula sulcata* Bronn, tissue weight peaks in August and coincides with the onset of spawning (Ansell, 1974). Reproductive effort may also change with age. *Nucula turgida*, which produces large eggs, adopts a "reckless" strategy in which older, larger individuals invest relatively greater energy in reproduction than young individuals, possibly to offset higher predation rates among large individuals (Wilson, 1988).

4.2 Gametes

Egg size in invertebrates is typically related to developmental mode; large, yolk-filled eggs are associated with lecithotrophic and brooded development, while small eggs are usually associated with planktotrophic development (Ockelmann, 1965). Among protobranchs, there is much greater latitude in the relationship between egg size and development. As far as it is known, protobranchs either brood their young internally or release lecithotrophic larvae, yet egg sizes vary widely in either case (see Gustafson and Reid, 1986), even among closely related species (Scheltema, 1972). In the genera *Silicula* and *Lametila* for instance, egg size predicts planktotrophic or lecithotrophic development but prodissoconch size indicates lecithotrophic or direct development (Allen and Sanders, 1973). *Solemya reidi*, which has one of the largest eggs for a protobranch (271 μm), is predicted to have direct development, yet it develops into a lecithotrophic larva (Gustafson and Reid, 1986). The deep-sea nuculid *Nuculoma similis* Rhind & Allen has an egg of similar size (270 μm) (Scheltema, 1972) (identified by Scheltema as *Nucula subovata* (Rhind and Allen, 1992)) but information on its development is lacking. *Microgloma pusilla*, on the other hand, has a small egg (85–90 μm) and direct development (Ockelmann and Warén, 1998).

Bivalve sperm are of the primitive type (Morse and Zardus, 1997), possessing a pointed acrosome, a large nucleus, a midpiece of mitochondria and a tail. Nuculid sperm are elongate with an extended nucleus (Franzén, 1983; Healy, 1989; Morse and Zardus, 1997), whereas sperm in *Yoldia limatula* are short and ovoid (Morse and Zardus, 1997). The brooded *Microgloma pusilla* has an elongate sperm with a unique bent acrosome, possibly a modification due to direct development (Ockelmann and Warén, 1998).

Karyotypes have been examined in a few protobranchs and may be important taxonomically. The diploid chromosome number in two solemyoids is 22 (Ieyama, 1982), 24 in a nuculid (Ieyama, 1992) and 38 in two malletiids and a sareptid (Thiriot-Quievreux *et al.*, 1987, 1991).

4.3. Reproductive periodicity

Reproduction in protobranchs is usually seasonal for shallow-water species but may be continuous or seasonal in the deep sea. In the northern hemisphere, coastal species generally breed in the fall and winter. Among nuculids, *Acila castrensis* breeds in November and December in the Pacific Northwest (Zardus and Morse, 1998), and *Nucula nucleus* (Linnaeus) during the fall in the Mediterranean (Chardy *et al.*, 1984). The sareptid *Yoldia limatula* reproduces from early to mid-summer near Prince Edward Island (Lewis *et al.*, 1982), while *Solemya velum* breeds in winter and early spring (Watzin, 1986). *Solemya reidi*, on the other hand, reproduces continuously (Gustafson *et al.*, 1987). Timing of reproduction may also vary regionally within a species. *Nuculoma tenuis* reproduces from September to December in the Scandinavian fjords (Valderhaug and Gray, 1984), from November to February in a Scottish loch (Harvey and Gage, 1995) and from October to January off the coast of Denmark (Christensen and Kanneworff, 1985). *Nucula turgida* spawns during fall to spring in the German Bight (Rachor, 1976), but during September in Dublin Bay (Davis and Wilson, 1983b, 1985).

Continuous or quasi-continuous reproduction is thought to be the norm for deep-sea protobranchs (Scheltema, 1972; Rokop, 1974, 1979). Indeed this has been found to be the case for *Nucula darella* Dall and *Nuculana pontonia* (Dall) (Rokop, 1974), *Tindaria cervola* (Rokop, 1979), *Deminucula atacellana* (Schenck) (Sanders and Hessler, 1969; Scheltema, 1972; Gage *et al.*, 1986; Tyler *et al.*, 1992) and *Malletia cuneata* (Gage *et al.*, 1986; Tyler *et al.*, 1992). However, at bathyal depths in the Rockall Trough, the sareptids *Ledella pustulosa* and *Yoldiella jeffreysi* (Hidalgo) have been found to breed seasonally (in October and January) (Lightfoot *et al.*, 1979; Gage *et al.*, 1986; Tyler *et al.*, 1992). Survival rates in continuously breeding species may not be similar throughout the year as indicated by a higher number of small-size-class juveniles in summer months in the Rockall Trough (Gage, 1994). Spawning may extend over several months but typically there are one or two distinct peaks. Differences in timing among populations probably correspond to variation in temperature and nutrient input. Phytoplankton production may be a major regulating mechanism for growth and reproduction in many species (Christensen and Kanneworff, 1985).

4.4. Larval development

The larval form characteristic of protobranchs is the pericalymma, which differs from the veliger larva of other bivalves (Zardus and Morse, 1998;

Zardus and Martel, 2002). The larval shell in the pericalymma develops beneath a thin external tissue, the test, which bears locomotory cilia (Figure 5). This larval type contrasts with the veliger, which swims with an apical velum that can be retracted into the external shell. With protection from the test, the larval shell or prodissoconch I (PI) is typically smooth in sculpture and a prodissoconch II (PII) stage (additional larval shell growth with distinct sculpture) does not form (Ockelmann, 1965). At settlement, growth of the sculptured and ridged dissoconch is demarcated by a clear boundary. *Nuculana grasslei* Allen, a hydrothermal vent species, is unusual in that the larval shell bears a reticulate sculpture unlike any other known bivalve (Berg and Van Dover, 1987; Allen, 1993). Some diminutive members of the Nuculidae also possess prodissoconchs with ridges or knobs (Moore, 1977; Gofas and Salas, 1996).

Larval development has been described for only a handful of protobranch species: two solemyaceans (Gustafson, 1986; Gustafson and Reid, 1986, 1988a, b; Gustafson and Lutz, 1992), a nuculanacean (Drew, 1897) and three nuculaceans (Drew, 1899a, b; Drew, 1901; Zardus and Morse, 1998). Limited observations of development in a few other nuculids and nuculaceans have also been made (Lebour, 1938; Trevallion, 1965). No

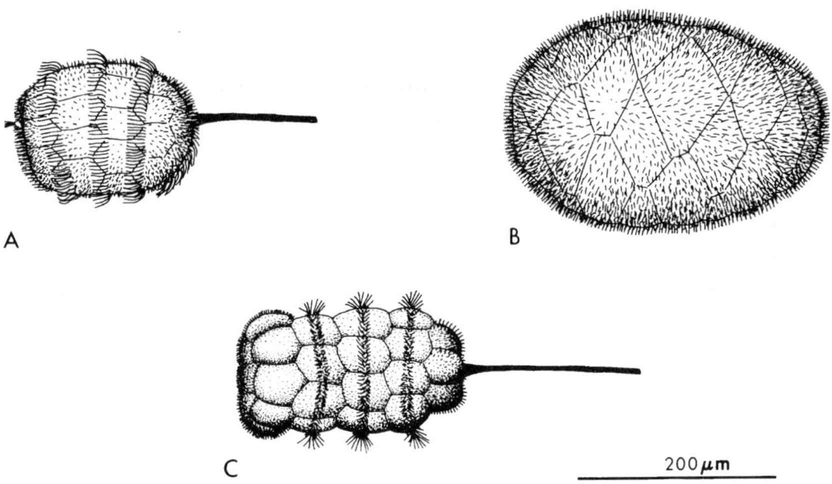

Figure 5 Pericalymma larvae of representatives from the three major protobranch superfamilies showing differences in form, ciliation, size and number of test cells. (A) *Acila castrensis* (redrawn with permission from the American Microscopical Society Inc., after Zardus and Morse, 1998). (B) *Solemya velum* (drawn with permission from the American Microscopical Society Inc., after Zardus and Morse, 1998). (C). *Yoldia limatula* (redrawn after Drew, 1897, 1899a, b). The actual size of the pericalymma of *Y. limatula* is not known but is estimated here from the ratio of egg size to larval length in known species (see Zardus and Morse, 1998 for details). Scale: 200 μm.

information has been obtained on development of any member of the Manzanellacea and the only deep-sea representative in which development has been observed is the pristiglomid *Microgloma pusilla* (Ockelmann and Warén, 1998). It develops directly by incubating a single embryo for over 9 months in a pouch of the ovary. The developing clam is probably provided with nourishment by the mother and is released as a juvenile at half the size of the parent.

Planktonic, lecithotrophic development is probably the most common developmental mode among shelf species of protobranchs; brooding and other forms of protected development may be more common in deep-sea species. The coastal nuculid *Nucula delphinodonta* is unusual in that the female produces a brood sac, attached externally to the shell in which lecithotrophic larvae are incubated (Drew, 1901). *Solemya reidi* develops from a free-swimming lecithotrophic larva (Gustafson and Reid, 1986), whereas *S. velum* embryos are encapsulated and attached to the substratum and remain protected until they emerge many days later as fully formed juveniles (Gustafson and Lutz, 1992). The Australian species *Nucula pusilla* Angas is reported to brood embryos (Bergmans, 1978). Post-embryonic shell growth also occurs during this period and a juvenile is released that is approximately half the size of the adult. The conclusion that Miocene fossil representatives of *N. proxima* incubated larvae (LaBarbera, 1974) is not supported by later work (Webb, 1987).

Reliance by solemyoids on chemolithoautotrophic symbionts for nutrition may start early. Using genetic markers it has been shown that bacterial symbionts in *Solemya reidi* are vertically transmitted during development (Cary, 1994). In *S. velum*, the symbionts are passed on in the oocytes, become sequestered in the pericalymma test and apparently colonize the gill filaments when the test is ingested by the larva at metamorphosis (Krueger *et al.*, 1996b).

Larval development in protobranchs ends with metamorphosis and shedding of the larval test. The larval period is short, from hours to days, but post-larval development may last for weeks. Post-larvae of *Nuculoma tenuis* grow slowly and are thought not to deposit-feed until the palps have developed (Mortimer, 1962). This is supported by the absence of distinct size classes in populations and the presence of individuals of less than 1.0 mm throughout the year (Harvey and Gage, 1995).

4.5. Growth and longevity

Due to cyclical fluctuations in food supply, growth in some protobranchs is seasonal (Rachor, 1976; Nakaoka, 1992a; Craig, 1994; Nakaoka and Matsui,

1994), producing bands corresponding to periodic growth in the shell (Hutchings and Haedrich, 1984; Gilkinson *et al.*, 1986). Dark bands in *Yoldia* species are thought to represent periods of slower calcification and are correlated with winter (Lewis *et al.*, 1982; Nakaoka, 1992b). Microgrowth patterns in the shell of several nuculids from shallow depths show sharply delimited bands corresponding to daily increments of tidal effects, whereas in bathyal species bands are of uniform thickness without sharp boundaries and lack seasonal clustering (Rhoads and Pannella, 1970).

Growth is most rapid in early stages and declines with age, often obscuring cohorts over time (Hutchings and Haedrich, 1984; Gage, 1994). Growth rates for eight species of *Nucula* were found to range from 0.38 to 2.0 mm per year but may be as high as 3.5 mm in the first year (Craig, 1994). *Nucula annulata* slows its growth dramatically with age as it invests more in gonadal output than somatic growth (Wilson, 1992).

Life span and age to maturity vary widely among species. A solemyid inhabiting seagrass beds is thought to grow rapidly, especially during the summer and at low density, with longevity estimated at 1 year (Rainer and Wadley, 1991). *Yoldia limatula* reaches 14–18 mm in length after 1 year and individuals 9 mm in length may be sexually mature (Lewis *et al.*, 1982). By contrast, *Y. notabilis* Yokoyama matures at 3–4 years old and may live for 13–17 years (Nakaoka, 1993). *Nucula turgida* also begins producing gametes at 3 years of age (Wilson, 1992) and lives for 7 (Rachor, 1976) to 10 years (Allen, 1953). The Arctic species *Y. eightsi* is limited to a yearly growing season of approximately 5 months, and large individuals 35 mm or longer may be 60–113 years old (Nolan and Clarke, 1993; Peck and Bullough, 1993). *Nucula nucleus* lives for approximately 10 years (Chardy *et al.*, 1984) and *Nuculana minuta* lives for 7 (Ansell *et al.*, 1978) or 8 years (Parulekar, 1984).

In the deep sea, great age and slow growth are not necessarily the norm, as previously thought (Gage, 1994), and may be on a par with shallow-water rates (Gage, 1992). Several deep-sea protobranchs have been found to live for 5–7 years (Gage *et al.*, 1986; Gage, 1992). Among the oldest perhaps is *Tindaria callistiformis*, which has been determined to reach maturity at 50 years. Individuals 8.4 mm in length may be up to 100 years old (Turekian *et al.*, 1975). However, this estimation of age rests on an assumed interval of shell ring deposition in this species that is most likely inflated (Gage, 1994).

Protobranchs tend to be relatively small and never reach the size of the largest lamellibranchs. The largest species are solemyoids of which *Solemya bartschii* Dall, a deep-water Philippine species, is the largest. Its calcareous shell can reach 19.1 cm in length and 24.0 cm including the extended periostracum (Dall, 1908). The smallest are nuculids, which on the continental shelf are generally in the 1.0 cm size range as adults;

however, a few species 1.0 mm or less in length are known (Moore, 1977; Gofas and Salas, 1996).

5. PHYSIOLOGY

5.1. Respiration

Hemocyanin is the common respiratory pigment of protobranchs (Morse *et al.*, 1986; Mangum *et al.*, 1987; Toulmond *et al.*, 1987; Taylor *et al.*, 1995; Angelini *et al.*, 1998; Sanders *et al.*, 1998) but otherwise does not occur in the Bivalvia. Common to gastropods and cephalopods, it varies in structure among the molluscan classes (Van Holde *et al.*, 1992). Hemoglobin is common among solemyoids and has been found in the hydrothermal vent species *Nuculana grasslei* (Allen, 1993). Several forms of hemocyanin occur among protobranchs, differing in oxygen-binding affinity and cooperativity (Mangum *et al.*, 1987). Though not as efficient as other bivalves in oxygen uptake at low oxygen tension (Wilson, 1981), some protobranchs can withstand anoxia for 2 weeks or more, far longer than any lamellibranch (Moore, 1931a; Taylor *et al.*, 1995). Oxygen consumption is highly variable among species (see Taylor *et al.*, 1995). In some species, oxygen consumption is higher at warmer temperatures and in larger individuals, reaching a maximum just before spawning (Davis and Wilson, 1985). In cold water, oxygen uptake has been found to be independent of temperature, which probably helps the organism to minimize energy loss during periods of limited food supply (Davenport, 1988b). *Yoldia limatula*, for instance, is inactive below 6°C. Activity rates, measured as expulsions of sediment from the siphons, steadily increase with temperature (Bender and Davis, 1984). *Nucula turgida* is metabolically more active in summer (Wilson and Davis, 1984), displaying a drop in metabolic activity associated with spawning (Ansell, 1974; Rachor, 1976; Ansell *et al.*, 1978) and a change in biochemical composition (Davis and Wilson, 1983a). Ctenidial pumping rates and hydrostatic pressures in the mantle cavity generated by pumping are lower in protobranchs than in other bivalves (Bernard and Noakes, 1990) as a result of their simple gill structure and decreased dependence on suspension feeding.

5.2. Salinity tolerance

Few protobranchs have been tested for their ability to withstand changes in salinity; however, most are probably not well adapted for salinities lower than 20‰. *Solemya velum*, which is able to withstand 15‰, may be an exception (Castagna and Chanley, 1973). *Portlandia aestuariorum* (Mosevich), an estuarine inhabitant, is apparently able to withstand very low salinities by remaining tightly closed for great lengths of time (Khlebovich, 1996). *Nucula proxima* is better able to withstand decreases in salinity at lower temperatures (Blake and Jeffries, 1971).

6. ECOLOGY

6.1. Habitat

Protobranchs, like other deposit-feeding bivalves, favor fine sediment (Carey *et al.*, 1984) and occur in sediment with a silt/clay fraction ranging from 16 to 75% (Sanders, 1956, 1958; Kinner *et al.*, 1974). Sediment grain size may be the most important factor influencing protobranch distributions. Non-overlapping distributions and zonation of four nuculids and a nuculanid in the Clyde Sea relative to sediment size suggests that the range of optimal grain size may be very narrow for some species (Allen, 1954). *Nucula annulata* and *N. proxima* are easily mistaken for each other and commonly occur in similar locales along the east coast of the United States; however, the former favors muddy sediment and the latter sandier substrata (Hampson, 1971; Howe *et al.*, 1988). *Nucula turgida* may be a species with a wide tolerance for sediment grain size, occurring in fine sand (125–150 μm) in which sufficient mud and particulate organic material is mixed (Davis and Wilson, 1983b; Creutzberg, 1986). *Yoldia limatula* and *Nucula proxima* distributions closely follow silt/clay fractions in Buzzards Bay, Massachusetts with little overlap between the species. *Yoldia limatula*, however, tolerates a wider range of grain sizes (Driscoll and Brandon, 1973). Among protobranchs, *Solemya* species are found in sediments with the highest silt/clay fraction (Driscoll and Brandon, 1973). In addition to size, sediment grain shape and composition may also determine the suitability of a substratum. Studies with artificial sediment composed of glass beads 150 and 250 μm in diameter indicated that when sediment is fine and dense, it can become too compact for efficient burrowing by protobranchs (Follum and Gray, 1987). On the other hand, low density sediments may be difficult for heavy organisms to inhabit and small protobranchs with thin shells such as *N. annulata* may be adapted for these conditions (Rhoads, 1974). Some nuculids appear to be phenotypically plastic, producing a lighter shell in the softest sediments and a heavier shell in harder substrata (Allen, 1954).

In general, protobranchs are shallow horizontal burrowers that tend to occur in level, relatively sheltered areas, away from turbulence (Rhoads, 1967). Nuculids, nuculanids and sareptids are important members of shelf communities, often dominating coastal assemblages (Ockelmann, 1958; Sanders, 1960; Rhoads, 1967; Skalkin and Tabunkov, 1969; Kinner *et al.*, 1974; Rachor, 1976; Davis and Wilson, 1983b; Chardy *et al.*, 1984; Valderhaug and Gray, 1984; Craig, 1994; Nakaoka and Matsui, 1994). Solemyids may be particularly important constituents of seagrass beds (Rainer and Wadley, 1991).

6.2. Bioturbation

Often numbering in the hundreds to thousands per square meter (Rhoads, 1963; Skalkin and Tabunkov, 1969; Zardus, in preparation), protobranchs are important bioturbators, redistributing and resuspending sediment into the water column at a greater rate than physical processes alone (Rhoads and Young, 1970; Bender and Davis, 1984). Continually processing sediment through its gut at a rate estimated at 2.17 g sediment/g tissue/day (Davis and Wilson, 1985), *Nucula turgida* may be a major processor of sediment organics in Dublin Bay, potentially reworking one quarter of the sewage sludge released into the bay annually (Davis and Wilson, 1985). This species may, however, be adversely affected by this pollutant as its distribution has been negatively correlated with sewage sludge dumping in the North Sea (Caspers, 1980). An average-sized *Yoldia limatula* (14.2 mm in shell length), annually reworks between 257 ml of wet mud (Rhoads, 1963) and 440 g of dry sediment (Bender and Davis, 1984). In Buzzards Bay, Massachusetts, *Y. limatula* was found capable of reworking the sediment deposited in the bay twice over in a year (Rhoads, 1967), and *N. annulata* from the same bay is known to have an even higher sediment processing rate (Rhoads, 1974).

Feeding activity by protobranchs loosens the sediment surface, decreases bulk density of sediment, increases turgidity, lowers critical erosion velocity and interjects fecal pellets into the substratum (Rhoads and Young, 1970). Mobile feeders such as nuculids rework the top few centimeters of sediment (Rhoads and Stanley, 1965; Rhoads, 1974) and can be responsible for the upward transport of a large volume of sediment in the top 2–3 cm (Shull, 2001). Because of their mode of conveyor belt feeding in which sediment processed at depth is ejected into the water column, sareptids may be the more significant bioturbators, especially in terms of resuspending sediment. *Yoldia eightsi* is particularly notable in this regard, ejecting greater volumes of sediment than any other species studied (Davenport, 1988a). *Yoldia* species also cause biogenic grading of sediment by removing the fine grained sediment fraction at the feeding depth and depositing it at the surface (Rhoads, 1967; Bender and Davis, 1984). In some locations, an opaque layer that is poor in organic material and high in mineral particles can be seen 2–3 cm below the sediment surface, which is the result of downward transport of particles by deposit feeders (Rhoads and Young, 1970) or the selective removal of organics at that depth (Bender and Davis, 1984). Though difficult to assay, the organic fraction of the sediment processed by the deposit feeders most likely remains bound up in fecal pellets at the feeding depth and may be reworked rather than ejected to the surface (Rhoads, 1967). Compaction of the sediment through pelletization by *Yoldia limatula* has been found to be minimal compared to several deposit-feeding polychaetes (Rhoads, 1967).

Bioturbation by protobranchs can also increase microbial (Yingst and Rhoads, 1980) or meiofaunal (Austen *et al.*, 1998) growth and so elevate food supply to deposit feeders. In mesocosm studies, *Nuculoma tenuis* was found to facilitate greatest meiofaunal diversity at intermediate densities (Austen *et al.*, 1998). In high abundance, protobranchs may have an amensalistic effect on other organisms (Zardus, 1998; Zardus, in preparation). Reworking of the sediment increases turbidity by destabilizing the sediment (Rhoads, 1973; Bender and Davis, 1984) and this can adversely affect sensitive taxa (Aller and Dodge, 1974). The trophic-group amensalism hypothesis posits that resuspension of sediment by deposit feeders inhibits suspension feeders by clogging up their filtering structures, burying or discouraging their newly settled larvae and preventing attachment of sessile epifauna (Rhoads and Young, 1970). Maurer *et al.* (1980) found that *Nucula proxima* was able to burrow up through 8–16 cm of sediment with mortality rates increasing with depth of burial; however, none was able to burrow to the surface through 32 cm of sediment.

6.3. Resource partitioning

Niche specialization in protobranchs is not well studied. Some protobranchs may be vertically stratified in the sediment, and indeed the bivalves *Nucula proxima*, *Macoma tenta* and *Yoldia limatula* may all co-exist through maintenance of different vertical distributions (Rhoads and Young, 1970; Levinton, 1977). Four sympatric nuculids and a nuculanid are also thought to burrow to different depths and at different rates, influencing the distribution of a commensal hydroid (Edwards, 1965). Depth-structuring of fossil assemblages of protobranchs and other deposit-feeding bivalves may have been similar to those of some present-day communities (Levinton, 1975). Studies with *N. proxima* indicate that spatial arrangements of individuals in populations are random and that they do not maintain territories to reduce feeding interference (Levinton, 1972a). However, interspecific interference competition has been demonstrated by X-radiograph studies in the laboratory between *Nucula–Yoldia* pairings and *Yoldia–Solemya* pairings. This mas in part explain their non overlapping distributions in the field (Levinton, 1977). In the deep Atlantic, 9 to 14 species of protobranchs occur at any given location and all are thought to have similar lifestyles (Allen and Sanders, 1996b). Under diminished food supply in a stable environment, deposit-feeders may evolve specializations that reduce competition (Levinton, 1972b). The enigma is how so many seemingly similar species can co-exist in this oligotrophic environment. A theory of resource partitioning by diet or feeding mode seems untenable but separation by depth horizon may be likely (Allen, 1978). The depth at which these species inhabit the sediment may also correspond to hindgut length, with deep-dwelling species possessing longer hindguts, a consequence of diminishing food resources (Allen, 1985).

6.4. Recruitment

Recruitment dynamics in protobranchs vary among species and habitats. In the deep sea, size frequencies of populations are typically skewed to larger individuals, presumably as a result of low fecundity and larval settlement, whereas shallow water populations are often skewed to younger individuals (Allen and Hannah, 1989). Some populations of deep-sea tindariids are dominated by large size classes but these species are very small and sampling may not have been adequate for juveniles (Sanders and Allen, 1977).

Low survivorship of recruits in shelf populations may be due in part to intrinsic factors, such as adult condition during spawning, which influences energy investment in propagules; however, extrinsic factors such as predation and climatic factors are probably more important (Wilson, 1992). For a population of *Nucula turgida*, spawning effort has been estimated at 1.1 million recruits annually and survivorship from this stage to appearance in the adult population at 1 in 10 000 (Wilson, 1992). In *Yoldia notabilis* Yokoyama, recruitment is sporadic and populations are probably maintained only through successful recruitment events, which occur about every 5 years (Nakaoka, 1993). A similar condition may exist in populations of *Y. eightsi* in the Antarctic, which exhibit bi-modal distributions of young and old individuals (Peck and Bullough, 1993). High densities of adults are often correlated with low numbers of conspecific juveniles (Harvey and Gage, 1995); however, contagious distributions of adults and juveniles have been found in *Nucula hartvigiana* Pfeiffer (Thrush *et al.*, 1989). Recruitment in this species, however, varies in relation to densities of the tellinid *Macomona liliana* Iredale, increasing when the habitat is muddy sand and decreasing when it is sand (Thrush *et al.*, 1996).

Population structure has not been well studied among protobranchs but it is thought to be influenced strongly by the short developmental time of the propagules (a few hours to 8 days) (Zardus and Morse, 1998), leading to low larval dispersal. In the deep sea, populations are also thought to be genetically impoverished as a result of high environmental stability. The latter idea has been challenged by the discovery of highly polymorphic loci for a number of proteins in two deep-sea protobranchs and other species (Gooch and Schopf, 1972). Studies of genetic variability in deep-sea protobranchs have shown population structure at small (100 km) and large (1000 km) scales despite the lack of obvious topographical or hydrographic impediments to gene flow (Chase *et al.*, 1998; Etter *et al.*, 1999; Zardus *et al.*, in preparation). In *Deminucula atacellana*, greater genetic differentiation was found with depth than with geographic distance (Zardus *et al.*, in preparation).

6.5. Predators, parasites and commensals

Protobranchs are prey to a number of benthic invertebrates. Nuculids and sareptids are eaten by mud-dwelling sea stars of the genus *Astropecten* (Hunt, 1925; Sierra *et al.*, 1978; Gemmell *et al.*, 1980), and crabs are heavy predators of smaller-sized *Yoldia notabilis* in Japan, exerting an influence on the optimal reproductive size of the clams (Nakaoka, 1994, 1996, 1998). Solemyids in seagrass beds of western Australia are an important part of the diet of juvenile rock lobsters (Rainer and Wadley, 1991). The enigmatic flatworm-like organism *Xenoturbella* is thought to feed on species of *Nucula* (Israelsson, 2000). Size may provide refuge from some predators. Adult members of the deep-sea genus *Microgloma* are minute and are preyed upon only by very young naticid gastropods (Ockelmann and Warén, 1998). Naticids prey on only the larger-sized *Nucula turgida* (Wilson, 1988) and young nuculanids in the deep sea (Hutchings and Haedrich, 1984). Bottom-feeding fish may be the largest group of predators on protobranchs. The siphons of *Yoldia* species are nipped off by some fish (Drew, 1899a), and the whole organism is consumed by flounder (Jewett and Feder, 1980; Lewis *et al.*, 1982), plaice (Tyler, 1972), ocean pout and cod (MacDonald *et al.*, 1982). Plaice and other flatfish also feed on species of *Nucula* (Hunt, 1925) and demersal fish in the deep sea eat young nuculanids (Hutchings and Haedrich, 1984). Though protobranchs are the dominant mollusks of the abyssal plain, an examination of stomach contents from 81 abyssal fish in the Gulf of Mexico found protobranchs to comprise an insignificant portion of the fish diet (Bright, 1970).

Several parasites of protobranchs have been reported. Trematodes have been found in both nuculids (Martorelli and Cremonte, 1998) and sareptids (Halton and McCrae, 1985), and are likely species specific with their hosts (see Køie, 1980). The bivalves serve as first intermediate hosts for the trematode sporocysts as they develop into cercariae. In *Malletia cuneata*, a possible metacercaria and an unidentified flatworm have been found in the proximity of the gonad and digestive gland. Parasite-like structures have also been found in the gonad or mantle cavity of *Yoldiella jeffreysi*, appearing as egg masses and partially destroying ovaries and testes (Tyler *et al.*, 1992).

Several kinds of organisms have been found living commensally with protobranchs. Most common are hydroids that live epifaunally on the valves of their hosts. The hydroid *Neoturris pileata* (Forskål) occurs exclusively on nuculids and prefers certain species over others. It most commonly occurs on *Nucula sulcata*, less so on *N. turgida*, sparingly on *N. nucleus* and rarely on *Nuculoma tenuis* (Edwards, 1965). Though *Nuculana minuta* and other hard-shelled organisms occur in the same habitat, they are never colonized. The burrowing habits of the nuculids vary and the hydroid, thought to feed infaunally, is associated with those species that burrow deeper and more frequently. Hydroids have also been found on the abyssal sareptids *Yoldiella lata* (Jeffreys), *Y. ella* Allen, Hannah & Sanders and *Y. curta* Verrill & Bush, as well as on the shelf species *Y. inconspicua inconspicua* Verrill & Bush (Allen *et al.*, 1995). Other records of commensals include a single pterobranch, possibly a species of *Cephalodiscus*, found attached to the valve of a solemyoid, *Acharax johnsoni* (Dall) collected at a depth of 3600 m (Knudsen, 1970), and a mantle-dwelling polychaete found in an undescribed solemyoid from a bathyal cold-seep in Sagami Bay, Japan (Miura and Laubier, 1990).

7. ZOOGEOGRAPHY

7.1. Global patterns

Protobranch bivalves are found in all the world's oceans and at all depths but they are primarily associated with the deep sea and low temperatures (Allen, 1978), a pattern that is also evident in tertiary faunas (Hickman, 1974). Lamellibranchs are the most important bivalves in continental shelf habitats but protobranchs become increasingly important with depth

(Figure 6), constituting 30% of the bivalve fauna from the shelf break to the slope, 50% from the slope to the rise and up to 80 or 90% out across the abyssal plain (Allen, 1971). Protobranchs make up 1–7% of the bivalve fauna in shallow tropical and warm temperate zones, 19% (on average) of the shallow temperate bivalve assemblage, 23% of Arctic bivalves, and only 13% of Antarctic bivalves (Nicol, 1972). Apart from worldwide and ocean-wide species lists (Clarke, 1962a; Knudsen, 1967, 1970; Allen and Sanders, 1996b), protobranchs have been cataloged in a number of regional faunas including the Gulf of Mexico (James, 1973), the Mediterranean (Sabelli *et al.*, 1992; Arduino *et al.*, 1995), the Adriatic (Hrs-Brenko and Legac, 1991), Britain (Bowden and Heppell, 1966; Tebble, 1976), Australia (Cotton, 1961, Beesley *et al.*, 1998; Lamprell and Healy, 1998), New Zealand (Dell, 1956, 1962; Powell, 1979), the Indo-Pacific (Prashad, 1932), the Arctic (Clarke, 1962b; Knudsen, 1985), the Antarctic (Powell, 1951; Clarke, 1961; Filatova and Vinogradova, 1974; Hain, 1990), the Eastern Pacific (Bernard, 1983; Coan *et al.*, 2000), Japan (Habe, 1951–52, 1977; Okutani, 1974), China (Bernard *et al.*, 1993), the Red Sea (Oliver, 1992), the Panamic-Pacific (Olsson, 1961) and the East China Sea (Xu, 1990).

The families Nuculidae and Nuculanidae are the dominant protobranch families on the continental shelf, followed by the Sareptidae and Solemyidae (Figure 7). Except for the solemyids, these families are also

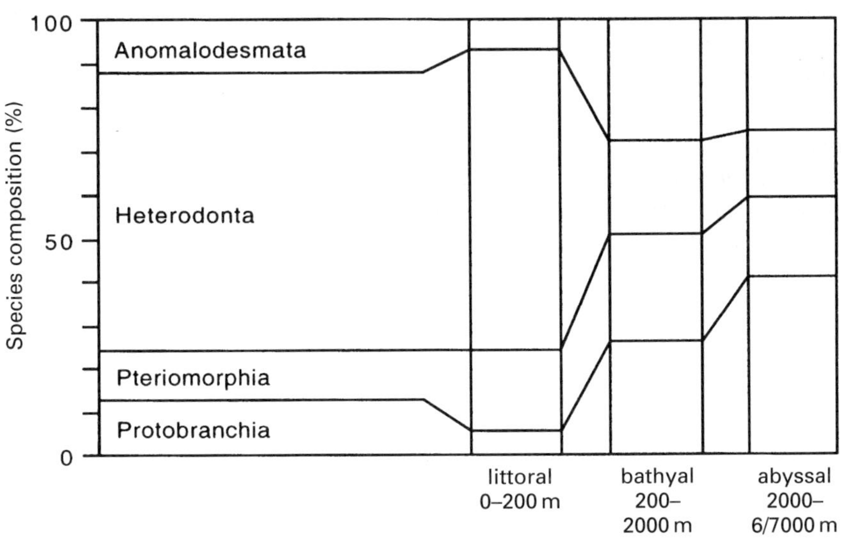

Figure 6 Percent species composition of the major bivalve subclasses by depth zone (redrawn with permission of the Western Society of Malacologists, after Hickman, 1974).

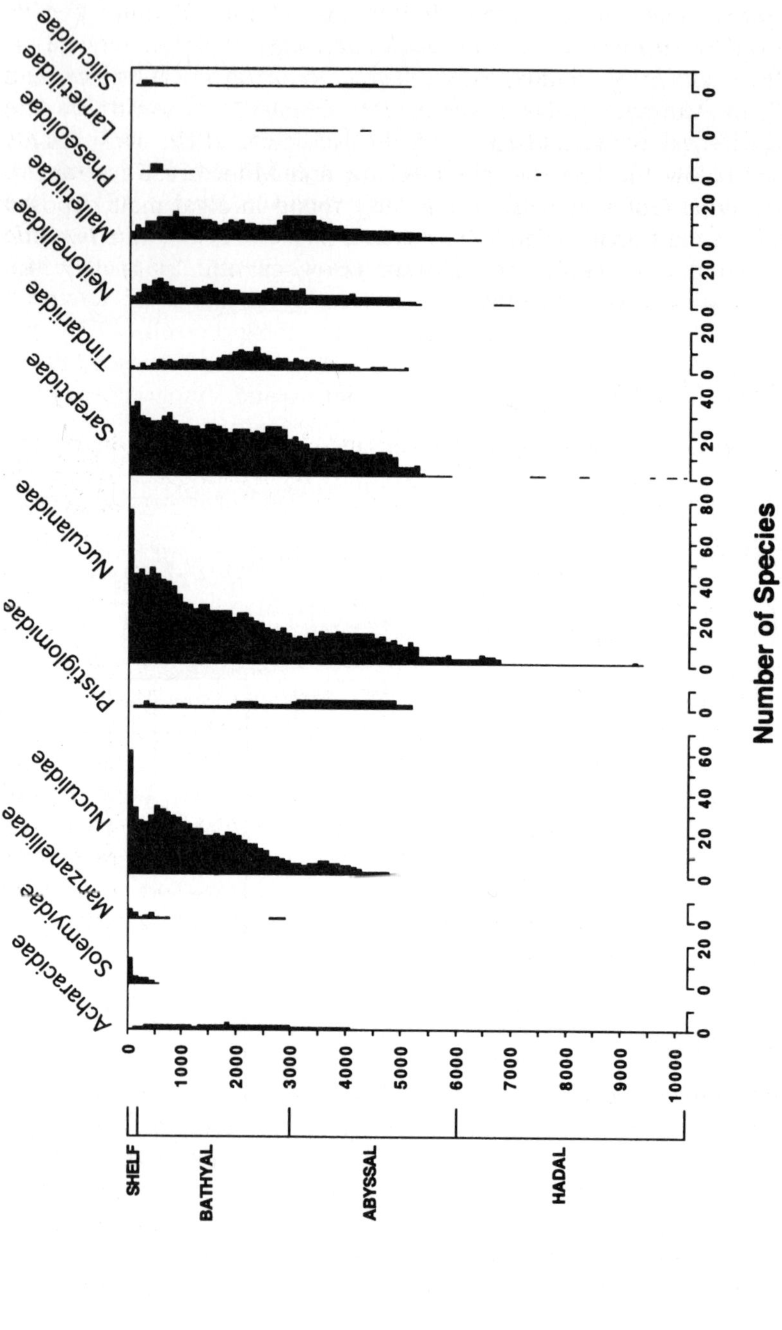

Figure 7 Depth distribution of 586 protobranch species by family in 100 m intervals. Compiled from reports in the literature (beach-washed shells not included).

well represented at bathyal and abyssal depths. The remaining protobranch families are primarily deep-sea species, and the bulk of all protobranch species occur below shelf depths. Some families, though containing few species, are represented over a great depth range (e.g. Siliculidae and Pristiglomidae). The deepest occurring protobranch, and indeed deepest known bivalve, is *Sarepta hadalis* Knudsen from 10 200 m in the Philippine Trench (Knudsen, 1970). Protobranchs of the shelf usually occur well below the low tide mark (Clark and Milne, 1955); however, *Nucula bicornis* Gofas & Salas is regularly found in algal mats exposed at low tide in the Canary Islands (Gofas and Salas, 1996). *Yoldia limatula* (Hunter and Brown, 1964) and *Solemya velum* (Smith, 1964) may also occur intertidally in soft sediments.

7.2. Cosmopolitanism

Malletia cuneata, an abyssal species, is the only truly cosmopolitan species, occurring at depths of from 1780 to 4797 m throughout the world (Knudsen, 1970; Sanders and Allen, 1985). *Ledella ultima* (Smith), also an abyssal species, is the most common and widespread protobranch in the Atlantic (Allen and Hannah, 1989) and may occur in other oceans as well (Allen and Sanders, 1996b). The success of *L. ultima* may be attributed to its long hindgut, which allows for increased digestive efficiency, and an inflated shell in older individuals, which allows for greater egg production (Allen and Hannah, 1989). In the Atlantic, many protobranch species are very widespread abyssally with large depth ranges (Allen and Sanders, 1996b). In the Pacific, there is a division in the abyssal fauna from east to west. Japan, for instance, completely lacks abyssal species that are found in the east Pacific, except for *Malletia cuneata* and the pan-Pacific *Bathyspinula calcar* (Dall, 1908), but does share species with the Indo-Malayan region (Okutani, 1974). This division is thought to be owing to the highly oligotrophic Central Pacific Basin acting as a faunal barrier (Knudsen, 1979).

7.3. Endemism

Endemism is high among protobranchs despite the widespread occurrence of many species. Up to 80% of the species of protobranchs in a basin may be endemic to that basin, with greater endemism at upper slope depths (Allen and Sanders, 1996b). The Argentine basin and several other southern Atlantic basins have the highest known endemism, as well as the highest species diversity (Allen and Sanders, 1996b). General species diversity, and diversity of bivalves as a whole, varies with latitude in the deep sea (Rex *et al.*, 1993, 1997). Protobranch diversity tends to be highest at low latitudes and declines toward the poles (Allen and Sanders, 1996b); however, along the continental shelf in the northeastern Pacific, protobranchs show no latitudinal diversity gradient (Roy *et al.*, 2000). High-latitude faunas, in general, tend to include more non-planktotrophic species (Pearse *et al.*, 1991; Hain and Arnaud, 1992) and this even distribution in the number of protobranch species along the northeastern Pacific shelf may be attributed to their developmental mode rather than to the latitudinal cline in available energy (Roy *et al.*, 2000). Similarly, there is a uniform distribution of protobranchs within the Arctic (MacGinitie, 1955; Carey *et al.*, 1984). *Yoldia limatula* is unusual in displaying an amphi-oceanic distribution pattern, occurring in the eastern Pacific and western Atlantic, but is absent from the Arctic (Ockelmann, 1954).

The vertical distribution of individual protobranch species can extend over several thousand meters in the deep sea (Knudsen, 1979) and varies with the species (Allen and Sanders, 1996b). The most distinct vertical change in taxonomic representation of protobranchs occurs between littoral and bathyal depths (Clarke, 1962a; Knudsen, 1979). On the continental shelf, protobranchs are represented almost entirely by only four families: Nuculidae, Nuculanidae, Sareptidae and Solemyidae (Allen, 1978), while the nine remaining protobranch families are exclusively bathyal or abyssal (Knudsen, 1979). It is thought that active, roaming deposit feeders, especially nuculids, have higher metabolic demands and thus are more common at depths where food is not limited (Allen, 1985). Though no distinct divisions exist between bathyal and abyssal species, abyssal species are not typically found in either Arctic or Antarctic faunas; although water temperatures would appear to be conducive to their survival (Knudsen, 1979).

The disparity between polar and abyssal protobranch faunas is somewhat enigmatic. The Antarctic, especially, is surprisingly depauperate and exhibits a high degree of endemism (Bouchet and Warén, 1979). The Nuculidae, which are common and abundant in continental faunas elsewhere in the world, are represented in the Antarctic by only a single species, *Nucula notobenthalis* Thiele, whereas the Nuculanidae are represented by at least five genera (Dell, 1972, 1990). The Arctic hosts high protobranch diversity, and maintains at least some linkage with the abyssal northern Pacific and Norwegian Sea (Knudsen, 1979). Several species of protobranchs are found in common between the Norwegian and Arctic Seas but none of these occur in the north Pacific, south of the Bering Straits (Bouchet and Warén, 1979). One species of Nuculidae and eight species of Nuculanidae occur in the abyssal Arctic (Knudsen, 1985). Five of these species are thought to be endemic to the Arctic (Knudsen, 1979).

7.4. Unusual habitats

Protobranchs may occur in extreme and unusual habitats including trenches, hydrothermal vents, hydrocarbon seeps and submarine caves. The hadal zone is the least studied zone of the deep sea; nevertheless, trenches have been found to exhibit high levels of endemism, often with endemic species occurring in only a single trench or trench system (Knudsen, 1979; Filatova and Shileyko, 1985; Vinogradova, 1997). The entire genus *Parayoldiella*, for instance, is hadal (Filatova, 1971; Filatova and Shileyko, 1985). Food levels in trenches may actually be favorable in the vicinity of continents where organic matter can accumulate. However, diversity in the trenches is lower than in adjacent abyssal plains, possibly owing to geological instability, limited area and the relatively short duration of trenches (Knudsen, 1979). Hydrothermal vents for the most part do not support protobranch populations except in the Guaymas Basin (in the Gulf of California), where *Nuculana grasslei* has been found dwelling in soft sediments beneath bacterial mats at the vent periphery (Allen, 1993). Of 24 known cold-water seep sites in the Atlantic and Pacific, solemyoids have been found at 10 of the Atlantic and 2 of the Pacific sites (reviewed in Sibuet and Olu, 1998). Several representatives of deep-water protobranch genera have also been found in shallow, sunless caves of southern Japan, including a manzanellid, *Huxleyia cavernicola* Hayami & Kase, a nuculid, *Pronucula insignis* Hayami & Kase and an unidentified solemyid (Hayami and Kase, 1993, 1996).

8. EVOLUTION

8.1. Early origins

Bivalves are thought to have arisen from rostroconch mollusks in the early Cambrian (Pojeta and Runnegar, 1976, 1985; Waller, 1998). The earliest known fossil bivalves are two genera, *Fordilla* and *Pojetaia,* from the late–early Cambrian (Pojeta *et al.*, 1973; Jell, 1980). Based on shell microstructure, *Fordilla* and *Pojetaia* are united in the family Fordillidae (Runnegar and Pojeta, 1992). Both genera are taxodont and *Pojetaia* compares well with praenuculids of the Ordovician (Pojeta and Runnegar, 1985) and nuculoids of the Devonian (Carter, 1990). Two later genera, *Pseudomyona* and *Tuarangia*, known from the middle Cambrian (Mackinnon, 1982; Berg-Madsen, 1987; Hinz-Schallreuter, 1995), are also thought to be early taxodonts (Cope, 1997). Protobranchs closely resembling modern-day representatives were widespread by the early Ordovician, as were bivalves from most other subclasses (Pojeta, 1971; Runnegar and Pojeta, 1992). Solemyoids first appear in the Ordovician (Starobogatov, 1992; Cope, 1996) or possibly as early as the late Cambrian (Allen, 1985).

Protobranchs are clearly separated from the rest of the Bivalvia, or Autobranchia, by a number of apomorphies (Waller, 1998), but whether they represent an ancestral, derived (Waller, 1990; Salvini-Plawen and Steiner, 1996) or polyphyletic condition (Adamkewicz *et al.*, 1997; Hoeh *et al.*, 1998) remains unknown. The pericalymma larva is unique to the Protobranchia and is probably a derived feature (Waller, 1998); however, lecithotrophic development is argued to be the primitive condition in mollusks (Chaffee and Lindberg, 1986). Conchological evidence from one of the oldest known bivalves, *Pojetaia runnegari* Jell, suggests this species possessed a large yolky egg (Waller, 1998). Lecithotrophic development is also common among other primitive mollusks including scaphopods, polyplacophorans and aplacophorans (Chaffee and Lindberg, 1986; Haszprunar *et al.*, 1995). Deposit feeding by use of labial palps is also thought to be the primitive trophic mode of bivalves (Yonge, 1939; Purchon, 1978), though it has been suggested that protobranchs were originally epifaunal suspension feeders that secondarily began deposit feeding (Stasek, 1965; Reid and Brand, 1986). The palp proboscides of protobranchs have also been compared with the feeding captaculum of scaphopods (Yonge, 1939, 1959; Dinamani, 1964), which may point to an altogether different origin for the group. Another apomorphic trait of the Protobranchia that suggests an origin far removed from other bivalves is the possession of hemocyanin as a respiratory pigment. In mollusks, hemocyanin is known to occur in cephalopods, polyplacophorans and gastropods. Protobranch hemocyanin most closely resembles that of the latter (Terwilliger *et al.*, 1988; Lambert *et al.*, 1995).

Molecular data concerning protobranch evolution are very sparse but the few studies to date suggest an early position for protobranchs in the Bivalvia, though not always one that places them as a monophyletic clade. A phylogeny using 18S rRNA from a limited number of bivalves groups the one solemyoid and one nuculid used in the study within the same clade (Campbell *et al.*, 1998). A separate analysis of 18S rRNA in the Mytilidae, using *Solemya reidi* as an outgroup, places *Solemya* as the most basal species in the phylogeny (Distel, 2000). A larger study looking at 18S rRNA from 28 bivalve species, including three protobranchs (*Solemya velum*, *Nucula proxima* and *Yoldia limatula*), and representing all subclasses and most orders, does not support the monophyly of the three protobranch species (Adamkewicz *et al.*, 1997). However, results from the most comprehensive cladistic analysis of the Bivalvia yet, incorporating both molecular and morphological characters, clusters protobranchs (including solemyoids) within two adjoining clades and places them basal to the entire class (Giribet and Wheeler, in press). Secondary structure of molecules is also a useful means of establishing evolutionary relaationships and identifying suprageneric motifs; however, among protobranchs the structure of only one molecule, the large ribosomal RNA subunit, in only one species, *Nucula nucleus*, has been examined to date (Chenuil *et al.*, 1997).

8.2. Radiation within the subclass

Pathways of evolution within the Protobranchia are uncertain but their earliest representative was most likely a nuculoid (Waller, 1998). The suggestion that protobranchs are derived from a *Nucinella*-like solemyoid ancestor (Allen, 1978, 1985) seems untenable given the later appearance of solemyoids in the fossil record, their highly specialized feeding and solid evidence that they evolved from the nuculoid-like Ctenodontidae (Pojeta, 1988; Waller, 1998). Evolution within the protobranchs has probably been constrained by feeding mechanisms and infaunal habit, and major radiations were most likely early, abrupt and of short duration, culminating in highly conserved forms (Allen and Hannah, 1986). The most dramatic differences appear among the Solemyoida. Symbiosis with sulfur-oxidizing bacteria has effected an enlargement of the gills, a reduction of the gut, a reduction of the palps, the partitioning of sulfidic and oxygenated water and the development of bacteriocytes (Reid and Brand, 1986). A critical point in their evolution must have involved a change from collecting food with the labial palps to filtering with the ctenidia, possibly assisted by symbionts, causing the ctenidia to greatly expand (Reid and Brand, 1986).

The more closely related superfamilies Nuculacea and Nuculanacea are separated on the basis of water flow into the mantle. Nuculanaceans switched from the more primitive state of anterior inhalant flow to posterior inhalant flow (Allen, 1985) along with developing inhalant siphons and elongating the shell. These modifications allowed feeding to become more independent of respiration and allowed for a greater diversification in form (Allen, 1985). The most speciose protobranchs are nuculanaceans, in the subfamily Yoldiellinae, with more representatives in this taxa than in any other protobranch family. Tindariids are among the more primitive nuculanaceans and possess an oval shape and no siphons. Generally considered more primitive, nuculaceans are, however, more advanced in terms of hindgut configuration, tending to have more loops than nuculanaceans (Allen, 1985).

A suggested unusual evolutionary radiation among the Proto-branchia concerns the enigmatic turbellarian-like organism *Xenoturbella*. Morphological aspects of its oogenesis (Israelsson, 1997), spermatogenesis (Israelsson, 2000) and larval development (Israelsson, 1999) have led to the speculation that it may in fact be a highly derived protobranch. Molecular evidence for such an interpretation has also been presented (Norén and Jondelius, 1997) but requires further validation, especially since *Nucula* species found in the digestive tract of *Xenoturbella* (Israelsson, 2000) could represent a source of cross-contaminating DNA. Recent ultrastructural evidence regarding cilia also casts doubt on a close relationship between this animal and protobranchs (Lundin and Schander, 2001).

Diversity in the Protobranchia peaked in the middle to late Ordovician with 30 genera known from the fossil record (Pojeta, 1978). Marked declines followed in the Silurian among solemyoids and in the Permian among nuculoids. Diversity in nuculoids rebounded in the Paleogene, but not to Ordovician levels (Starobogatov, 1992), with a few recent radiations represented by genera such as *Adrana* and *Acila* (Ockelmann and Warén, 1998). Protobranch diversity in recent times is highest in the deep sea and protobranchs are an important faunal element of the deep sea with diversities intermediate to other abundant deep-sea taxa (Sanders and Grassle, 1971). These bivalves are well suited for the deep sea and may have originated there or alternatively may have been relegated to it (Knudsen, 1979). Fossil evidence suggests that protobranchs became less abundant in warm shelf waters during the Oligocene, possibly as a result of competition from deposit-feeding tellinid bivalves (Nicol, 1972). On the other hand, species diversity differences between the continental shelf and deep sea may be merely a result of the great disparaity in area between the two habitats (pers. comm. Carole Hickman). For a detailed review of the evolution of the deep-sea forms see Allen (1978).

9. CONCLUSIONS

Though it is uncertain where the Protobranchia fit in the evolution of the Bivalvia, they are a unique and undoubtedly natural group as evidenced by a number of synapomorphies. Analogous larval forms, shared respiratory pigments, similar digestion and specific correspondence in shell microstructure, gill structure and other morphological aspects, strongly support this conclusion. Among the oldest of bivalves, probably occurring from the late–early or middle Cambrian, they may represent an ancestral form of the Bivalvia or an early divergence. Alternatively their derivation may be paraphyletic to the Bivalvia and represent a convergence in form.

Their feeding and digestion are key elements of their biology and may account for their longevity in the fossil record. Highly modified labial palps, which efficiently collect and extensively sort sediment, combined with extracellular digestion allow them to exploit refractory food sources and dominate oligotrophic sediments that may be less available to other taxa. In conjunction with lecithotrophic development, which allows development in a nutrient-poor water column, protobranchs are especially well adapted for living in the deep sea where they predominate. Their disproportionate representation in shallow water may be owing to competition from lamellibranch bivalves or limitations of habitat size.

Ecologically, protobranchs can be important bioturbators, especially when in high abundance in shallow-water sediments, redistributing sediment by grain size and influencing community assembly. In the deep sea, it is an enigma that numerous species of protobranchs are able to co-exist in the absence of any obvious niche specialization. Perhaps subtle differences in feeding mode, feeding depth and possibly diet allow resource partitioning among them. Much remains, however, to be learned concerning the ecology and evolution of the deep-sea benthos.

Geographic and genetic distributions of protobranchs indicate that despite physical stability and homogeneity in the deep sea, their dispersal and gene flow can be restricted. Striking levels of isolation are especially seen in the Antarctic protobranch fauna and in oceanic trenches. Levels of endemism are also fairly high in oceanic basins; however, within the Atlantic Ocean there are a number of protobranch species that occur in many basins. Not as well sampled and much larger in size, the Pacific Ocean appears to exhibit great isolation between east and west, with only two species of protobranchs in common between them.

ACKNOWLEDGEMENTS

Thanks are due to Dr Kenneth J. Boss and other members of the Mollusk Department of the Museum of Comparative Zoology, Harvard University for making available extensive library resources. Professor John Allen contributed important advice and comments that helped direct this project. An anonymous reviewer made valuable recommendations which greatly improved the manuscript. This work was supported by a grant from the National Science Foundation (OCE-9811925) to Drs R. Etter and M. Rex.

REFERENCES

Adamkewicz, S. L., Harasewych, M. G., Blake, J., Saudek, D. and Bult, C. J. (1997). A molecular phylogeny of the bivalve mollusks. *Molecular Biology and Evolution* **14**, 619–629.

Allen, J. A. (1953). Observations on *Nucula turgida* Marshall and *N. moorei* Winckworth. *Journal of the Marine Biological Association of the United Kingdom* **31**, 515–528.

Allen, J. A. (1954). A comparative study of the British species of *Nucula* and *Nuculana*. *Journal of the Marine Biological Association of the United Kingdom* **33**, 457–472.

Allen, J. A. (1971). Evolution and functional morphology of the deep water protobranch bivalves of the Atlantic. *In* "Proceedings of the Joint Oceanographic Assembly" (M. Uda, ed.), pp. 251–253. Japan Society for the Promotion of Science, Tokyo.

Allen, J. A. (1973). The adaptations of the bivalves of the Atlantic abyssal plain. *Proceedings of the Challenger Society* **4**, 148–251.

Allen, J. A. (1978). Evolution of the deep sea protobranch bivalves. *Philosophical Transactions of the Royal Society of London* B **284**, 387–401.

Allen, J. A. (1983). The ecology of deep-sea molluscs. *In* "Ecology, The Mollusca, Volume 6" (W. D. Russell-Hunter, ed.) pp. 29–75. Academic Press, Orlando, FL.

Allen, J. A. (1985). The recent Bivalvia: their form and evolution. *In* "Evolution, The Mollusca, Volume 10" (E. R. Trueman and M. R. Clarke, eds), pp. 337–399. Academic Press, New York, NY.

Allen, J. A. (1992). The evolution of the hindgut of the deep-sea protobranch bivalves. *American Malacological Bulletin* **9**, 187–191.

Allen, J. A. (1993). A new deep-water hydrothermal species of *Nuculana* (Bivalvia: Protobranchia) from the Guaymas Basin. *Malacologia* **35**, 141–151.

Allen, J. A. and Hannah, F. J. (1986). A reclassification of the recent genera of the subclass Protobranchia (Mollusca: Bivalvia). *Journal of Conchology* **32**, 225–249.

Allen, J. A. and Hannah, F. J. (1989). Studies on the deep sea Protobranchia: The subfamily Ledellinae (Nuculanidae). *Bulletin of the British Museum of Natural History (Zoology)* **55**, 123–171.

Allen, J. A. and Sanders, H. L. (1969). *Nucinella serrei* Lamy (Bivalvia: Protobranchia), a monomyarian solemyid and possible living actinodont. *Malacologia* **7**, 381–396.

Allen, J. A. and Sanders, H. L. (1973). Studies on deep-sea Protobranchia (Bivalvia); the families Siliculidae and Lametilidae. *Bulletin of the Museum of Comparative Zoology, Harvard University* **145**, 263–309.

Allen, J. A. and Sanders, H. L. (1982). Studies on the deep sea Protobranchia (Bivalvia); the subfamily Spinulinae (family Nuculanidae). *Bulletin of the Museum of Comparative Zoology, Harvard University* **150**, 1–30.

Allen, J. A. and Sanders, H. L. (1996a). Studies on the deep-sea Protobranchia (Bivalvia): the family Neilonellidae and the family Nuculanidae. *Bulletin of the Natural History Museum of London (Zoology)* **62**, 101–132.

Allen, J. A. and Sanders, H. L. (1996b). The zoogeography, diversity and origin of the deep-sea protobranch bivalves of the Atlantic: the epilogue. *Progress in Oceanography* **38**, 95–153.

Allen, J. A., Sanders, H. L. and Hannah, F. J. (1995). Studies on the deep-sea Protobranchia (Bivalvia); the subfamily Yoldiellinae. *Bulletin of the Natural History Museum of London (Zoology)* **61**, 11–90.

Aller, R. C. and Dodge, R. E. (1974). Animal–sediment relations in a tropical lagoon Discovery Bay, Jamaica. *Journal of Marine Research* **32**, 209–230.

Anderson, A. E., Childress, J. J. and Favuzzi, J. A. (1987). Net uptake of CO_2 driven by sulphide and thiosulphate oxidation in the bacterial symbiont-containing clam *Solemya reidi*. *Journal of Experimental Biology* **133**, 1–31.

Anderson, A. E., Felbeck, H. and Childress, J. J. (1990). Aerobic metabolism is maintained in animal tissue during rapid sulfide oxidation in the symbiont-containing clam *Solemya reidi*. *Journal of Experimental Zoology* **256**, 130–134.

Angelini, E., Salvato, B., Di-Muro, P. and Beltramini, M. (1998). Respiratory pigments of *Yoldia eightsi*, an Antarctic bivalve. *Marine Biology* **131**, 15–23.

Ansell, A. D. (1974). Seasonal changes in biochemical composition of the bivalve *Nucula sulcata* from the Clyde Sea area. *Marine Biology* **25**, 101–108.

Ansell, A. D., Parulekar, A. H. and Allen, J. H. (1978). On the rate of growth of *Nuculana minuta* (Müller) (Bivalvia; Nuculanidae). *Journal of Molluscan Studies* **44**, 71–82.

Arduino, G., Locatelli, B., Orlando, F. and Repetto, G. (1995). "Catalogo Illustrato delle Conchiglie Marine del Mediterraneo". Associazione Amici del Museo "Federico Eusebio", Alba.

Atkins, D. (1936). On the ciliary mechanisms and interrelationships of lamellibranchs. Part I. New observations on sorting mechanisms. *Quarterly Journal of Microscopical Science* **79**, 181–308.

Atkins (1937a). On the ciliary mechanisms and interrelationships of lamellibranchs. Part III. Types of lamellibranch gills and their food currents. *Quarterly Journal of Microscopical Science* **79**, 375–421.

Atkins, D. (1937b). On the ciliary mechanisms and interrelationships of lamellibranchs. Part II. Sorting devices on the gills. *Quarterly Journal of Microscopical Science* **79**, 339–373.

Austen, M. C., Widdicombe, S. and Villano-Pitacco, N. (1998). Effects of biological disturbance on diversity and structure of meiobenthic nematode communities. *Marine Ecology Progress Series* **174**, 233–246.

Bayne, B. L. and Newell, R. C. (1983). Physiological energetics of marine molluscs. *In* "Physiology. Part 1, The Mollusca, Volume 4" (A. S. M. Saleuddin and K. M. Wilbur, eds), pp. 407–515. Academic Press, New York, NY.

Beedham, G. E. and Owen, G. (1965). The mantle and shell of *Solemya parkinsoni* (Protobranchia: Bivalvia). *Proceedings of the Zoological Society of London* **145**, 405–430.

Beesley, P. L., Ross, G. J. B. and Wells, A., eds (1998). Mollusca: The Southern Synthesis. Part A. CSIRO Publishing, Melbourne.

Bender, K. and Davis, W. R. (1984). The effect of feeding by *Yoldia limatula* on bioturbation. *Ophelia* **23**, 91–100.

Berg, C. J. and Van Dover, C. L. (1987). Benthopelagic macrozooplankton communities at and near deep-sea hydrothermal vents in the eastern Pacific Ocean and the Gulf of California. *Deep-Sea Research* **34**, 379–401.

Berg-Madsen, V. (1987). *Tuarangia* from Bornholm (Denmark) and similarities in Baltoscandian and Australasian late Middle Cambrian faunas. *Alcheringa* **11**, 245–259.

Bergmans, W. (1978). Taxonomic revision of recent Australian Nuculidae (Mollusca: Bivalvia) except *Ennucula* Iredale, 1931. *Records of the Australian Museum* **31**, 673–736.

Bernard, F. R. (1980). A new *Solemya* s. str. from the Northeastern Pacific (Bivalvia: Cryptodonta). *Venus (Japanese Journal of Malacology)* **39**, 17–23.

Bernard, F. R. (1983). "Catalogue of the living Bivalvia of the eastern Pacific Ocean: Bering Strait to Cape Horn". Department of Fisheries and Oceans, Ottawa.

Bernard, F. R. and Noakes, D. J. (1990). Pumping rates, water pressures, and oxygen use in eight species of marine bivalve molluscs from British Columbia. *Canadian Journal of Fisheries and Aquatic Science* **47**, 1302-1306.

Bernard, F. R., Cai, Y.-Y. and Morton, B. (1993). "Catalogue of the living bivalve molluscs of China". University Press, Hong Kong.

Blake, N. J. and Jeffries, H. P. (1971). The structure of an experimental infaunal community. *Journal of Experimental Marine Biology and Ecology* **6**, 1–14.

Bouchet, P. and Warén, A. (1979). The abyssal molluscan fauna of the Norwegian Sea and its relation to other faunas. *Sarsia* **64**, 211–243.

Bowden, J. and Heppell, D. (1966). Revised list of British Mollusca. I. Introduction; Nuculacea–Ostreacea. *Journal of Conchology* **26**, 99–124.

Bright, T. J. (1970). Food of deep-sea bottom fishes. *In* "Contributions on the biology of the Gulf of Mexico, Texas A&M University Oceanographic Studies, Volume 1" (W. E. Pequegnat and F. A. Chace, Jr., eds), pp. 245–252. Gulf Publishing Co., Houston, TX.

Brooks, W. K. (1875). On an organ of special sense in the lamellibranchiate genus *Yoldia*. *Proceedings of the American Association for the Advancement of Science* **23**, 80–82.

Campbell, D. C., Hoekstra, K. J. and Carter, J. G. (1998). 18S ribosomal DNA and evolutionary relationships within the Bivalvia. *In* "Bivalves: an eon of evolution" (P. A. Johnston and J. W. Haggart, eds), pp. 75–85. University of Calgary Press, Calgary, Alberta.

Carey, A. G., Jr., Scott, P. H. and Walters, K. R. (1984). Distributional ecology of shallow southwestern Beaufort Sea (Arctic Ocean) bivalve Mollusca. *Marine Ecology Progress Series* **17**, 125–134.

Carter, J. G. (1990). Evolutionary significance of the shell microstructure in the Palaeotaxodonta, Pteriomorphia and Isofilibranchia (Bivalvia, Mollusca). *In* "Skeletal biomineralization: patterns, processes and evolutionary trends, Volume 1" (J. G. Carter, ed.), pp. 135–296. Van Nostrand Reinhold, New York, NY.

Carter, J. G., Lawrence, D. R. and Sanders, H. L. (1990). Shell microstructural data for the Bivalvia. Part II. Orders Nuculoida and Solemyoida. *In* "Skeletal Biomineralization: Patterns, Processes and Evolutionary Trends, Volume 1" (J. G. Carter, ed.), pp. 303–319. Van Nostrand Reinhold, New York, NY.

Cary, S. C. (1994). Vertical transmission of a chemoautotrophic symbiont in the protobranch bivalve, *Solemya reidi*. *Molecular Marine Biology and Biotechnology* **3**, 121–130.

Caspers, H. (1980). Long-term changes in benthic fauna resulting from sewage sludge dumping into the North Sea. *Progress in Water Technology* **12**, 461–479.

Castagna, M. and Chanley, P. E. (1973). Salinity tolerance of some marine bivalves from inshore and estuarine environments in Virginia waters on the western mid-Atlantic coast. *Malacologia* **12**, 47–96.

Cavanaugh, C. M. (1980). Symbiosis of chemoautotrophic bacteria and marine invertebrates. *Biological Bulletin, Marine Biological Laboratory, Woods Hole* **159**, 457.

Cavanaugh, C. M. (1983). Symbiotic chemoautotrophic bacteria in marine invertebrates from sulphide-rich habitats. *Nature* **302**, 58–61.

Cavanaugh, C. M. (1985). Symbioses of chemoautotrophic bacteria and marine invertebrates from hydrothermal vents and reducing sediments. *Bulletin of the Biological Society of Washington* **6**, 373–388.

Cavanaugh, C. M., Abbott, M. S. and Veenhuis, M. (1988). Immuno-chemical localization of ribulose-1,5-bisphosphate carboxylase in the symbiont containing gills of *Solemya velum* (Bivalvia: Mollusca). *Proceedings of the National Academy of Sciences USA* **85**, 7786–7789.

Chaffee, C. and Lindberg, D. R. (1986). Larval biology of early Cambrian molluscs: The implications of small body size. *Bulletin of Marine Science* **39**, 536–549.

Chardy, P., Guillaumont, B. and Hamon, D. (1984). Population dynamics of *Nucula nucleus* (Bivalvia, Protobranchiata) in the Cap of Flamanville (Manche). *Oceanologica Acta* **7**, 103–112.

Chase, M. R., Etter, R. J., Rex, M. A. and Quattro, J. M. (1998). Bathymetric patterns of genetic variation in a deep-sea protobranch bivalve, *Deminucula atacellana*. *Marine Biology* **131**, 301–308.

Cheng, I.-J. and Lopez, G. R. (1991). Contributions of bacteria and sedimentary organic matter to the diet of *Nucula proxima*. *Ophelia* **34**, 157–170.

Chenuil, A., Solignac, M. and Bernard, M. (1997). Evolution of the large-subunit ribosomal RNA binding site for protein L23/25. *Molecular Biology and Evolution* **14**, 578–588.

Christensen, H. and Kanneworff, E. (1985). Sedimenting phytoplankton as major food source for suspension and deposit feeders in the Øresund. *Ophelia* **24**, 223–244.

Clark, R. B. and Milne, A. (1955). The sublittoral fauna of two sandy bays on the Isle of Cumbrae, Firth of Clyde. *Journal of the Marine Biological Association of the United Kingdom* **34**, 161–180.

Clarke, A. H., Jr. (1961). Abyssal mollusks of the South Atlantic Ocean. *Bulletin of the Museum of Comparative Zoology, Harvard University* **125**, 345–387.

Clarke, A. H., Jr. (1962a). Annotated list and bibliography of the abyssal marine molluscs of the world. *National Museum of Canada, Bulletin* **181**.

Clarke, A. H., Jr. (1962b). Arctic archibenthal and abyssal molluscs – II. Molluscs dredged from drifting station Charlie (Alpha II). *National Museum of Canada, Bulletin* **185**, 90–109.

Coan, E. V., Scott, P. V. and Bernard, F. R. (2000). "Bivalve Seashells of Western North America". Santa Barbara Museum of Natural History, Santa Barbara, CA.

Conway, N. M. and Capuzzo, J. M. (1990). The use of biochemical indicators in the study of trophic interactions in animal–bacteria symbioses: *Solemya velum*, a case study. *In* "Trophic Relationships in the Marine Environment. Proceedings of the 24th European Marine Biology Symposium" (M. Barnes and R. N. Gibson, eds), pp. 553–564. Aberdeen University Press, Aberdeen.

Conway, N. M. and McDowell–Capuzzo, J. E. (1991). Incorporation and utilization of bacterial lipids in the *Solemya velum* symbiosis. *Marine Biology* **108**, 277–291.

Conway, N. M. and McDowell-Capuzzo, J. E. (1992). High taurine levels in the *Solemya velum* symbiosis. *Comparative Biochemistry and Physiology* **102B**, 175–185.

Conway, N. M., Capuzzo, J. M. and Fry, B. (1989). The role of endosymbiotic bacteria in the nutrition of *Solemya velum*: evidence from a stable isotope analysis of endosymbionts and hosts. *Limnology and Oceanography* **34**, 249–255.

Conway, N. M., Howes, B. L., McDowell-Capuzzo, J. E., Turner, R. D. and Cavanaugh, C. M. (1992). Characterization and site description of *Solemya borealis* (Bivalvia: Solemyidae), another bivalve–bacteria symbiosis. *Marine Biology* **112**, 601–613.

Cope, J. C. W. (1996). The early evolution of the Bivalvia. *In* "Origin and Evolutionary Radiation of the Mollusca" (J. D. Taylor, ed.), pp. 361–370. Oxford University Press, London.

Cope, J. C. W. (1997). The early phylogeny of the class Bivalvia. *Paleontology* **40**, 713–746.

Cotton, B. C. (1961). "South Australian Mollusca". W.L. Hawkes, Government Printer, Adelaide.

Cowan, I. M. (1968). The interrelationships of certain Boreal and Arctic species of *Yoldia* Möller, 1842. *The Veliger* **11**, 51–58.

Cox, L. R. (1959). The geological history of the Protobranchia and dual origin of taxodont Lamellibranchia. *Proceedings of the Malacological Society of London* **33**, 200–209.

Cox, L. R. (1960). Thoughts on the classification of the Bivalvia. *Proceedings of the Malacological Society of London* **34**, 60–88.

Craig, N. I. (1994). Growth of the bivalve *Nucula annulata* in nutrient-enriched environments. *Marine Ecology Progress Series* **104**, 77–90.

Creutzberg, F. (1986). Distribution patterns of two bivalve species (*Nucula turgida*, *Tellina fabula*) along a frontal system in the Southern North Sea. *Netherlands Journal of Sea Research* **20**, 305–311.

Dall, W. H. (1908). A gigantic *Solemya* and a new *Vesicomya*. *Nautilus* **22**, 61–63.

Davenport, J. (1988a). The feeding mechanism of *Yoldia* (=*Aequiyoldia*) *eightsi* (Couthouy). *Proceedings of the Royal Society of London* B **232**, 431–442.

Davenport, J. (1988b). Oxygen consumption and ventilation rate at low temperatures in the antarctic protobranch bivalve mollusc *Yoldia* (=*Aequiyoldia*) *eightsi* (Couthouy). *Comparative Biochemistry and Physiology* **90**A, 511–513.

Davis, J. P. and Wilson, J. G. (1983a). Seasonal changes in tissue weight and biochemical composition of the bivalve *Nucula turgida* in Dublin Bay with reference to gametogenesis. *Netherlands Journal of Sea Research* **17**, 84–95.

Davis, J. P. and Wilson, J. G. (1983b). The population structure and ecology of *Nucula turgida* (Leckenby and Marshall) in Dublin Bay. *Progress in Underwater Sciences* **8**, 53–60.

Davis, J. P. and Wilson, J. G. (1985). The energy budget and population structure of *Nucula turgida* in Dublin Bay. *Journal of Animal Ecology* **54**, 557–571.

Dell, R. K. (1956). "Archibenthal Mollusca of New Zealand". Dominion Museum, Wellington.

Dell, R. K. (1962). Additional archibenthal Mollusca from New Zealand. *Records of the Dominion Museum* **4**, 67–76.

Dell, R. K. (1972). Antarctic benthos. *Advances in Marine Biology* **10**, 1–216.

Dell, R. K. (1990). Antarctic Mollusca with special reference to the fauna of the Ross Sea. *Royal Society of New Zealand Bulletin* **27**, 1–311.

Dinamani, P. (1964). Feeding in *Dentalium conspicuum*. *Proceedings of the Malacological Society of London* **36**, 1–5.

Distel, D. L. (2000). Phylogenetic relationships among Mytilidae (Bivalvia): 18S rRNA data suggest convergence in mytilid body plans. *Molecular Phylogenetics and Evolution* **15**, 25–33.

Distel, D. L., Felbeck, H. and Cavanaugh, C. M. (1994). Evidence for phylogenetic congruence among sulfur-oxidizing chemoautotrophic bacterial endosymbionts and their bivalve hosts. *Journal of Molecular Evolution* **38**, 533–542.

Doeller, J. E., Kraus, D. W., Colacino, J. W. and Wittenberg, J. B. (1988). Gill hemoglobin may deliver sulfide to bacterial symbionts of *Solemya velum* (Bivalvia, Mollusca). *Biological Bulletin, Marine Biological Laboratory, Woods Hole* **175**, 388–396.

Drew, G. A. (1897). Notes on the embryology, anatomy, and habits of *Yoldia limatula*, Say. *Johns Hopkins University Circulars* **17**, 11–14.

Drew, G. A. (1899a). The anatomy, habits, and embryology of *Yoldia limatula*, Say. *Memoirs from the Biological Laboratory of the Johns Hopkins University* **4**, 1–37.

Drew, G. A. (1899b). Some observations on the habits, anatomy and embryology of members of the Protobranchia. *Anatomischer Anzeiger* **15**, 493–519.

Drew, G. A. (1900). Locomotion in *Solenomya* and its relatives. *Anatomischer Anzeiger* **17**, 257–266.

Drew, G. A. (1901). The life-history of *Nucula delphinodonta* (Mighels). *Quarterly Journal of Microscopical Science* **44**, 313–392.

Driscoll, E. G. (1964). Accessory muscle scars, an aid to protobranch orientation. *Journal of Paleontology* **38**, 61–66.

Driscoll, E. G. and Brandon, D. E. (1973). Mollusc–sediment relationships in northwestern Buzzards Bay, Massachusetts, USA. *Malacologia* **12**, 13–46.

Edwards, C. (1965). The hydroid and the medusa *Neoturris pileata*. *Journal of the Marine Biological Association of the United Kingdom* **45**, 443–458.

Eisen, J. A., Smith, S. W. and Cavanaugh, C. M. (1992). Phylogenetic relationship of chemoautotrophic bacterial symbionts of *Solemya velum* Say (Mollusca: Bivalvia) determined by 16S rRNA gene sequence analysis. *Journal of Bacteriology* **174**, 3416–3421.

Embley, R. W., Eittreim, S. L., McHugh, C. H., Normark, W. R., Rau, G. H., Hecker, B., DeBevoise, A. E., Greene, H. G., Ryan, W. B. F., Harrold, C. and Baxter, C. H. (1990). Geological setting of chemosynthetic communities in the Monterey Fan Valley system. *Deep-Sea Research* **37A**, 1651–1667.

Etter, R. J., Rex, M. A., Chase, M. C. and Quattro, J. M. (1999). A genetic dimension to deep-sea biodiversity. *Deep-Sea Research* **46**, 1095–1099.

Felbeck, H. (1983). Sulfide oxidation and carbon fixation by the gutless clam *Solemya reidi*: an animal–bacteria symbiosis. *Journal of Comparative Physiology* B **152**, 3–11.

Filatova, Z. A. (1971). On some mass species of bivalve molluscs from the ultra-abyssal zone of the Kurile-Kamchatka Trench. *Trudy Instituta Okeanologiya Akademia Nauk SSSR* **92**, 46–60 (in Russian).

Filatova, Z. A. and Shileyko, A. A. (1985). Composition, morphology and distribution of the ultra-abyssal genus *Parayoldiella* (Bivalvia, Protobranchia). *Trudy Zoologicheskogo Instituta* **135**, 76–94 (in Russian).

Filatova, Z. A. and Vinogradova, N. G. (1974). Bottom fauna of the South Atlantic deep-sea trenches. Biological investigations in the Atlantic sector of the Antarctic Ocean. *Trudy Instituta Okeanologiya Akademia Nauk SSSR* **98**, 141–156 (in Russian).

Fisher, C. R. (1990). Chemoautotrophic and methanotrophic symbioses in marine invertebrates. *CRC Reviews in Aquatic Sciences* **2**, 399–436.

Fisher, C. R. and Childress, J. J. (1986). Translocation of fixed carbon from symbiotic bacteria to host tissues in the gutless bivalve *Solemya reidi*. *Marine Biology* **93**, 59–68.

Follum, O. A. and Gray, J. S. (1987). Nitrogenous excretion by the sediment-living bivalve *Nucula tenuis* from the Oslofjord, Norway. *Marine Biology* **96**, 355–358.

Franc, A. (1960). Classe des bivalves. *In* "Traité de Zoologie, Volume 5, part 2" (P.-P. Grassé, ed.), pp. 1845–2164. Masson et Cie, Paris.

Franzén, Å. (1983). Ultrastructural studies of spermatozoa in three bivalve species with notes on evolution of elongated sperm nucleus in primitive spermatozoa. *Gamete Research* **7**, 199–214.

Frey, R. W. (1967). The lebensspuren of some common marine invertebrates near Beaufort, North Carolina. I. Pelecypod burrows. *Journal of Paleontology* **42**, 570–574.

Fuiman, L. A., Gage, J. D. and Lamont, P. A. (1999). Shell morphology of the deep sea protobranch bivalve *Ledella pustulosa* in the Rockall Trough, north-east Atlantic. *Journal of the Marine Biological Association of the United Kingdom* **79**, 661–671.

Gage, J. D. (1992). Benthic secondary production in the deep sea. *In* "Deep-sea Food Chains and the Global Carbon Cycle, NATO ASI Series C: Mathematical and Physical Sciences, Volume 360" (G. T. Rowe and V. Pariente, eds), pp. 199–216. Kluwer Academic Publishers, Boston, MA.

Gage, J. D. (1994). Recruitment ecology and age structure of deep-sea invertebrate populations. *In* "Reproduction, Larval Biology, and Recruitment of the Deep-sea Benthos" (C. M. Young and K. J. Eckelbarger, eds), pp. 223–242. Columbia University Press, New York, NY.

Gage, J. D., Tyler, P. A., Davies, G. and Harvey, R. (1986). Life-history studies on deep-sea protobranch bivalves from the Rockall Trough (NE Atlantic). *In* "Ninth International Malacological Congress" (D. Heppell, ed.), p. 27. Unitas Malacologica, Edinburgh.

Galliher, E. W. (1931). Notes on excrement. *Micropaleontology Bulletin* **3**, 11–12.

Gemmell, J., Hertz, C. M. and Myers, B. W. (1980). Seastar predation on mollusks in the San Felipe Bay area, Baja California, Mexico. *Festivus* **12**, 24–55.

Gilkinson, K. D., Hutchings, J. A., Oshel, P. E. and Haedrich, R. L. (1986). Shell microstructure and observations on internal banding patterns in the bivalves *Yoldia thraciaeformis* Storer, 1838, and *Nuculana pernula* Muller, 1779 (Nuculanidae), from a deep-sea environment. *The Veliger* **29**, 70–77.

Giribet, G. and Wheeler, W. (in press). On bivalve phylogeny: a high-level analysis of Bivalvia (Mollusca) based on combined morphology and DNA sequence data. *Invertebrate Biology*.

Gofas, S. and Salas, C. (1996). Small Nuculidae (Bivalvia) with functional primary hinge in the adults. *Journal of Conchology* **35**, 427–435.

Gooch, J. L. and Schopf, T. J. M. (1972). Genetic variability in the deep sea: relation to environmental variability. *Evolution* **26**, 545–552.

Gustafson, R. G. (1986). Development, morphogenesis, and putative symbiont transmission in the pericalymma larva of *Solemya reidi* (Bivalvia, Cryptodonta, Solemyidae). PhD thesis. University of Victoria, Victoria, BC, Canada.

Gustafson, R. G. and Lutz, R. A. (1992). Larval and early post-larval development of the protobranch bivalve *Solemya velum* (Mollusca: Bivalvia). *Journal of the Marine Biological Association of the United Kingdom* **72**, 383–402.

Gustafson, R. G. and Reid, R. G. B. (1986). Development of the pericalymma larva of *Solemya reidi* (Bivalvia: Cryptodonta: Solemyidae) as revealed by light and electron microscopy. *Marine Biology* **93**, 411–427.

Gustafson, R. G. and Reid, R. G. B. (1988a). Association of bacteria with larvae of the gutless protobranch bivalve *Solemya reidi* (Cryptodonta: Solemyidae). *Marine Biology* **97**, 389–401.

Gustafson, R. G. and Reid, R. G. B. (1988b). Larval and post-larval morphogenesis in the gutless protobranch bivalve *Solemya reidi* (Cryptodonta: Solemyidae). *Marine Biology* **97**, 373–387.

Gustafson, R. G., Gustafson, B. D. and Reid, R. G. B. (1987). Continuous reproduction in the protobranch bivalve *Solemya reidi* (Cryptodonta: Solemyidae). *The Veliger* **29**, 367–373.

Habe, T. (1951–52). "Genera of Japanese Shells (Pelecypods and Scaphopoda)", Volumes 1 and 2. Tokyo.

Habe, T. (1977). "Systematics of Mollusca in Japan. Bivalvia and Scaphopoda". Hokurikukanu, Japan.

Hain, S. (1990). The benthic seashells (Gastropoda and Bivalvia) of the Weddell Sea, Antarctica. *Reports on Polar Research* **70**, 1–181.

Hain, S. and Arnaud, P. M. (1992). Notes on the reproduction of high-Antarctic molluscs from the Weddell Sea. *Polar Biology* **12**, 303–312.

Halton, D. W. and McCrae, J. M. (1985). Development of the tegument and alimentary tract in a digenetic trematode, *Fellodistomum fellis*. *Parasitology* **90**, 193–204.

Halton, D. W. and Owen, G. (1968). The fine structure and histochemistry of the gastric cuticle of the protobranchiate bivalve, *Nucula sulcata* Bronn. *Proceedings of the Malacological Society of London* **38**, 71–81.

Hampson, G. R. (1971). A species pair of the genus *Nucula* (Bivalvia) from the eastern coast of the United States. *Proceedings of the Malacological Society of London* **39**, 333–342.

Harvey, R. and Gage, J. D. (1995). Reproduction and recruitment of *Nuculoma tenuis* (Bivalvia: Nuculoida) from Loch Etive, Scotland. *Journal of Molluscan Studies* **61**, 409–419.

Haszprunar, G. (1985). On the anatomy and fine-structure of a peculiar sense organ in *Nucula* (Bivalvia, Protobranchia). *The Veliger* **28**, 52–62.

Haszprunar, G., Salvini-Plawen, L. v. and Rieger, R. M. (1995). Larval planktotrophy – A primitive trait in the Bilateria? *Acta Zoologica* **76**, 141–154.

Hayami, I. and Kase, T. (1993). Submarine cave Bivalvia from the Ryukyu Islands: systematics and evolutionary significance. *University of Tokyo University Museum Bulletin* **35**, 1–133.

Hayami, I. and Kase, T. (1996). Characteristics of submarine cave bivalves in the northwestern Pacific. *American Malacological Bulletin* **12**, 59–65.

Healy, J. M. (1989). Spermiogenesis and spermatozoa in the relict bivalve genus *Neotrigonia*: relevance to trigonioid relationships, particularly Unionoidea. *Marine Biology* **103**, 73–85.

Heath, H. (1937). The anatomy of some protobranch mollusks. *Mémoires du Musée Royal D'Histoire Naturelle de Belgique* **12**, 1–26.

Hickman, C. S. (1974). Characteristics of bathyal mollusk faunas in the Pacific Coast Tertiary. *Annual Report of the Western Society of Malacologists* **7**, 41–50.

Hickman, C. S. (1984). Composition, structure, ecology, and evolution of six Cenozoic deep-water mollusk communities. *Journal of Paleontology* **58**, 1215–1234.

Hinz-Schallreuter, I. (1995). Muscheln (Pelecypoda) aus dem Mittelkambrium von Bornholm. *Geschiebekunde aktuell* **11**, 71–84.

Hirasaka, K. (1927). Notes on *Nucula*. *Journal of the Marine Biological Association of the United Kingdom* **14**, 629–645.

Hoeh, W. R., Black, M. B., Gustafson, R. G., Bogan, A. E., Lutz, R. A. and Vrijenhoek, R. C. (1998). Testing alternative hypotheses of *Neotrigonia* (Bivalvia: Trigonioida) phylogenetic relationships using cytochrome C oxidase subunit I DNA sequences. *Malacologia* **40**, 267–278.

Howe, S., Maurer, D. and Leathem, W. A. (1988). Secondary production of benthic molluscs from the Delaware Bay and coastal area. *Estuarine, Coastal & Shelf Science* **26**, 81–94.

Hrs-Brenko, M. and Legac, M. (1991). A review of bivalve species in the eastern Adriatic Sea 1. Protobranchia (Solemyidae, Nuculidae, Nuculanidae). *Acta Adriatica* **32**, 655–670.

Hunt, O. D. (1925). The food of the bottom fauna of the Plymouth fishing grounds. *Journal of the Marine Biological Association of the United Kingdom* **13**, 560–599.

Hunter, W. R. and Brown, S. C. (1964). Phylum Mollusca. *In* "Keys to Marine Invertebrates of the Woods Hole Region" (R. I. Smith, ed.), pp. 129–152. Spaulding Co., Boston, MA.

Hutchings, J. A. and Haedrich, R. L. (1984). Growth and population structure in two species of bivalves (Nuculanidae) from the deep sea. *Marine Ecology Progress Series* **17**, 135–142.

Hylleberg, J. (1975). Selective feeding by *Abarenicola pacifica* with notes on *Abarenicola vagabunda* and a concept of gardening in lugworms. *Ophelia* **14**, 113–137.

Ieyama, H. (1982). Karyotypes in two species of the Solemyidae (Bivalvia, Cryptodonta). *Venus (Japanese Journal of Malacology)* **40**, 232–236.

Ieyama, H. (1992). Chromosomes of two species in two families of Palaeotaxodonta (Bivalvia). *Venus (Japanese Journal of Malacology)* **51**, 115–120.

Israelsson, O. (1997). *Xenoturbella*'s molluscan relatives . . . and molluscan embryogenesis. *Nature* **390**, 32.

Israelsson, O. (1999). New light on the enigmatic *Xenoturbella* (phylum uncertain): ontogeny and phylogeny. *Proceedings of the Royal Society of London* B **266**, 835–841.

Israelsson, O. (2000). *Xenoturbella*. PhD thesis. Stockholm University, Stockholm.

James, B. M. (1973). Systematics and biology of the deep-water Palaeotaxodonta (Mollusca: Bivalvia) from the Gulf of Mexico. PhD thesis, Texas A&M University, College Station, Texas.

Jell, P. A. (1980). Earliest known pelecypod on earth – a new Early Cambrian genus from South Australia. *Alcheringa* **4**, 233–239.

Jewett, S. C. and Feder, H. M. (1980). Autumn food of adult starry flounders, *Platichthys stellatus,* from the northeastern Bering Sea and the southeastern Chukchi Sea. *Journal du Conseil* **39**, 7–14.

Kellogg, J. L. (1915). Ciliary mechanisms of Lamellibranchs with descriptions of anatomy. *Journal of Morphology* **26**, 625–701.

Khlebovich, V. V. (1996). *Portlandia aestuariorum* (Mossewitsch, 1928) [Bivalvia, Nuculanidae] in its relation to salinity. *Ruthenica* **6**, 167–169.

Kinner, P., Maurer, D. and Leathem, W. A. (1974). Benthic invertebrates in Delaware Bay: animal–sediment associations of the dominant species. *Internationale Revue der gesamten Hydrobiologie* **59**, 685–701.

Knudsen, J. (1967). The deep-sea Bivalvia. *Scientific Reports of the John Murray Expedition* **11**(3), 237–343.

Knudsen, J. (1970). The systematics and biology of abyssal and hadal Bivalvia. *Galathea Reports* **2**, 7–241 and 220 plates.

Knudsen, J. (1979). Deep-sea bivalves. *In* "Pathways in Malacology" (S. van der Spoel, A. C. van Bruggen and J. Lever, eds), pp. 195–224. Scheltema & Holkema, Utrecht.

Knudsen, J. (1985). Abyssal Mollusca of the Arctic Ocean. *Journal of Conchology* **32**, 97–107.

Køie, M. (1980). On the morphology and life-history of *Steringotrema pagelli* (van Beneden, 1871) Odhner, 1911 and *Fellodistomum fellis* (Olsson, 1868) Nicoll, 1909 [syn. *S. ovacutum* (Lebour, 1908) Yamaguti, 1953] (Trematoda, Fellodistomidae). *Ophelia* **19**, 215–236.

Kraus, D. W. (1995). Heme proteins in sulfide-oxidizing bacteria/mollusc symbioses. *American Zoologist* **35**, 112–120.

Krueger, D. M., Gallager, S. M. and Cavanaugh, C. M. (1992). Suspension feeding on phytoplankton by *Solemya velum*, a symbiont-containing clam. *Marine Ecology Progress Series* **86**, 145–151.

Krueger, D. M., Gustafson, R. G. and Cavanaugh, C. M. (1996a). Vertical transmission of chemoautotrophic symbionts in the bivalve *Solemya velum* (Bivalvia: Protobranchia). *Biological Bulletin, Marine Biological Laboratory, Woods Hole* **190**, 195–202.

Krueger, D. M., Dubilier, N. and Cavanaugh, C. M. (1996b). Chemoautotrophic symbiosis in the tropical clam *Solemya occidentalis* (Bivalvia: Protobranchia): ultrastructural and phylogenetic analysis. *Marine Biology* **126**, 55–64.

Kuznetsov, A. P. and Ohta, S. (1989). Common and morphometric characteristics of several Protobranchia bivalve molluscs inhabiting various depth zones in Suruga Bay (Japan). *Doklady Akademii Nauk USSR (Biological Science)* **306**, 1012–1015.

Kuznetsov, A. P. and Shileiko, A. A. (1984). Gutless Protobranchia (Bivalvia). *Nauchnye Doklady vysshei Shkoly Biologicheskie Nauki* 1984 (2), 39–49 (in Russian).

Kuznetsov, A. P., Ohta, S. and Endow, K. (1991). Morphofunctional consequences of bacterial symbiotrophy in *Solemya* (*Petrasma*) *pusilla* (Protobranchia, Bivalvia) from Sagami Bay (central Japan). *Izvestiya Akademii Nauk SSSR, ser. Biologicheskaya* 1991 (6), 895–903 (in Russian). Translation in *Biology Bulletin of the Academy of Science of the USSR* **17**, 587–594.

LaBarbera, M. (1974). Larval and post-larval development of five species of Miocene bivalves (Mollusca). *Journal of Paleontology* **48**, 256–277.

Lambert, O., Taveau, J.-C., Boisset, N. and Lamy, J. N. (1975). Three-dimensional reconstruction of the haemocyanin of the protobranch bivalve mollusc *Nucula hanleyi* from frozen-hydrated specimens. *Archives of Biochemistry and Biophysics* **319**, 231–243.

Lamprell, K. L. and Healy, J. M. (1998). "Bivalves of Australia". Backhuys Publishers, Leiden.

Lebour, M. V. (1938). Notes on the breeding of some lamellibranchs from Plymouth, and their larvae. *Journal of the Marine Biological Association of the United Kingdom* **23**, 119–144.

Lee, R. W., Robinson, J. J. and Cavanaugh, C. M. (1999). Pathways of inorganic nitrogen assimilation in chemoautotrophic bacteria–marine invertebrate symbioses: expressions of host and symbiont glutamine synthetase. *Journal of Experimental Biology* **202**, 289–300.

Levinton, J. S. (1972a). Spatial distribution of *Nucula proxima* Say (Protobranchia): an experimental approach. *Biological Bulletin, Marine Biological Laboratory, Woods Hole* **143**, 175–183.

Levinton, J. S. (1972b). Stability and trophic structure in deposit-feeding and suspension-feeding communities. *American Naturalist* **106**, 472–486.

Levinton, J. S. (1977). Ecology of shallow-water deposit-feeding communities, Quisset Harbor, Massachussetts. *In* "Ecology of Marine Benthos" (B. C. Coull, ed.), pp. 191–227. University of South Carolina, Columbia, SC.

Levinton, J. S. and Banbach, R. K. (1975). A comparative study of Silurian and Recent deposit-feeding communities. *Paleobiology* **1**, 97–124.

Levinton, J. S., Ward, J. E. and Thompson, R. J. (1996). Biodynamics of particle processing in bivalve molluscs: models, data, and future directions. *Invertebrate Biology* **115**, 232–242.

Lewis, J. B., Saleh, S., Reiswig, H. M. and Lalli, C. M. (1982). Growth, production and biomass of the burrowing protobranch mollusc *Yoldia limatula* in the Bideford River, Prince Edward Island, Canada. *Marine Biology* **70**, 173–179.

Lightfoot, R. H., Tyler, P. A. and Gage, J. D. (1979). Seasonal reproduction in deep-sea bivalves and brittlestars. *Deep-Sea Research* **26A**, 967–973.

Liljedahl, L. (1984). *Janeia silurica*, a link between nuculoids and solemyoids (Bivalvia). *Paleontology* **27**, 693–698.

Lopez, G. R. (1980). The availability of microorganisms attached to sediment as food for some marine deposit-feeding molluscs, with notes on microbial detachment due to the crystalline style. *In* "Marine Benthic Dynamics" (K. Tenore and B. C. Coull, eds), pp. 387–406. South Carolina University Press, Columbia, S.C.

Lopez, G. R. and Cheng, I.-J. (1983). Synoptic measurements of ingestion rate, ingestion selectivity, and absorption efficiency of natural foods in the deposit-feeding molluscs *Nucula annulata* (Bivalvia) and *Hydrobia toteni* (Gastropoda). *Marine Ecology Progress Series* **11**, 55–62.

Lopez, G. R. and Levinton, J. S. (1987). Ecology of deposit-feeding animals in marine sediments. *The Quarterly Review of Biology* **62**, 235–260.

Lopez, G. R., Tantichodok, P. and Cheng, I.-J. (1989). Radiotracer methods for determining utilization of sedimentary organic matter by deposit feeders. *In* "Ecology of Marine Deposit Feeders" (G. R. Lopez, G. L. Taghon and J. S. Levinton, eds), pp. 149–170. Springer-Verlag, New York, NY.

Lundin, K. and Schander, C. (2001). Ciliary ultrastructure of protobranchs (Molluscs, Bivalvia). *Invertebrate Biology* **120**, 350–357.

MacDonald, J. S., Waiwood, K. G. and Green, R. H. (1982). Rates of digestion of different prey in Atlantic cod (*Gadus morhua*), ocean pout (*Macrozoarces americanus*), winter flounder (*Pseudopleuronectes americanus*) and American plaice (*Hippoglossoides platessoides*). *Canadian Journal of Fisheries and Aquatic Science* **39**, 651–659.

MacGinitie, G. E. (1955). Distribution and ecology of the marine invertebrates of Point Barrow, Alaska. *Smithsonian Miscellaneous Collections* **128**, 1–201.

Mackinnon, D. I. (1982). *Tuarangia paparua* n. gen. and n. sp., a late Middle Cambrian pelecypod from New Zealand. *Journal of Paleontology* **56**, 589–598.

Mangum, C. P., Miller, K. I., Scott, J. L., Van Holde, K. E. and Morse, M. P. (1987). Bivalve hemocyanin: structural, functional, and phylogenetic relationships. *Biological Bulletin, Marine Biological Laboratory, Woods Hole* **173**, 205–221.

Martorelli, S. R. and Cremonte, F. (1998). A proposed three-host life history of *Monascus filiformis* (Rudolphi, 1819) (Digenea: Fellodistomidae) in the southwest Atlantic Ocean. *Canadian Journal of Zoology* **76**, 1198–1203.

Maurer, D., Keck, R. T., Tinsman, J. C. and Leathem, W. A. (1980). Vertical migration and mortality of benthos in dredged material – part 1: Mollusca. *Marine Environmental Research* **4**, 299–319.

Maxwell, P. A. (1988). Comments on "A reclassification of the recent genera of the subclass Protobranchia (Mollusca: Bivalvia)" by J. A. Allen and F. J. Hannah (1986). *Journal of Conchology* **33**, 85–96.

Métivier, B. and von Cosel, R. (1993). *Acharax alinae* n. sp., Solemyidae (Mollusca: Bivalvia) géante du bassin de Lau. *Comptes Rendus de l'Académie des Sciences*. Serie III, Sciences de la vie **316**, 229–237.

Meyhöfer, E., Morse, M. P. and Robinson, W. E. (1985). Podocytes in bivalve molluscs: morphological evidence for ultrafiltration. *Journal of Comparative Physiology* B **156**, 151–161.

Miura, T. and Laubier, L. (1990). Nautiliniellid polychaetes collected from the Hatsushima cold-seep site in Sagami Bay, with descriptions of new genera and species. *Zoological Science* **7**, 319–325.

Moore, D. R. (1977). Small species of Nuculidae (Bivalvia) from the tropical western Atlantic. *Nautilus* **91**, 119–128.

Moore, H. B. (1931a). The muds of the Clyde Sea area. III, Chemical and physical conditions; rate and nature of sedimentation; and fauna. *Journal of the Marine Biological Association of the United Kingdom* **17**, 325–358.

Moore, H. B. (1931b). The systematic value of a study of Mollusca faeces. *Proceedings of the Malacological Society of London* **19**, 281–290.

Moore, H. B. (1931c). The specific identification of faecal pellets. *Journal of the Marine Biological Association of the United Kingdom* **17**, 359–365.

Morris, N. J. and Fortey, R. A. (1976). The significance of *Tironucula* gen. nov. to the study of bivalve evolution. *Journal of Paleontology* **50**, 701–709.

Morse, E. S. (1913). Observations on living *Solenomya* (*velum* and *borealis*). *Biological Bulletin, Marine Biological Laboratory, Woods Hole* **25**, 261–281.

Morse, E. S. (1919). Observations on living lamellibranchs of New England. *Proceedings of the Boston Society of Natural History* **35**, 139–196.

Morse, M. P. (1987). Comparative functional morphology of the bivalve excretory system. *American Zoologist* **27**, 737–746.

Morse, M. P. and Meyhöfer, E. (1990). Ultrastructural studies on the heart–kidney complex of three species of protobranch bivalve molluscs. *In* "The Bivalvia. Proceedings of a Memorial Symposium in Honour of Sir Charles Maurice Yonge (1899–1986), Edinburgh, 1986" (B. J. Morton, ed.), pp. 223–235. Hong Kong University Press, Hong Kong.

Morse, M. P. and Zardus, J. D. (1997). Bivalvia. *In* "Mollusca II, Microscopic Anatomy of Invertebrates, Volume 6A" (F. W. Harrison and A. J. Kohn, eds), pp. 7–118. Wiley-Liss, Inc., New York, NY.

Morse, M. P., Meyhöfer, E., Otto, J. J. and Kuzirian, A. M. (1986). Hemocyanin respiratory pigment in bivalve mollusks. *Science* **231**, 1302–1304.

Mortimer, J. E. (1962). A comparative study of post-larval feeding mechanisms in the Bivalvia. PhD thesis. University of Glasgow.

Morton, B. J. (1996). The evolutionary history of the Bivalvia. *In* "Origin and Evolutionary Radiation of the Mollusca" (J. D. Taylor, ed.), pp. 337–359. Oxford University Press, Oxford.

Nakaoka, M. (1992a). Spatial and seasonal variation in growth rate and secondary production of *Yoldia notabilis* in Otsuchi Bay, Japan, with reference to the influence of food supply from the water column. *Marine Ecology Progress Series* **88**, 215–223.

Nakaoka, M. (1992b). Age determination and growth analysis based on external shell rings of the protobranch bivalve *Yoldia notabilis* Yokoyama in Otsuchi Bay, northeastern Japan. *Benthos Research* **43**, 53–66.

Nakaoka, M. (1993). Yearly variation in recruitment and its effect on population dynamics in *Yoldia notabilis* (Mollusca: Bivalvia), analyzed using projection matrix model. *Research in Population Ecology* **35**, 199–213.

Nakaoka, M. (1994). Size-dependent reproductive traits of *Yoldia notabilis* (Bivalvia: Protobranchia). *Marine Ecology Progress Series* **114**, 129–137.

Nakaoka, M. (1996). Size-dependent survivorship of the bivalve *Yoldia notabilis* (Yokoyama, 1920): the effect of crab predation. *Journal of Shellfish Research* **15**, 355–362.

Nakaoka, M. (1998). Optimal resource allocation of the marine bivalve *Yoldia notabilis*: the effects of size-limited reproductive capacity and size-dependent mortality. *Evolutionary Ecology* **12**, 347–361.

Nakaoka, M. and Matsui, S. (1994). Annual variation in the growth rate of *Yoldia notabilis* (Bivalvia: Nuculanidae) in Otsuchi Bay, northeastern Japan, analyzed using shell microgrowth patterns. *Marine Biology* **119**, 397–404.

Newell, N. D. (1969). Classification of Bivalvia. *In* "Part N, Treatise on invertebrate paleontology, Volume 1, Mollusca 6, Bivalvia", pp. N205–N224. Geological Society of America and University of Kansas, Lawrence, KA.

Nicol, D. (1972). Geologic history of deposit-feeding pelecypods. *Nautilus* **86**, 11–15.

Nolan, C. P. and Clarke, A. (1993). Growth in the bivalve *Yoldia eightsi* at Signy Island, Antarctica, determined from internal shell increments and calcium-45 incorporation. *Marine Biology* **117**, 243–250.

Norén, M. and Jondelius, U. (1997). *Xenoturbella*'s molluscan relatives. *Nature* **390**, 31–32.

Ockelmann, K. W. (1954). On the interrelationship and the zoogeography of northern species of *Yoldia* Möller, s. str. (Mollusca, Fam. Ledidae). *Meddelelser om Grønland* **107**, 32.

Ockelmann, K. W. (1958). The zoology of East Greenland: marine Lamellibranchiata. *Meddelelser om Grønland* **122**, 1–256.

Ockelmann, K. W. (1965). Developmental types in marine bivalves and their distribution along the Atlantic coast of Europe. *In* "Proceedings of the First European Malacological Congress" (L. R. Cox and J. F. Peake, eds), pp. 25–35. Conchological Society of Great Britain and Ireland and the Malacological Society of London, London.

Ockelmann, K. W. and Warén, A. (1998). Taxonomy of and biological notes on the bivalve genus *Microgloma*, with comments on protobranch nomenclature. *Ophelia* **48**, 1–24.

Okutani, T. (1968). Bathyal and abyssal Mollusca trawled from Sagami Bay and the south off Boso Peninsula by the R/V Soyo-Maru, 1965–1967. *Bulletin of the Tokai Regional Fisheries Research Laboratory* **56**, 7–56.

Okutani, T. (1974). Review and new records of abyssal and hadal molluscan fauna in Japanese and adjacent waters. *Venus (Japanese Journal of Malacology)* **33**, 23–39.

Oliver, P. G. (1992). "Bivalved seashells of the Red Sea". Verlag Christa Hemmen & National Museum of Wales, Weisbaden, Germany and Cardiff, Wales.

Olsson, A. A. (1961). "Mollusks of the Tropical Eastern Pacific: Panamic-Pacific Pelecypoda". Paleontological Research Institution, Ithaca, NY.

Orton, J. H. (1912). The mode of feeding of *Crepidula*, with an account of the current-producing mechanism in the mantle cavity, and some remarks on the mode of feeding in gastropods and lamellibranchs. *Journal of the Marine Biological Association of the United Kingdom* **9**, 444–478.

Orton, J. H. (1913). The ciliary mechanisms on the gill and the mode of feeding in *Amphioxus*, ascidians, and *Solenomya togata*. *Journal of the Marine Biological Association of the United Kingdom* **10**, 19–49.

Owen, G. (1956). Observations on the stomach and digestive diverticula of the Lamellibranchia II. The Nuculidae. *Quarterly Journal of Microscopical Science* **97**, 541–568.

Owen, G. (1959). The ligament and digestive system in the taxodont bivalves. *Proceedings of the Malacological Society of London* **33**, 215–223.

Owen, G. (1961). A note on the habits and nutrition of *Solemya parkinsoni* (Protobranchia: Bivalvia). *Quarterly Journal of Microscopical Science* **102**, 15–21.

Owen, G. (1973). The fine structure and histochemistry of the digestive diverticula of the protobranchiate bivalve *Nucula sulcata*. *Proceedings of the Royal Society of London Series B* **183**, 249–264.

Owen, G. (1978). Classification and bivalve gill. *Philosophical Transactions of the Royal Society of London* B **284**, 377–385.

Parulekar, A. H. (1984). Studies on growth and age of bivalves from temperate and tropical estuarine ecosystems. *Indian Journal of Marine Sciences* **13**, 193–195.

Pearse, J. S., McClintock, J. B. and Bosch, I. (1991). Reproduction of Antarctic benthic marine invertebrates: tempos, modes and timing. *American Zoologist* **31**, 65–80.

Peck, L. S. and Bullough, L. W. (1993). Growth and population structure in the infaunal bivalve *Yoldia eightsi* in relation to iceberg activity at Signy Island, Antarctica. *Marine Biology* **117**, 235–241.

Pelseneer, P. (1888). Report on the anatomy of the deep-sea Mollusca collected by H.M.S. Challenger in the years 1873–76. *Challenger Reports, Zoology*: *Anomura, Deep-Sea Mollusca,* Phoronis buskii *and Tunicata* **27**, 1–42 and 44 plates.

Pelseneer, P. (1889). Sur la classification phylogénetique des pélécypodes (communication preliminaire). *Bulletin de la Societe Scientifique de France et de Belgique* **20**, 27–52.

Phillips, N. W. (1984). The role of different microbes and substrates as potential suppliers of specific, essential nutrients to marine detritivores. *Bulletin of Marine Science* **35**, 283–298.

Pojeta, J., Jr. (1971). Review of Ordovician pelecypods. *United States Geological Survey Professional Papers* **695**, 1–46.

Pojeta, J., Jr. (1978). The origin and early taxonomic diversification of pelecypods. *Philosophical Transactions of the Royal Society of London* B **284**, 225–246.

Pojeta, J., Jr. (1988). The origin and Paleozoic diversification of solemyoid pelecypods. *New Mexico Bureau of Mines and Mineral Resources Memoir* **44**, 201–271.

Pojeta, J., Jr. and Runnegar, B. (1976). The paleontology of rostroconch mollusks and the early history of the phylum Mollusca. *United States Geological Survey Professional Papers* **968**, 84.

Pojeta, J., Jr. and Runnegar, B. (1985). The early evolution of diasome molluscs. *In* "Evolution, The Mollusca, Volume 10" (E. R. Trueman and M. R. Clarke, eds), pp. 295–336. Academic Press, New York, NY.

Pojeta, J., Jr., Runnegar, B. and Kriz, J. (1973). *Fordilla troyensis* Barrande: the oldest known pelecypod. *Science* **180**, 66–68.

Powell, A. W. B. (1951). Antarctic and subantarctic Mollusca: Pelecypoda and Gastropoda. *Discovery Reports* **26**, 47–196.

Powell, A. W. B. (1979). "New Zealand Mollusca: Marine, Land and Freshwater Shells". Collins, Auckland.

Powell, M. A. and Somero, G. N. (1985). Sulfide oxidation occurs in the animal tissue of the gutless clam, *Solemya reidi*. *Biological Bulletin, Marine Biological Laboratory, Woods Hole* **169**, 164–181.

Powell, M. A. and Somero, G. N. (1986). Hydrogen sulfide oxidation is coupled to oxidative phosphorylation in mitochondria of *Solemya reidi*. *Science* **233**, 563–566.

Prashad, B. (1932). The Lamellibranchia of the Siboga Expedition. Systematic part. II. Pelecypoda (exclusive of the Pectinidae). *In* "Siboga-Expeditie. Mollusca III. Lamellibranchia, Volume 53c", pp. 1–353. E. J. Brill Ltd, Leiden.

Purchon, R. D. (1956). The stomach in the Protobranchia and Septibranchia (Lamellibranchia). *Proceedings of the Zoological Society of London* **127**, 511–525.

Purchon, R. D. (1978). An analytical approach to a classification of the Bivalvia. *Philosophical Transactions of the Royal Society of London* B **284**, 425–436.

Purchon, R. D. (1987). The stomach in the Bivalvia. *Philosophical Transactions of the Royal Society of London* B **316**, 183–276.

Rachor, E. (1976). Structure, dynamics and productivity of a population of *Nucula nitidosa* (Bivalvia, Protobranchia) in the German Bight. *Bericht der deutschen wissenschaftlichen Kommission für Meeresforschung* **24**, 296–331.

Rainer, S. F. and Wadley, V. A. (1991). Abundance, growth and production of the bivalve *Solemya* sp., a food source for juvenile rock lobsters in a seagrass community in Western Australia. *Journal of Experimental Marine Biology and Ecology* **152**, 201–223.

Reid, R. G. B. (1980). Aspects of the biology of a gutless species of *Solemya* (Bivalvia: Protobranchia). *Canadian Journal of Zoology* **58**, 386–393.

Reid, R. G. B. (1990). Evolutionary implications of sulphide-oxidizing symbioses in bivalves. *In* "The Bivalvia: Proceedings of a Memorial Symposium in Honour of Sir Charles Maurice Yonge (1899–1986), Edinburgh, 1986" (B. J. Morton, ed.), pp. 127–140. Hong Kong University Press, Hong Kong.

Reid, R. G. B. and Bernard, F. R. (1980). Gutless bivalves. *Science* **208**, 609–610.

Reid, R. G. B. and Brand, D. G. (1986). Sulfide-oxidizing symbiosis in lucinaceans: implications for bivalve evolution. *The Veliger* **29**, 3–24.

Reid, R. G. B. and Brand, D. G. (1987). Observations on Australian Solemyidae. *Journal of the Malacological Society of Australia* **8**, 41–50.

Rex, M. A., Stuart, C. T., Hessler, R. R., Allen, J. A., Sanders, H. L. and Wilson, G. D. F. (1993). Global-scale latitudinal patterns of species diversity in the deep-sea benthos. *Nature* **365**, 636–639.

Rex, M. A., Etter, R. J. and Stuart, C. T. (1997). Large-scale patterns of species diversity in the deep-sea benthos. *In* "Marine biodiversity: patterns and processes" (R. F. G. Ormond, J. D. Gage and M. V. Angel, eds), pp. 94–121. Cambridge University Press, Cambridge.

Rhind, P. M. and Allen, J. A. (1992). Studies on the deep-sea Protobranchia (Bivalvia): the family Nuculidae. *Bulletin of the British Museum of Natural History (Zoology)* **58**, 61–93.

Rhoads, D. C. (1963). Rates of sediment reworking by *Yoldia limatula* in Buzzards Bay, Massachusetts, and Long Island Sound. *Journal of Sedimentary Petrology* **33**, 723–727.

Rhoads, D. C. (1967). Biogenic reworking of intertidal and subtidal sediments in Barnstable Harbor and Buzzards Bay, Massachusetts. *Journal of Geology* **75**, 461–476.

Rhoads, D. C. (1973). The influence of deposit-feeding benthos on water turbidity and nutrient recycling. *American Journal of Science* **273**, 1–22.

Rhoads, D. C. (1974). Organism–sediment relations on the muddy sea floor. *Oceanography and Marine Biology, an Annual Review* **12**, 263–300.

Rhoads, D. C. and Pannella, G. (1970). The use of molluscan growth patterns in ecology and paleoecology. *Lethaia* **3**, 143–161.

Rhoads, D. C. and Stanley, D. J. (1965). Biogenic graded bedding. *Journal of Sedimentary Petrology* **35**, 956–963.

Rhoads, D. C. and Young, D. K. (1970). The influence of deposit-feeding organisms on sediment stability and community structure. *Journal of Marine Research* **28**, 150–178.

Ridewood, W. G. (1903). On the structure of the gills of the Lamellibranchia. *Philosophical Transactions of the Royal Society of London* B **195**, 147–284.

Rokop, F. J. (1974). Reproductive patterns in the deep-sea benthos. *Science, NY* **186**, 743–745.

Rokop, F. J. (1979). Year round reproduction in deep-sea bivalve molluscs. *In* "Reproductive Ecology of Marine Invertebrates" (S. E. Stancyk, ed.), pp. 189–198. University of South Carolina Press, Columbia, SC.

Roy, K., Jablonski, D. and Valentine, J. W. (2000). Dissecting latitudinal diversity gradients: functional groups and clades of marine bivalves. *Proceedings of the Royal Society of London* B **267**, 293–299.

Runnegar, B. and Pojeta, J., Jr. (1985). Origin and diversification of the Mollusca. *In* "Evolution, The Mollusca, Volume 10" (E. R. Trueman and M. R. Clarke, eds), pp. 1–57. Academic Press, New York, NY.

Runnegar, B. and Pojeta, J., Jr. (1992). The earliest bivalves and their Ordovician descendants. *American Malacological Bulletin* **9**, 117–122.

Sabelli, B., Giannuzzi-Savelli, R. and Bedulli, D. (1992). "Catalogo Annotato dei Molluschi Marini del Mediterraneo", 3 volumes, Societa Italiana di Malacologia. Libreria Naturalistica Bolognese, Bologna.

Salvini-Plawen, L. v. and Steiner, G. (1996). Synapomorphies and plesiomorphies in higher classification of Mollusca. *In* "Origin and Evolutionary Radiation of the Mollusca" (J. D. Taylor, ed.), pp. 29–51. Oxford University Press, Oxford.

Sanders, H. L. (1956). Oceanography of Long Island Sound, 1952–1954. X. The biology of marine bottom communities. *Bulletin of the Bingham Oceanographic Collection* **15**, 345–414.

Sanders, H. L. (1958). Benthic studies in Buzzards Bay I. Animal–sediment relationships. *Limnology and Oceanography* **3**, 245–258.

Sanders, H. L. (1960). Benthic studies in Buzzards Bay III. The structure of the soft-bottom community. *Limnology and Oceanography* **5**, 138–153.

Sanders, H. L. and Allen, J. A. (1973). Studies on deep-sea Protobranchia (Bivalvia); prologue and the Pristiglomidae. *Bulletin of the Museum of Comparative Zoology, Harvard University* **145**, 237–262.

Sanders, H. L. and Allen, J. A. (1977). Studies on the deep sea Protobranchia (Bivalvia); the family Tindariidae and the genus *Pseudotindaria*. *Bulletin of the Museum of Comparative Zoology, Harvard University* **148**, 23–59.

Sanders, H. L. and Allen, J. A. (1985). Studies on deep-sea Protobranchia (Bivalvia): the family Malletiidae. *Bulletin of the British Museum of Natural History (Zoology)* **49**, 195–238.

Sanders, H. L. and Grassle, J. F. (1971). The interactions of diversity, distribution and mode of reproduction among major groupings of the deep-sea benthos. *In* "Proceedings of the Joint Oceanographic Assembly" (M. Uda, ed.), pp. 260–262. Japan Society for the Promotion of Science, Tokyo.

Sanders, H. L. and Hessler, R. R. (1969). Diversity and composition of abyssal benthos. *Science* **166**, 1034.

Sanders, H. L., Hessler, R. R. and Hampson, G. R. (1965). An introduction to the study of deep-sea benthic faunal assemblages along the Gay Head–Bermuda transect. *Deep-Sea Research* **12**, 845–867.

Sanders, N., Childress, J. J. and McMahon, B. R. (1998). Oxygen transport by the hemocyanin of the protobranch mollusc *Solemya reidi. Marine Biology* **131**, 293–299.

Scarlato, O. A. and Starobogatov, Y. I. (1985). General evolutionary patterns and the system of the class Bivalvia. *Special Occasional Publications, Department of Mollusks, Harvard University* **5**, 1–67.

Schaefer, K. (2000). The adoral sense organ in protobranch bivalves (Mollusca): comparative fine structure with special reference to *Nucula nucleus. Invertebrate Biology* **119**, 188–214.

Scheltema, R. S. (1972). Reproduction and dispersal of bottom dwelling deep-sea invertebrates: a speculative summary. *In* "Barobiology and the Experimental Biology of the Deep Sea" (R. W. Brauer, ed.), pp. 58–66. N. Carolina Sea Grant Program, University of N. Carolina, Chapel Hill, NC.

Schenck, H. G. (1934). Classification of nuculid pelecypods. *Bulletin of the Royal Museum of Natural History, Belgium* **20**, 1–74.

Schenck, H. G. (1936). Nuculid bivalves of the genus *Acila. Geological Society of America Special Papers* **4**, 1–149.

Self, R. F. L. and Jumars, P. A. (1988). Cross-phyletic patterns of particle selection by deposit feeders. *Journal of Marine Research* **46**, 119–143.

Shull, D. H. (2001). Transition-matrix model of bioturbation and radionuclide diagenesis. *Limnology and Oceanography* **46**, 905–916.

Sibuet, M. and Olu, K. (1998). Biogeography, biodiversity and fluid dependence of deep-sea cold-seep communities at active and passive margins. *Deep-Sea Research* Part II **45**, 517–567.

Sierra, A., Garcia, L. and Lloris, D. (1978). Trofismo y competencia alimentaria en asteroideos de la bahía de Almería. *Investigacion Pesquera (Barcelona)* **42**, 485–499.

Skalkin, V. A. and Tabunkov, V. D. (1969). *Nuculana pernula* (Müller) (Taxodonta, Nuculanidae) of the south-east coast of Sakhalin: its biology, distribution and stock. *Zoologicheskii Zhurnal* **48**, 1147–1155.

Smith, R. I., ed. (1964). Keys to Marine Invertebrates of the Woods Hole Region. Spaulding Co., Boston, MA.

Stanley, S. M. (1970). Relation of shell form to life habits of the Bivalvia (Mollusca). *Geological Society of America Memoirs* **125**, 296.

Starobogatov, Y. I. (1992). Morphological basis for phylogeny and classification of Bivalvia. *Ruthenica* **2**, 1–25.

Stasek, C. R. (1961). The ciliation and function of the labial palps of *Acila castrensis* (Protobranchia, Nuculidae), with an evaluation of the role of the protobranch organs of feeding in the evolution of the Bivalvia. *Proceedings of the Zoological Society of London* **137**, 511–538.

Stasek, C. R. (1965). Feeding and particle-sorting in *Yoldia ensifera* (Bivalvia: Protobranchia), with notes on other nuculanids. *Malacologia* **2**, 349–366.

Stempell, W. (1898). Beiträge zur Kenntniss der Nuculiden. *Zoologische Jahrbücher* Supplement **4**, 339–430.

Stenta, M. (1909). Über ein neues mantelrandorgan bei *Leda commutata*. *Zoologische Jahrbücher* **35**, 154–157.

Stoll, E. (1939). *Malletia gigantea* (Smith) ein beitrag zur Kenntnis der protobranchiaten Muscheln. *Acta Zoologica* **20**, 339–438.

Suess, E., Carson, B., Ritger, S. D., Moore, J. C., Jones, M. L., Kulm, L. D. and Cochrane, G. R. (1985). Biological communities at vent sites along the subduction zone off Oregon. *Bulletin of the Biological Society of Washington* **6**, 475–484.

Sun, M.-Y., Aller, R. C., Lee, C. and Wakeham, S. G. (1999). Enhanced degradation of algal lipids by benthic macrofaunal activity: effect of *Yoldia limatula*. *Journal of Marine Research* **57**, 775–804.

Tabita, F. R. (1988). Molecular and cellular regulation of autotrophic carbon dioxide fixation in microorganisms. *Microbiological Reviews* **52**, 155–189.

Taylor, A. C., Davenport, J. and Allen, J. A. (1995). Anoxic survival, oxygen consumption and haemocyanin characteristics in the protobranch bivalve *Nucula sulcata* Bronn. *Comparative Biochemistry and Physiology* **112A**, 333–338.

Taylor, J. D. (1973). The structural evolution of the bivalve shell. *Paleontology* **16**, 519–534.

Taylor, J. D., Kennedy, W. J. and Hall, A. (1969). The shell structure and mineralogy of the Bivalvia. Introduction. Nuculacea – Trigonacea. *Bulletin of the British Museum of Natural History (Zoology)* Supplement **3**, 1–125 and 129 plates.

Tebble, N. (1976). "British Bivalve Seashells: a Handbook for Identification", 2nd edition. Royal Scottish Museum, Edinburgh.

Terwilliger, N. B., Terwilliger, R. C., Meyhöfer, E. and Morse, M. P. (1988). Bivalve hemocyanins – a comparison with other molluscan hemocyanins. *Comparative Biochemistry and Physiology* **89B**, 189–195.

Thiriot-Quievreux, C., Soyer, J. and Bouvy, M. (1987). Etude des chromosomes du bivalve protobranche *Malletia sabrina* Hedley, 1916. *Vie et Milieu* **37**, 175–180.

Thiriot-Quievreux, C., Albert, P. and Soyer, J. (1991). Karyotypes of five subantarctic bivalve species. *Journal of Molluscan Studies* **57**, 59–70.

Thrush, S. F., Hewitt, J. E. and Pridmore, R. D. (1989). Patterns in the spatial arrangements of polychaetes and bivalves in intertidal sandflats. *Marine Biology* **102**, 529–535.

Thrush, S. F., Hewitt, J. E., Pridmore, R. D. and Cummings, V. J. (1996). Adult/juvenile interactions of infaunal bivalves: contrasting outcomes in different habitats. *Marine Ecology Progress Series* **132**, 83–92.

Toulmond, A., Jouin, C. and de Frescheville, J. (1987). Hemocyanin of the protobranch bivalve mollusc *Nucula hanleyi* Winckworth. *Comparative Biochemistry and Physiology* **88B**, 71–74.

Trevallion, A. C. (1965). A study of detritus feeding bivalve molluscs and an investigation on detritus. PhD thesis. University of Southampton, Southampton.

Trueman, E. R. (1952). Observations on the ligament of *Nucula*. *Proceedings of the Malacological Society of London* **29**, 201–205.

Turekian, K. K., Cochran, J. K., Kharkar, D. P., Cerrato, R. M., Vaisnys, J. R., Sanders, H. L., Grassle, J. F. and Allen, J. A. (1975). Slow growth of a deep-sea clam determined by ^{228}Ra chronology. *Proceedings of the National Academy of Sciences USA* **72**, 2829–2832.

Tyler, A. V. (1972). Food resource division among northern, marine, demersal fishes. *Journal of the Fisheries Research Board of Canada* **29**, 997–1003.

Tyler, P. A., Harvey, R., Giles, L. A. and Gage, J. D. (1992). Reproductive strategies and diet in deep-sea nuculanid protobranchs (Bivalvia: Nuculoidea) from the Rockall Trough. *Marine Biology* **114**, 571–580.

Tyler, P. A., Campos-Creasey, L. S. and Giles, L. A. (1994). Environmental control of quasi-continuous and seasonal reproduction in deep-sea benthic invertebrates. *In* "Reproduction, Larval Biology, and Recruitment of the Deep-sea Benthos" (C. M. Young and K. J. Eckelbarger, eds), pp. 158–178. Columbia University Press, New York, NY.

Valderhaug, V. A. and Gray, J. S. (1984). Stable macrofauna community structure despite fluctuating food supply in subtidal soft sediments of Oslofjord, Norway. *Marine Biology* **82**, 307–322.

Van Holde, K. E., Miller, K. I. and Lang, W. H. (1992). Molluscan hemocyanins: structure and function. *In* "Advances in Comparative and Environmental Physiology. Blood and Tissue Oxygen Carriers. Volume 13" (C. P. Mangum, ed.), pp. 257–300. Springer-Verlag, Berlin.

Vetter, R. D. (1991). Symbiosis and the evolution of novel trophic strategies: thiotrophic organisms at hydrothermal vents. *In* "Symbiosis as a Source of Evolutionary Innovation: Speciation and Morphogenesis" (L. Margulis and R. Fester, eds), pp. 219–245. MIT Press, Cambridge, MA.

Vinogradova, N. G. (1997). Zoogeography of the abyssal and hadal zones. *Advances in Marine Biology* **32**, 325–387.

Vlés, F. (1905). Sur un nouvel organe sensitif de *Nucula nucleus* L. *Bulletin de la Société Zoologique de France* **30**, 88–90.

Vokes, H. E. (1956). Notes on the Nucinellidae (Pelecypoda) with the description of a new species from the Eocene of Oregon. *Journal of Paleontology* **30**, 652–671.

Waller, T. R. (1990). The evolution of ligament systems in the Bivalvia. *In* "The Bivalvia – Proceedings of a Memorial Symposium in Honour of Sir Charles Maurice Yonge, Edinburgh, 1986" (B. Morton, ed.), pp. 49–71. Hong Kong University Press, Hong Kong.

Waller, T. R. (1998). Origin of the molluscan class Bivalvia and a phylogeny of major groups. *In* "Bivalves: An Eon of Evolution – Paleobiological Studies Honoring Norman D. Newell" (P. A. J. J. W. Haggart, ed.), pp. 1–45. University of Calgary Press, Calgary.

Watzin, M. C. (1986). Larval settlement into marine soft-sediment systems: interactions with the meiofauna. *Journal of Experimental Marine Biology and Ecology* **98**, 65–113.

Webb, C. M. (1987). Post-larval development of the bivalves *Nucula turgida*, *Venus striatula*, *Spisula subtruncata* and *S. elliptica* (Mollusca: Bivalvia) (with reference to the late larva). *Journal of the Marine Biological Association of the United Kingdom* **67**, 441–459.

Wilson, J. G. (1981). Temperature tolerance of circatidal bivalves in relation to their distribution. *Journal of Thermal Biology* **6**, 279–286.

Wilson, J. G. (1988). Resource partitioning and predation as a limit to size in *Nucula turgida* (Leckenby & Marshall). *Functional Ecology* **2**, 63–66.

Wilson, J. G. (1992). Age-specific energetics of reproduction in *Nucula turgida* (Leckenby and Marshall) a bivalve with lecithotrophic larval development. *Invertebrate Reproduction and Development* **22**, 275–279.

Wilson, J. G. and Davis, J. P. (1984). The effect of environmental variables on the oxygen consumption of the protobranch bivalve *Nucula turgida* (Leckenby and Marshall). *Journal of Molluscan Studies* **50**, 73–77.

Wilson, J. G. and Shelley, C. (1986). The distribution of *Nucula turgida* (Bivalvia: Protobranchia) from Dublin Bay, Ireland, and the effect of sediment organic content. *Journal of the Marine Biological Association of the United Kingdom* **66**, 119–130.

Xu, F. (1990). The Bivalvia in the deep-water area of the East China Sea. *Studia Marina Sinica* **31**, 185–193.

Yingst, J. Y. and Rhoads, D. C. (1980). The role of bioturbation in the enhancement of bacterial growth rates in marine sediments. *In* "Marine Benthic Dynamics" (K. R. Tenore and B. C. Coull, eds), pp. 407–421. University of South Carolina Press, Columbia, SC.

Yonge, C. M. (1939). The protobranchiate Mollusca; a functional interpretation of their structure and evolution. *Philosophical Transactions of the Royal Society of London* B **230**, 79–147.

Yonge, C. M. (1959). The status of the Protobranchia in the bivalve Mollusca. *Proceedings of the Malacological Society of London* **33**, 210–214.

Zardus, J. D. (1998). The larval development and autecology of the protobranch bivalve *Acila castrensis*. PhD thesis. Northeastern University, Boston, MA.

Zardus, J. D. (in preparation). Community assembly in a soft-sediment habitat dominated by a deposit-feeding bivalve.

Zardus, J. D. and Martel, A. L. (2001). Phylum Mollusca: Bivalvia. *In* "Atlas of Marine Invertebrate Larvae" (C. M. Young, M. A. Sewell and M. E. Rice, eds), pp. 289–325. Academic Press, London.

Zardus, J. D. and Morse, M. P. (1998). Embryogenesis, morphology and ultrastructure of the pericalymma larva of *Acila castrensis* (Bivalvia: Protobranchia: Nuculoida). *Invertebrate Biology* **117**, 221–244.

Zardus, J. D., Chase, M. R., Etter, R. J. and Rex, M. A. (in preparation). Bathymetric and geographic population structure in *Deminucula atacellana*, a widely distributed deep-sea bivalve.

Shelled Opisthobranchs

Paula M. Mikkelsen

Division of Invertebrate Zoology, American Museum of Natural History, Central Park West at 79th Street, New York, NY 10024-5192, USA
e-mail: mikkel@amnh.org

ADVANCES IN MARINE BIOLOGY VOL. 42
ISBN 0-12-026142-1

In his contributions to the monographic series "Manual of Conchology", Henry Pilsbry reviewed the subgroup Tectibranchiata, comprising those opisthobranch snails that (at least primitively) still possess a shell (Pilsbry, 1894–1896). Exemplified by the Cephalaspidea (bubble shells), others included in this group at Pilsbry's time and since were Anaspidea (sea hares) and the shelled members of Notaspidea (side-gilled slugs) and Sacoglossa (leaf slugs). Pilsbry (and others since his time) considered tectibranchs to be the "root stock" from which more advanced gastropods such as Nudibranchia and Pulmonata were derived. Tectibranch systematics is firmly based on conchology and most species were originally described from empty shells. However, soft-anatomical characters were acknowledged quite early on as equally important in tectibranchs, due to the reduction of their shells and their evolutionary proximity to unshelled gastropods. Today, Tectibranchiata is not recognized as a natural taxon although the word "tectibranch" (like "prosobranch" and "mesogastropod") continues in vernacular use. Shelled opisthobranchs have been redistributed among various taxa, including several new ones – the unresolved basal opisthobranchs (Architectibranchia) and the "lower Heterobranchia", an enigmatic and currently much-studied group of families considered basal to all of Euthyneura (Opisthobranchia and landsnails (Pulmonata)). Despite their polyphyletic status, shelled opisthobranchs remain important subjects in evolutionary studies of gastropods – as the most basal members of nearly every opisthobranch clade and as organisms with mosaic combinations of primitive and derived features within evolutionary "trends" (e.g., loss of the shell, detorsion, concentration of the nervous system, ecological specialization, etc.). Although they play a pivotal role, the shelled opisthobranchs have received minimal attention in more comprehensive gastropod studies, often relegated to token representatives at the derived end of prosobranchs or at the basal end of nudibranchs. The choice of this representative in a larger study is critical if its morphology and/or molecules are to adequately exemplify a larger group.

This review explores the shelled opisthobranchs, including their history, current status and presumed synapomorphies, and emphasizes the importance of anatomical data to our current understanding of these "transitional" forms. A synthetic phylogenetic analysis, based on a combination of characters used in four published phylogenies involving tectibranchs, shows the current state of our knowledge and emphasizes areas for future study. The results indicate that Opisthobranchia, Cephalaspidea and Sacoglossa are monophyletic taxa, and that Acteon, *the traditional basal opisthobranch, is convincingly a lower heterobranch. In most of the resulting cladograms, Anaspidea formed a monophyletic group with Cephalaspidea, as did pleurobranchoidean Notaspidea with Nudibranchia (the latter recently named as Nudipleura Wägele and Willan, 2000).*

1. INTRODUCTION

> *In the present state of conchology, all systematic work on Tectibranchs is of a tentative and superficial character . . . it remains for those who have opportunity to observe living examples to properly classify many of the species.* (Pilsbry, 1895 (in 1894–1896):242)

The implementation of modern systematic techniques during the past 20 years, including revitalized anatomical studies and refined molecular techniques employed with now-requisite phylogenetic analyses, have greatly improved our understanding of mollusks as functioning, evolving organisms. Whole animal studies of morphology, behavior, diet, population ecology, etc. have focused attention away from the charismatic external shells that have kept various taxonomic groups united for over a century. The results have suggested that our dogmatic shell-based beliefs about relationships could be flawed and deserve closer scrutiny.

Prosobranch gastropods have received the bulk of phylogenetic attention in recent decades (Ponder, 1988; Bieler, 1992; Ponder and Lindberg, 1997), leading us from the simple triad of Archaeo-, Meso- and Neogastropoda to a still-fluid complex of five to eight major clades, some of which are still unnamed. Prosobranchia itself is now considered paraphyletic and is no longer accepted as a valid taxon, although the term "prosobranch" is still widely used.

On the more derived side of Gastropoda is Heterobranchia, the name now applied to Euthyneura (= Opisthobranchia (sea slugs) plus Pulmonata (land snails)) plus an unresolved, presumably basal group of families collectively called "lower heterobranchs" (or less frequently, Heterostropha or Allogastropoda). Lower Heterobranchs (including the families Rissoellidae, Omalogyridae, Cimidae, Xylodisculidae, Cornirostridae, Hyalogyrinidae, Architectonicidae, Mathildidae, Omalaxidae, Pyramidellidae, Amathinidae, Ebalidae, Tjaernoeidae, Valvatidae, Glacidorbidae and Orbitestellidae; Warén *et al.*, 1993; Millard, 1997) are currently receiving a great deal of attention from systematic malacologists, especially with regard to minute species with nondescript shells but interesting anatomy (e.g., Warén, 1994; Bieler *et al.*, 1998). Heterobranchia as a whole has received much less revisionary attention than Prosobranchia, with most recent analyses focusing within the larger clades, i.e., among Opisthobranchia (e.g., Thollesson, 1998; Wägele, 1998) or among Pulmonata (e.g., Emberton *et al.*, 1990; Emberton, 1991; Wade *et al.*, 2000).

Study of the overlap of Prosobranchia and Opisthobranchia has been minimal, in part because the derived nature of heterobranchs versus the more plesiomorphic shelled snails makes the comparison and coding of morphological characters especially difficult. Typically, whenever authors have attempted to include both groups, only a few token representatives serve as cladistic outgroups or exemplars of the "other clade". For example, Ponder and Lindberg's (1997) gastropod phylogeny treated the morphology of 32 higher-level prosobranch groups with two species of opisthobranchs. On a smaller scale, Haszprunar (1986) examined 49 genus-level taxa of gastropods for 12 fine-structural characters of the osphradium, a chemosensory organ; of these taxa, two were lower heterobranchs and another two were opisthobranchs. From the heterobranch perspective, Wägele (1998) investigated a number of specialized organs in 32 species of sea slug opisthobranchs (Nudibranchia) analyzed with four exemplars from other opisthobranch groups. Wägele and Willan's (2000) nudibranch phylogeny used 28 taxa of Nudibranchia and two shelled notaspideans.

Two recent molecular studies have investigated phylogenetic relationships within Gastropoda (Yoon and Kim, 2000; Dayrat *et al.,* 2001). Although both professed to emphasize Euthyneura, one (Yoon and Kim, 2000) included only two euthyneuran taxa that were not pulmonates. The other study (Dayrat *et al.,* 2001) was far more balanced, with one prosobranch as outgroup, 17 opisthobranchs and 14 pulmonates.

In all of the previous morphological examples, plus Yoon and Kim's (2000) molecular study, the representative "other taxa" belong to Tectibranchiata Cuvier, 1812, also known as Tectibranchia or shelled opisthobranchs. In simplified terms, tectibranchs are sea slugs (by virtue of their anatomical organization) that also bear a shell. This brief definition implies (or in many cases states outright) that tectibranchs "still" bear a shell, in the presumed evolutionary trend from shelled prosobranch to unshelled sea slug. Pilsbry (1895 in 1894–1896:134) exemplified this approach in characterizing Tectibranchia as the "synthetic type, from which in the remote past, the Nudibranchiata and Pulmonata have no doubt been derived". Well into the twentieth century, tectibranchs continued to be depicted as the rootstock of evolutionary trees (Boettger, 1955: figure 1; Morton, 1963: figure 2; Schmekel, 1985: figure 17; Clark *et al.*, 1990: figure 1; Poulicek *et al.*, 1991: figure 1).

A number of tectibranchs (used here in the same vernacular sense as "prosobranchs") are extremely well studied and reasonably well recognized in invertebrate zoology (Figure 1). The taxon Tectibranchia, however, has not been considered monophyletic or a natural grouping since the 1960s, although it or its vernacular form has been used in print as late as the 1990s as well as on a wide variety of current World-Wide Web sites (e.g., Schiøtte, 1992; World-Wide Web sites by Bibliomania, Marine Life Information Network for Britain and Ireland, British Marine Life Study Society, International Council for Exploration of the

Sea). Understandably, yet unfortunately, there are few active systematists specializing in this no-longer-recognized group. In addition, gastropodologists have been (and still are) traditionally polarized, focusing either on shelled gastropods (i.e., prosobranchs) or on unshelled sea slugs (usually nudibranchs), but rarely on both. To some extent, these specialties have biased their approaches toward either the shell or the soft body. For example, at the turn of the previous century, Henry Pilsbry and William Healey Dall, American malacologists who published on a wide variety of shelled animals, described many tectibranch species largely on the basis of shell characters. During the same period, Rudolf Bergh and Jules Guiart (German and French, respectively, both opisthobranch specialists) provided scores of pages of detailed anatomy for many of the same taxa.

Although phylogenetic methods using morphology were once considered problematic to impossible in opisthobranch studies (citing "rampant parallelism"; Gosliner and Ghiselin, 1984), successful cladistic analyses have recently been produced for most of the major clades (Notaspidea – Willan, 1987; Cephalaspidea – Mikkelsen, 1996; Sacoglossa – Jensen, 1996 and Mikkelsen, 1998a; Nudibranchia – Wägele and Willan, 2000). While there have been no attempts to analyze Opisthobranchia as a whole, each of these smaller analyses has prompted or confirmed changes to previous ideas through in-depth anatomical study of their focus group. One such change in tectibranch classification has taken a remarkably long time to become widely accepted. The transfer of externally shelled *Akera* from Cephalaspidea to Anaspidea (associated with internally shelled and unshelled forms, e.g., *Aplysia* and *Bursatella*, respectively) was first suggested by Guiart (1901), resurrected by Morton and Holme (1955), but was not fully accepted until after a series of confirming studies in the 1970s and 1980s (Morton, 1972; Thompson and Seaward, 1986, 1989). The most recent phylogenetic work on Anaspidea, a molecular investigation by Medina and Walsh (2000), retested this position yet again and confirmed *Akera*'s anaspidean nature for what one assumes is the last time. A similar shift, moving externally shelled *Volvatella, Ascobulla* and, most recently, *Cylindrobulla* to the Sacoglossa (associated with unshelled forms like *Elysia*; Marcus, 1982; Mikkelsen, 1996, 1998a) is finding acceptance more readily (e.g., Wägele and Willan, 2000).

While Tectibranchia is no longer recognized, tectibranch genera or species frequently take part in comparative gastropod studies and phylogenetic analyses. Which tectibranch is chosen as an outgroup or exemplar is critical because its morphology and/or molecules must exemplify the larger group. In light of today's revisions, it can be uncertain whether a chosen tectibranch will perform its task adequately or in the way intended by the author(s). For these reasons, it seems prudent to review the history

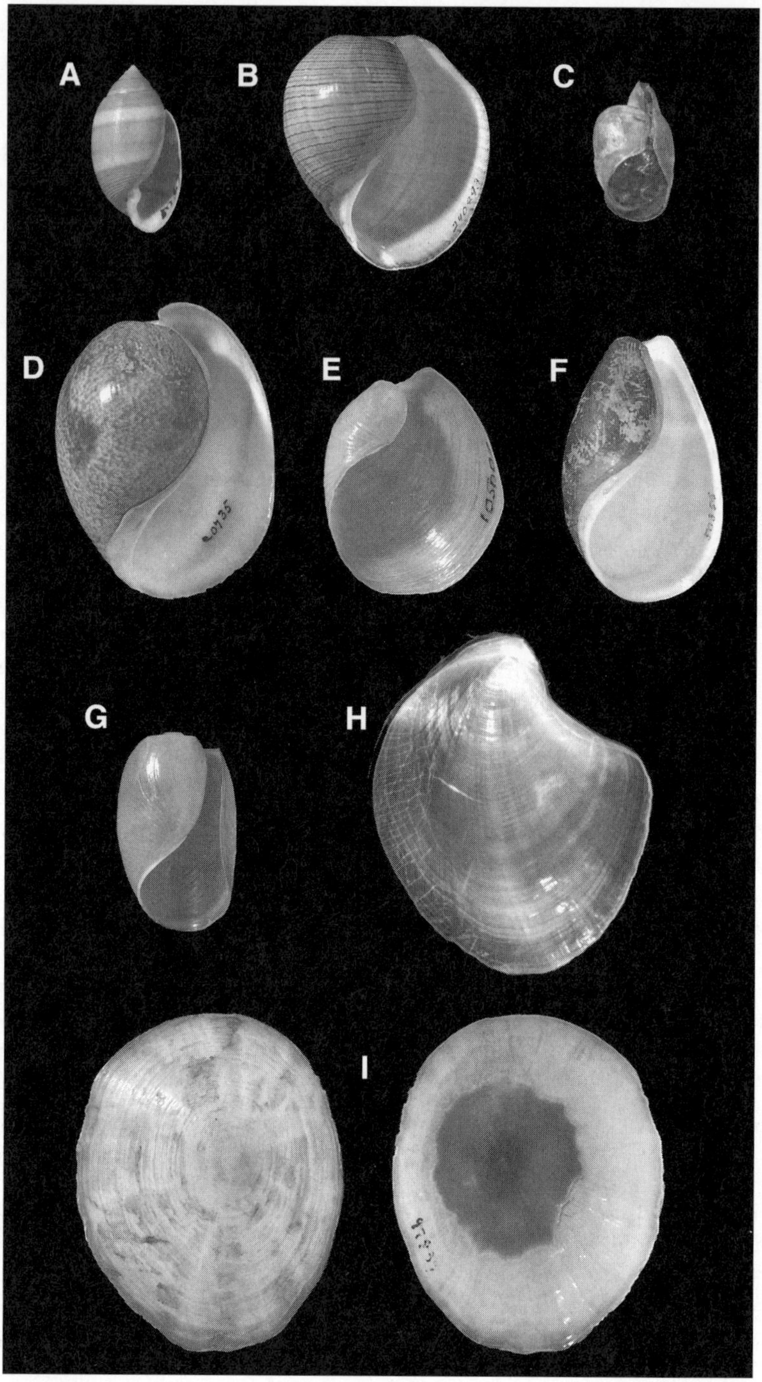
A
B
C
D
E
F
G
H
I

and current knowledge of shelled opisthobranchs, to assist future analysts in the choice of exemplars and to stimulate additional interest in this "borderline" area of malacology.

2. COMPONENT TAXA AND THEIR PRESUMED SYNAPOMORPHIES

By way of orientation, the component taxa in Tectibranchia and the characters defining them must be briefly defined. Starting at the highest level, opisthobranch gastropods are almost exclusively marine hermaphroditic snails; unlike prosobranchs, their nervous systems are at least partially euthyneurous (i.e., with an uncrossed visceral nerve loop) and the gill lies posterior to the heart (*opisthen* Gr. behind + *branchos* Gr. gill).

Tectibranchs (*tectus* L. cover) are opisthobranchs in which the gill is covered by a shell (as opposed to nudibranchs; *nudus* L. naked). This shell is heterostrophic, that is, the protoconch (larval shell) coils in the opposite direction to that of the adult shell (teleoconch). For the purposes of this discussion, the primary tectibranch taxa are Architectibranchia, Cephalaspidea, Anaspidea, Sacoglossa and Notaspidea. The taxonomic ranks of these taxa (whether subclass or order) are ill defined and inconsistent; these arbitrary ranks are immaterial for the purposes of this discussion and will not be used. The informal category of "shelled opisthobranchs" also technically includes the holoplanktonic Thecosomata or pteropods (e.g., *Cavolinia, Creseis*), but these are highly modified for a planktonic existence and are not discussed here with the otherwise benthic Tectibranchia. Two relatively small groups,

Figure 1 Tectibranch shells of some of the more widely familiar taxa. A. *Acteon tornatilis* (Linné, 1758) (Lower Heterobranchia; 20.9 mm, AMNH 50670, Great Britain). B. *Hydatina physis* (Linné, 1758) (Architectibranchia; 49.9 mm, AMNH 240893, Queensland, Australia). C. *Volvulella vigourouxi kawamurai* Habe, 1946 (Sacoglossa; 16.3 mm, AMNH 232659, Guam). D. *Bulla ampulla* Linné, 1758 (Cephalaspidea; 67.1 mm, AMNH 20735, Indian Ocean). E. *Philine aperta* (Linné, 1767) (Cephalaspidea; 37.6 mm, AMNH 105008, South Africa). F. *Scaphander lignarius* (Linné, 1758) (Cephalaspidea; 57.1 mm, AMNH 50356, Great Britain). G. *Akera soluta* (Gmelin, 1791) (Anaspidea; 31.1 mm, AMNH 135079, Zanzibar). H. *Aplysia dactylomela* Rang, 1828 (Anaspidea; shell 58.0 mm maximum diameter, from animal ca. 20 cm in length, FMNH 290367, Florida Keys). I. *Umbraculum umbraculum* (Lightfoot, 1786) (Notaspidea; dorsal and ventral views, 83.7 mm, AMNH 97837, India).

Acochlidiacea and Runcinacea, are poorly understood and of uncertain affinities to the remaining groups and will also not be covered.

In the sections that follow, tectibranch taxa (i.e., those with a shell present in the adult) will be emphasized over unshelled taxa in those groups that include both. Commonly encountered synonyms are provided (indicated by '+') to assist the nonspecialist. For each group, an exemplar genus has been selected that can adequately represent that group in a larger analysis. Anatomical drawings are provided for each of these genera, depicting diagrammatically the gross morphology and the arrangement of the major organ systems. More detailed discussion of the individual characters is provided in Section 3.

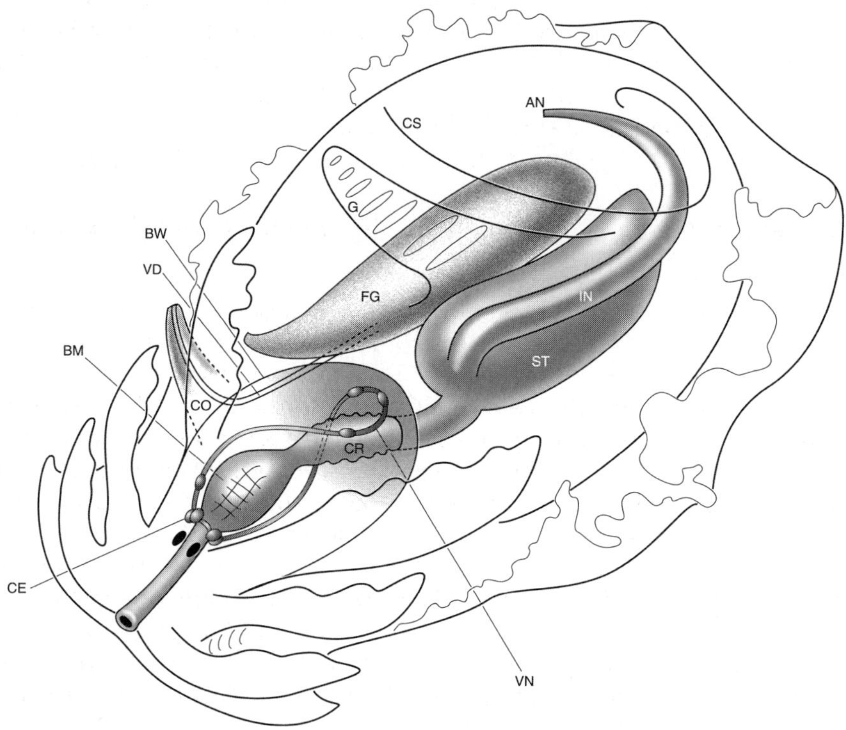

Figure 2 Diagrammatic anatomy of *Hydatina* (Architectibranchia). AN, anus; BM, buccal mass (containing radula); BW, wall of anterior body cavity; CE, circumesophageal nerve ring; CO, copulatory organ; CR, crop; CS, ciliated strips; EX, exterior ciliated groove; F, posterior limit of foot; FG, female gland mass; G, gill; GZ, gizzard; HO, Hancock's Organ; IN, intestine; OP, operculum; PF, posterior foot; ST, stomach; VD, vas deferens; VN, visceral nerve loop.

2.1. Architectibranchia (*Hydatina,* Figure 2)

Architectibranchia is probably paraphyletic, without known synapomorphies. As of this date, it serves as a "holding place" for basal opisthobranchs removed from other redefined clades. A number of gastropod plesiomorphies presently unite them: a streptoneurous nervous system, a gonoduct located within the mantle cavity, absence of a crushing gizzard in the esophagus and the absence of a lateral external ciliated groove for transport of eggs and/or sperm. Two trends have also characterized architectibranchs (that is, have presumably occurred within the confines of this taxon): reduction of the stomach and rotation of the mantle cavity opening from anterior to lateral (Haszprunar, 1985b; Salvini-Plawen, 1988, 1991a, b; Mikkelsen, 1996). The family Acteonidae, when not allocated to Cephalaspidea, is placed here by most authors (e.g., Gosliner, 1994; Rudman and Willan, 1998). However, accumulating evidence is strong to support its further removal into the "lower heterobranchs" (see Section 5). No comparative analyses have so far investigated the interrelationships of architectibranch taxa.

Synapomorphies: none currently identified.

Member taxa: Bullinidae, Aplustridae (+ Hydatinidae), Ringiculidae.

2.2. Cephalaspidea (+ Bullomorpha, Tectibranchia s.s.) (*Bulla* and *Cylichna,* Figures 3–4)

"Bubble snails" are the quintessential shelled opisthobranchs and Cephalaspidea has effectively replaced Tectibranchia in most non-cladistic evolutionary schemes. All members retain shells as adults and most are involute, that is, with the coiled, pointed spire of the shell covered by subsequent whorls. The original diagnostic character for the group, the shovel-shaped cephalic shield, with its sessile eyes and posterior tentacular flaps, is likely ecophenotypic for their characteristic burrowing lifestyle and is arguably too variable to code phylogenetically (Mikkelsen, 1996).

Synapomorphies (Mikkelsen, 1996): three hardened plates in an esophageal gizzard, ciliated strips (for water circulation in the mantle cavity) with a flexure near the lateral mantle opening, the ganglionic nerve ring located (secondarily) anterior to the buccal mass (or pharynx, thus "prepharyngeal") and the genital ganglion located on the visceral nerve loop (rather than on a separate nerve).

Member taxa: two superfamilies are recognized:

- Bulloidea (Figure 3), with Bullidae (+ Bullariidae) and Haminoeidae (+ Atyidae, Smaragdinellidae); characterized by exogyrous ciliated strips and an autosperm storage bulb at the end of the cephalic copulatory organ.

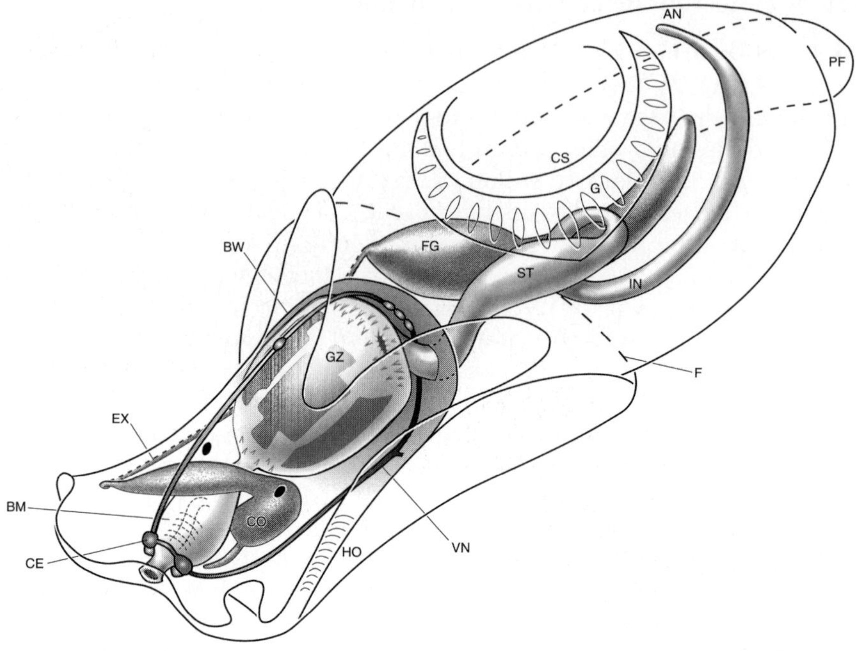

Figure 3 Diagrammatic anatomy of *Bulla* (Cephalaspidea: Bulloidea). Abbreviations as in Figure 2.

- Philinoidea (Figure 4), with Philinidae, Cylichnidae (+Scaphandridae, Acteocinidae), Retusidae (+Volvulidae), Gastropteridae and Aglajidae (+Dorididae); characterized by a unique radular bauplan (bilobed rachidian tooth, sickle-shaped lateral teeth), eyes that are deeply set and directed ventrolaterally and the absence (presumably loss) of spines accompanying the gizzard plates. Calcification of the gizzard plates occurs in some members of this clade. Shells are internal in members of a few families (Philinidae, Aglajidae, Gastropteridae). Aglajids are nearly bilaterally symmetrical, converging in external appearance with nudibranchs. Radular teeth are absent in retusids and all but one aglajid. Thollesson (1999) also found this group (represented by *Philine* and *Scaphander*) to be monophyletic by mitochondrial DNA (16S rRNA) analysis. Philinoidea (represented by *Philine, Scaphander* and *Cylichna*) was not, however, monophyletic in an analysis of 28S rRNA sequences by Dayrat *et al.* (2001).

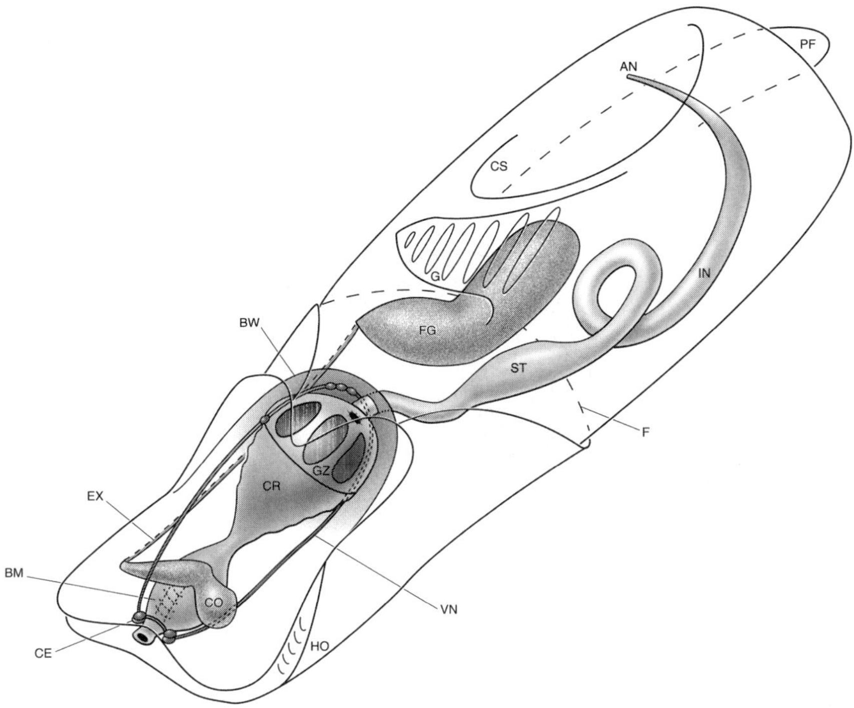

Figure 4 Diagrammatic anatomy of *Cylichna* (Cephalaspidea: Philinoidea). Abbreviations as in Figure 2.

2.3. Anaspidea (+ Aplysiomorpha) (*Akera*, Figure 5)

"Sea hares" are relatively large-bodied snails with wide, muscular parapodia that can be used for brief periods of swimming. The best-known member *Aplysia*, widely utilized as an invertebrate model in neurological or behavioral research (e.g., Wayne and Block, 1992; Brezina *et al.*, 1996), has an internalized, barely calcified, plate-like shell (Figure 1H); other members (e.g., *Bursatella, Dolabrifera*) lack shells as adults. The externally shelled *Akera* was the first of the traditional cephalaspids to be reassigned to another group as a result of holistic study of the animal's anatomy and behavior (Thompson and Seaward, 1989). Anaspidea formed a monophyletic group with the holoplanktonic pteropods (Gymnosomata and Thecosomata) in the 28S rRNA molecular phylogeny by Dayrat *et al.* (2001).

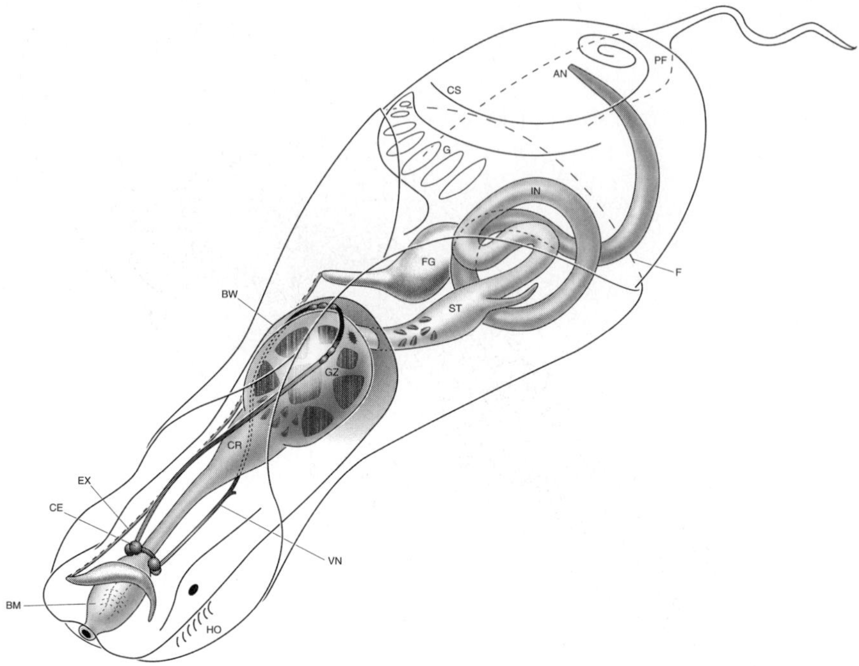

Figure 5 Diagrammatic anatomy of *Akera* (Anaspidea). Abbreviations as in Figure 2.

Synapomorphies (Mikkelsen, 1996): two unique characters of the digestive system (presence of a stomach caecum (also present in pteropods) and a filter chamber between the gizzard and stomach), supported by details of the nervous sytem. Features of the spermatozoa and of larval development are also unique. Characters given in support of Anaspidea by Schmekel (1985: figure 17) are primarily symplesiomorphies.

Member taxa: Akeridae (+Aceridae), Aplysiidae, Dolabriferidae, Notarchidae.

2.4. Sacoglossa (+ Ascoglossa) (*Cylindrobulla*, Figure 6)

"Leaf slugs" are suctorial feeders, often encountered on *Caulerpa* or other marine chlorophytes, and are often themselves green in color. Many species exhibit kleptoplasty, the retention of functional chloroplasts from

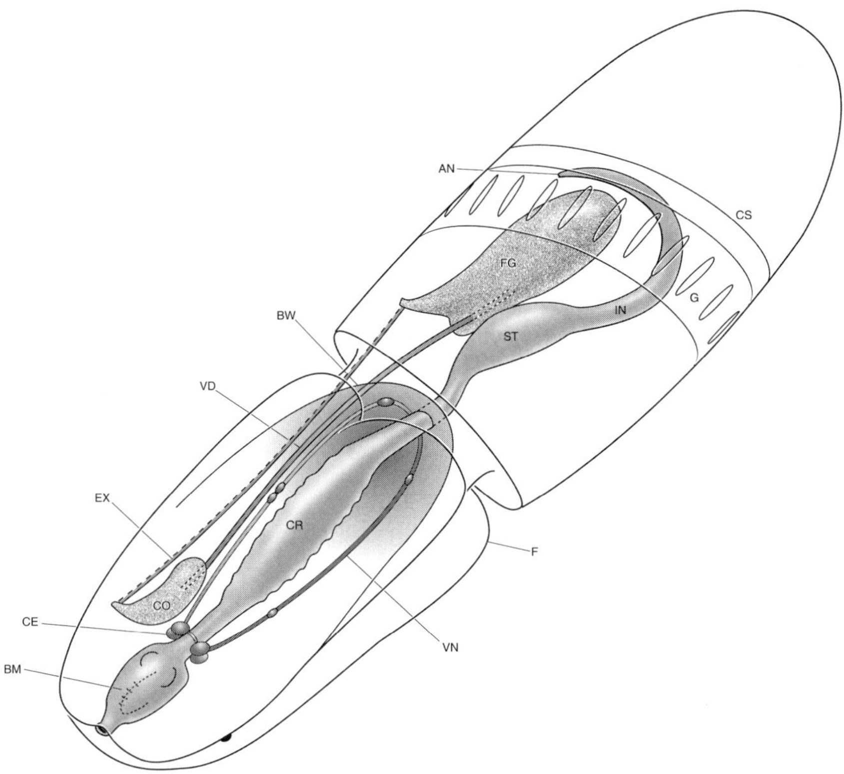

Figure 6 Diagrammatic anatomy of *Cylindrobulla* (Sacoglossa). Abbreviations as in Figure 2.

their prey, a feature considered plesiomorphic for Sacoglossa by Clark *et al.* (1990). Although long characterized by dagger-shaped radular teeth (used in piercing algal cells), that feature is no longer synapomorphic since *Cylindrobulla* (with quadrate teeth) was reallocated to Sacoglossa. The famous "bivalved gastropods" (Juliidae) are also classified here; these are a classic example of convergence in shell shape, so perfect that the first examples were originally described as clams (Kay, 1968). Shelled sacoglossans comprise a monophyletic group according to the most recent phylogenetic analyses based on morphology (Jensen, 1996; reanalyzed by Mikkelsen, 1998a) and molecules (Dayrat *et al.*, 2001).

Synapomorphies: a uniserial radula, an ascus, pharyngeal pouches (associated with suctorial feeding) and laterally directed eyes (Mikkelsen, 1996, 1998a).

Member taxa: two superfamilies are recognized:

- Oxynoidea, with Oxynoidae (+ Lobigeridae), Lobigeridae, Juliidae, Volvatellidae (+ Ascobullidae, Arthessidae) and Cylindrobullidae; characterized by a lamellate gill, a female genital papilla and shell adductor muscle(s) in some taxa.
- Plakobranchoidea (for included families, see Jensen, 1996) is characterized by the absence (loss) of the shell. For the purposes of this discussion, plakobranchoids are not considered tectibranchs.

2.5. Notaspidea (+ Umbraculomorpha, Pleurobranchomorpha) (*Tylodina*, Figure 7)

The "side-gilled sea slugs" form a relatively small taxonomic group, including both shelled and unshelled forms. In the most basal taxa, the shell is limpet-shaped and external. Other notaspids have an internal shell

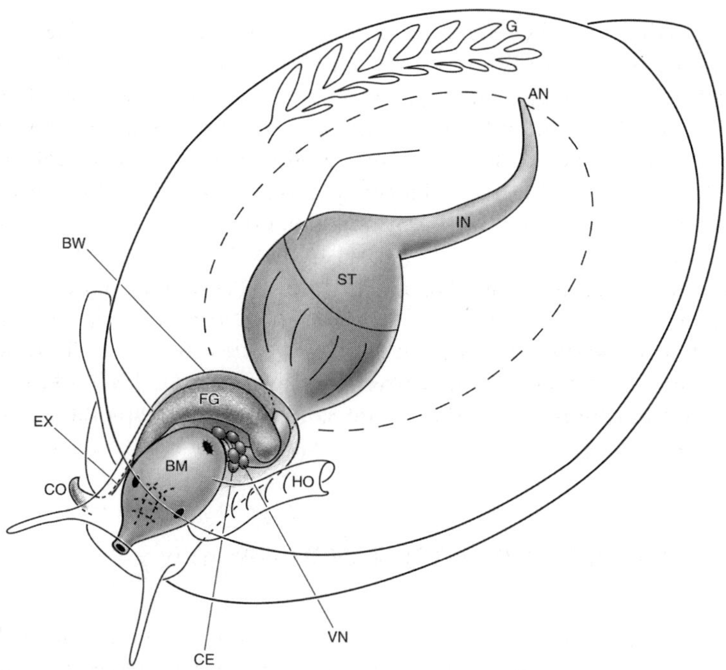

Figure 7 Diagrammatic anatomy of *Tylodina* (Notaspidea). Abbreviations as in Figure 2.

(usually ear-shaped, although that of *Tomthompsonia* is coiled) or lack the shell entirely. As in many tectibranchs, the soft bodies of notaspids far surpass their shells as sources of characters. This is particularly evident in the group's most flamboyant member, *Umbraculum*, whose large, yellow–orange, pustulose body contrasts with its plain, whitish, disk-shaped shell, prompting superlative-laden descriptions from conchologists about the living animal (Hartley, 1964; Tarr, 1989). The group is rather well studied anatomically and phylogenetically. It has long been believed to show close affinity to Nudibranchia (Guiart, 1901; Ghiselin, 1966; Gosliner, 1981; Wägele and Willan, 2000). Notaspidea did not form a monophyletic group in the 28S rRNA molecular phylogeny by Dayrat *et al.* (2001).

Synapomorphies (Willan, 1987): a bipinnate gill on the right side of the animal (not homologous with the ctenidium of prosobranchs nor with the plicatidium of cephalaspids and anaspids (Morton, 1972)), longitudinally slit rhinophores, a broad oral veil between the anterior tentacles and the anus located at the rear of the gill.

Member taxa: the four recognized families are organized into two superfamilies (elevated to ordinal level by Schmekel, 1985):

- Umbraculoidea (+Tylodinoidea), including Umbraculidae and Tylodinidae, with a limpet-like shell perched on top of the body and a cuticularized thickened ring just inside the mouth (Willan, 1987). Schmekel (1985) included fusion of the pleural, parietal and esophageal ganglia on each side (resulting in six ganglia in the circumesophageal ring, but only one on the visceral loop) as a synapomorphy of this group.
- Pleurobranchoidea, including Pleurobranchidae and Pleurobranchaeidae, the latter of which is entirely unshelled (non-tectibranch). Pleurobranchoidea have an internal or internalized shell, more-or-less rectangular in shape, rhinophores with fused bases, a pedal gland, a median buccal gland that presumably secretes an acidic fluid into the gut, and a retractible, ungrooved penis. Willan (1987) identified no synapomorphies for the shelled pleurobranchoideans (i.e., Pleurobranchidae).

3. CHARACTERS AND EVOLUTIONARY PATHWAYS

The literature is replete with reference to "evolutionary trends" in shelled opisthobranchs, within the context of the hypothetical transition from shelled prosobranch to unshelled nudibranch. The starting point for discussion of such trends is often a hypothetical ancestor, comprising the combination of all plesiomorphic character states. Graphically this combination has been presented as the Hypothetical Ancestral Opisthobranch (HAO) by Gosliner (1978: 124, reproduced only slightly modified in 1981: figure 4, and again in 1994: figure 2, as "general opisthobranch"). Although earlier authors (e.g., Fretter and Graham, 1954; Thompson, 1976) considered *Acteon* as the closest living "ancestral form", Gosliner's HAO is based more closely on the cephalaspid family Ringiculidae.

No attempt is made here to review fully the morphology of each organ system in each opisthobranch taxon. Attention is focused on those features of particular relevance to tectibranch morphology and systematics, and the character states are provided mainly as they exist in the exemplar taxa as basal and representative of each group. For more general information, and that on unshelled opisthobranchs, the reader is referred to the excellent reviews of opisthobranch morphology presented by Hoffmann (1932–1939), Gosliner (1994) and Rudman and Willan (1998), and to the detailed anatomical works of Eveline and Ernst Marcus (e.g., 1967a), Rudman (e.g., 1971, 1972a, b, c) and Wägele (e.g., 1998; also Wägele and Willan, 2000). In summary of this section, Table 1 presents characters and suggested character states for the selected exemplar taxa that are potentially useful in phylogenetic analysis. For completeness, this table also includes *Acteon*, as the traditionally most basal opisthobranch (Figure 8), and Gosliner's HAO.

3.1. Shell

The presence of an external shell has been the single most important character uniting the tectibranchs within Opisthobranchia. In acknowledgement of its "reduction" in various ways (i.e., size, thickness, sculpture; Mikkelsen, 1998b), the shell is now recognized as secondary to the underlying soft anatomy for supra-specific classification and phylogenetic studies. In part consequently, the tectibranchs are no longer considered a unified group, based on the presence of that shell.

Despite their diminished importance across higher taxonomic levels, tectibranch shell characters are still of considerable value at the species level. This is in part an historical artifact of original descriptions, which in malacology were routinely exclusively shell-based well into the 1960s; therefore, most tectibranch species names are inexorably linked to a usually dried, usually empty, type specimen shell. Critical re-examination of a taxonomic group can sometimes reveal additional shell characters that are distinct at levels not previously recognized. For example, morphology of the protoconch in *Acteocina* spp. (Cephalaspidea: Philinoidea) was

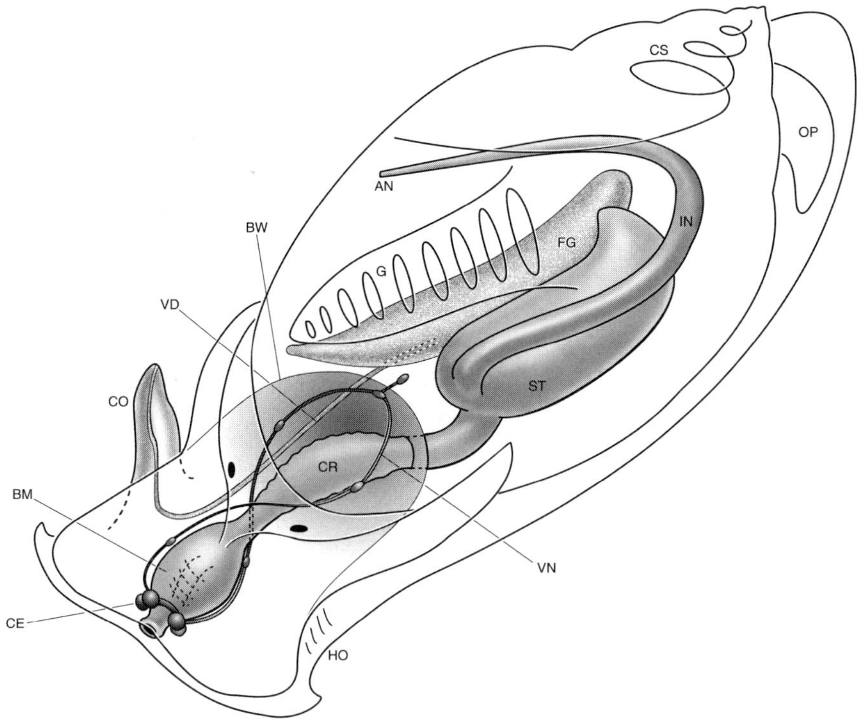

Figure 8 Diagrammatic anatomy of *Acteon*. Abbreviations as in Figure 2.

unrecognized until the mid-1980s (Mikkelsen and Mikkelsen, 1984) when two distinct morphologies were described, adding a valuable species-level shell character with implications for inferring developmental mode (Mikkelsen, 1985).

Although a tectibranch can usually be identified to species level on the basis of its shell, generic and familial assignment in most cases requires anatomical features (sometimes minimal, e.g., radular morphology). Species of *Ascobulla* and *Cylindrobulla* (Sacoglossa) or of *Acteocina*, *Tornatina* and *Retusa* (Cephalaspidea) have remarkably similar shell morphologies, but their soft parts differ to an extent requiring placement in different families (Marcus, 1977; Mikkelsen, 1998a).

While the classic definition of an opisthobranch shell includes heterostrophy, Willan (1987) claimed that notaspidean shells are secondarily orthostrophic (i.e., not heterostrophic) associated with "detorsion". However, this determination is not precise. The protoconch of a tectibranch notaspidean such as *Umbraculum* is certainly sinistral

Table 1 Tectibranch characters, as represented in the exemplar taxa.

Characters	HAO	*Acteon*	*Hydatina* Architectibranchia	*Bulla* Cephalaspidea: Bulloidea	*Cylichna* Cephalaspidea: Philinoidea	*Akera* Anaspidea	*Cylindrobulla* Sacoglossa	*Tylodina* Notaspidea
Shell spire	exposed	exposed	involute	involute	involute	exposed	involute	exposed
Complete withdrawal into shell	possible	possible	possible	possible	possible	possible	possible	not possible
Shell thickness	thick	thick	thin	thick	thick	thin	thin	thin
Operculum	present	present	absent	absent	absent	absent	absent	absent
Cephalic shield	present	present	absent?	present	present	present	present	absent
Parapodia	absent	absent	present	present	present	present	absent	absent
Foot length	short	long	long	short	short	long	short	long
Posterior foot	absent	absent	absent	present	present	present	absent	absent
Hancock's organs	present	present	present	present	present	present	present	absent
Rhinophoral organs	absent	absent	absent	absent	absent	absent	absent	present
Oral tentacles	absent	absent	present	present	absent	absent	absent	present
Propodial tentacles	absent	present	present	absent	absent	absent	absent	absent
Mantle opening	anterior	anterior	lateral	lateral	lateral	lateral	lateral	lateral
Gill	"plicate ctenidium" = plicatidium	plicatidium	plicatidium	plicatidium	plicatidium	plicatidium	lamellate gill	bipectinate gill
Ciliated strips	??	blunt	blunt	exogyrous	flexed	blunt	blunt	absent?
Adductor muscle	absent	absent	absent	absent	absent	present	present	absent
Jaws	present	absent	present	present	present	present	absent	absent
Oral cuticle	??	denticled	smooth	smooth	smooth	smooth	smooth, thickened	smooth, thickened

Radular width	broad	broad	broad	broad	narrow	broad	narrow (single row)	broad
Ascus	absent	absent	absent	absent	absent	absent	present	absent
Pharyngeal pouches	absent	absent	absent	absent	absent	absent	present	absent
Crop	present	present	present	absent	present	present	present	absent
Gizzard/gizzard plates	absent	absent	absent	present	present	present	absent	absent
Gizzard spines	n/a	n/a	n/a	present	absent	present	n/a	n/a
Stomach	flow-through	with chamber	with chamber	flow-through	flow-through	flow-through	flow-through	flow-through
Stomach caecum	absent	absent	absent	absent	absent	present	absent	absent
Typhlosole	??	short	absent	absent	short	absent	long	??
Position of anus	anterior	anterolateral	posterolateral	posterolateral	posterolateral	posterolateral	posterolateral	posterior
Copulatory organ	non-retractile	non-retractile	non-retractile	retractile	retractile	retractile	retractile	non-retractile
Internal sperm duct	absent	present	present	absent	absent	absent	present	present?
External ciliated groove	present	absent	absent	present	present	present	present	present
"Prostate"	lining external groove	surrounding internal duct	surrounding internal duct	copulatory organ	copulatory organ	absent	surrounding internal duct	surrounding internal duct?
Sperm sacs	proximal RS distal "BC"	proximal RS and GG, in tandem	proximal RS distal GG	proximal RS distal GG	proximal RS distal GG	proximal RS distal GG	distal RS absent GG distal BC (novel)	proximal RS distal "BC"
Circumesophageal nerve ring	pre-pharyngeal	pre-pharyngeal	pre-pharyngeal	pre-pharyngeal	pre-pharyngeal	post-pharyngeal	post-pharyngeal	post-pharyngeal
Cerebral/pleural ganglia	separate	fused	fused	separate	separate	separate	fused	separate
Visceral nerve loop	streptoneurous	streptoneurous	slightly streptoneurous	slightly streptoneurous	slightly streptoneurous	slightly streptoneurous	slightly streptoneurous	euthyneurous (concentrated)
Radular nerve	??	??	separate	fused	fused	fused	separate	fused
Genital ganglion	??	off visceral ganglion	off visceral ganglion	on visceral loop	on visceral loop	on visceral loop	off visceral ganglion	??

The HAO (Hypothetical Ancestral Opisthobranch) is from Gosliner (1994: figure 2). The following characters are common to all included taxa and have been omitted from the table: shell present, shell external, notum absent, radula present. BC, bursa copulatrix; GG, gametolytic gland; RS, receptaculum seminis.

(Willan, 1987; personal observation), while the animal is dextral (evidenced by the position of certain anatomical features, e.g., the anus on the right side), thus hyperstrophy of the protoconch is confirmed. The direction of teleoconch coiling, however, is severely obscured by the extreme expansion of the body whorl into a plate or limpet shape. Thus determination of a change in coiling (heterostrophy), or not, is very uncertain, and no larval development studies have been published that would settle the case. On the other hand, *Tomthompsonia spiroconchalis* Wägele and Hain, 1991, the only notaspid with a spirally coiled (albeit internal) shell, was recently determined to be heterostrophic (Wägele and Hain, 1991), illustrating what may be the plesiomorphic condition in the group.

Significant numbers of Recent tectibranchs are still only known from empty shells. For example, in the author's own research database, Bullidae includes 349 presumably valid species and, of these, anatomical data of some sort (sometimes as minimal as the radula alone) are known for only six species (<2%). Aplustridae, with 21 valid species, fares better with five species (23%) with some known anatomy, and Philinidae is still more complete with 40 of 139 valid species (29%) known anatomically. Still, this is a remarkably low statistic for a Recent taxonomic group. This figure would presumably be higher for tectibranchs with internal shells that are less likely to be described as empty, e.g., *Aplysia*; this factor may contribute to the higher value cited here for Philinidae, although empty shells of *Philine* have certainly been collected. Standards of today's malacology strongly encourage new species descriptions of shelled opisthobranchs to include basic anatomy (gross body form, radular and gizzard plate morphology, copulatory organ) at least to the extent that is necessary for generic and family placement. Those that have not been so described may be synonymized once anatomical data are gathered (e.g., *Haminoea taylorae* Petuch, 1987, versus *H. elegans* (Gray, 1825); Martínez and Ortea, 1997). Studies of living populations also quite often reveal species distinctions based on soft-body color or reproductive mode that were undetectable or unrecognized from empty shells or wet-preserved specimens. For example, lecithotrophic *Oxynoe azuropunctata* Jensen, 1980, from the Florida Keys, is distinguished from planktotrophic *O. antillarum* Mörch, 1863, from the Caribbean by body color, number and size of integumental papillae and by the green algal species on which it lives and feeds (Jensen, 1980).

The question arises: in a group based on soft anatomical characters, how do we properly classify species known only from shells and that we cannot (and perhaps never can) dissect? Malacology is replete with such examples, fossil and Recent, and no solution is fully satisfactory. Provisional placement is perhaps the best alternative, albeit one that is strongly speculative in some cases. The eminent opisthobranch anatomist Eveline Marcus (1977) solved this problem another way, while considering the genus *Acteocina* (Cephalaspidea: Cylichnidae) based on the type species *A. wetherilli* Lea, 1833, from the New Jersey Tertiary. Because of close shell morphologies and because the radula of *A. wetherilli* could never be known, *Acteocina* could not positively be placed in Cylichnidae (with radula) or in Retusidae (without radula). Her recommendation was to restrict the entire genus *Acteocina* to those species, like its type *A. wetherilli*, whose anatomy would never be known, and to use the next available genus (in this case *Utriculastra*) for those that could be properly placed. This is an admittedly extreme and artificial solution; other systematists have not generally practiced it. (This case was ultimately circumvented, when the fossil *A. wetherilli* was synonymized with the Recent species *Volvaria canaliculata* Say, 1826, thus fixing the anatomical definition of *Acteocina*; see Mikkelsen and Mikkelsen, 1984.)

Extant tectibranch groups with relatively robust shells (e.g., Acteonidae, Ringiculidae, Bullidae, Cylichnidae) have rather good fossil records (remarkably good for opisthobranchs) and support a large literature base. Most of these families reach back into the Jurassic Period (Tracey *et al.*, 1993), and have extensive Tertiary fossil records. Four families of extinct tectibranchs (Cylindrobullinidae, Ceritellidae, Actaeonellidae, Trochactaeonidae) lived during the Jurassic and Cretaceous Periods; the earliest recognizable heterobranchs are from the mid-Devonian (Bandel, 1994). They have all been presumed allied to Acteonidae, by virtue of a suite of similar shell characters. They have heterostrophic protoconchs, spires either involute or with tall-exposed whorls, thick walls and columellar folds (Sohl and Kollmann, 1985; Bandel, 1994). Muscle insertion scars left on the shells have led paleontologists to conclude that the soft parts could not be fully retracted into the shells. Some formations have included specimens preserved in burrows that agree well with those described for burrowing cephalaspideans; other species show traces of shell wear that have been interpreted as indicative of an epifaunal lifestyle. Although the heterostrophic protoconchs leave little doubt that these fossil forms are heterobranchs, further determination of which clade they belong to might not be possible under the present anatomical criteria employed in the study of tectibranchs. From the shell forms illustrated in the literature (e.g., Sohl and Kollmann, 1985), these taxa could equally be placed in Acteonidae ("lower heterobranchs"), Cylichnidae or Retusidae (Cephalaspidea).

Not surprisingly, the unshelled opisthobranchs, such as nudibranchs or plakobranchoidean sacoglossans, have virtually no fossil record (Tracey *et al.*, 1993). Especially thin-shelled groups (e.g., Philinidae, Akeridae) are represented by so few specimens that each new discovery literally warrants a separate publication, e.g., the aplysiid *Floribella aldrichi* (Dall, 1890) from the Miocene of the Dominican Republic (Geiger and Jung, 1996). In the latter example, re-examination of shell morphology compared to Recent material was sufficient to move the species and genus from Philinidae in Cephalaspidea, to the Anaspidea. Such rarity or absences in the fossil record are less an indication of low abundance or paleontological age than they are of poor fossilization of fragile or soft-bodied forms.

Reduction or loss of the shell is perhaps the most frequently cited evolutionary trend for shelled opisthobranchs (e.g., Thompson, 1976; Poulicek *et al.*, 1991; Rudman and Willan, 1998). All except very few opisthobranchs (even the unshelled forms) possess larval shells before hatching or until metamorphosis into benthic juveniles; this is solid ontogenetic evidence to support the presence of a shell as plesiomorphic. Loss of the shell is widely acknowledged as having happened independently in parallel many times in Opisthobranchia (Gosliner and Ghiselin, 1984). Phylogenetic studies have shown this to be probable and have contributed to the break-up of Tectibranchia as a recognized taxon. By today's taxonomic definitions, Anaspidea, Sacoglossa and Notaspidea each include both shelled and unshelled forms (Table 2). Of the Anaspidea, members of the genera *Bursatella* (Notarchidae) and *Dolabrifera* (Dolabriferidae) lack an adult shell. In Notaspidea, members of the monophyletic family Pleurobranchaeidae all lack adult shells (Willan, 1987; herein). The shells of members of Pleurobranchidae are internal and usually auriform (coiled only in *Tomthompsonia*; Wägele and Hain, 1991); in *Berthellina*, the shell is "occasionally absent" (Willan, 1983). In Sacoglossa, current classifications recognize two monophyletic groups: Oxynoidea, where the adult shell is present, and Plakobranchoidea, which lack adult shells (Jensen, 1996; Mikkelsen, 1998a). Wägele and Willan (2000) considered absence (loss) of the shell a synapomorphy of Nudibranchia.

All evidence to date suggests that internalization of the shell preceded its evolutionary loss (Gosliner and Ghiselin, 1984). However, there do seem to be several steps or stages of internalization. Internalized shells are here defined as largely covered by soft tissue, temporarily by expansion–contraction or permanently, but never completely (i.e., still open to the surrounding water); internalized shells can still be fully exposed when the animal retracts or is preserved. Internal shells, then, are permanently beneath tissue layers and are not fully exposed upon retraction. As an example, most members of Aplysiidae (Anaspidea) have a plate-like internalized shell open to the surrounding water through a small foramen, while members of *Bursatella* lack an adult shell

Table 2 Taxonomic history of Tectibranchia relative to the unshelled Nudibranchia.[a]

Pilsbry (1894–1896)	Thiele (1931 in 1929–1935)	Odhner in Franc (1968)	Current classification
Tectibranchiata	Pleurocoela	Cephalaspidea	"Lower Heterobranchia"
Cephalaspidea	Cephalaspidea	*Acteon*	*Acteon*
Acteon	*Acteon*	*Hydatina*	
Hydatina	*Hydatina*	*Bulla*	Architectibranchia
Bulla	*Bulla*	*Scaphander*	*Hydatina*
Scaphander	*Scaphander*	*Philine*	Cephalaspidea
Philine	*Philine*	*Diaphana*	*Bulla*
Diaphana	*Diaphana*	*Akera*	*Scaphander*
Akera	*Akera*	Runcinacea	*Philine*
Anaspidea	*Runcina*	*Runcina*	?*Diaphana*
Aplysia	Anaspidea	Anaspidea	?*Runcina*
Oxynoe	*Aplysia*	*Aplysia*	Anaspidea[b]
Notaspidea[b]	Sacoglossa[b]	Sacoglossa[b]	*Aplysia*
Umbraculum	*Oxynoe*	*Oxynoe*	*Akera*
Runcina	Acoela	Notaspidea[b]	Sacoglossa[b]
Sacoglossa[c]	Notaspidea[b]	*Umbraculum*	*Oxynoe*
Nudibranchia[c]	*Umbraculum*	(four groups of "nudibranchs" at this rank)[c]	Notaspidea[b]
	Nudibranchia[c]		*Umbraculum*
			Nudibranchia[c]

[a]Representative genera included for most groups; Acochlidiacea and pteropod groups excluded; currently accepted taxonomic names used throughout; [b]both shelled and unshelled taxa; [c]unshelled taxa only. See text for references on today's classification.

completely; although no phylogenetic analysis has been published for Anaspidea, internalization preceding loss is implied by this pattern. In Notaspidea, phylogenetic reanalysis clearly shows shell internalization to precede loss; the internal shell of *Bathyberthella* (Pleurobranchidae) is completely uncalcified (Willan, 1983, 1987) suggesting a further intermediate state. A similar succession is seen in heterobranch land snails, from shelled snails, to slugs with small, thin internal shells, to unshelled slugs (Smith and Stanisic, 1998).

Internal or heavily internalized shells almost invariably have an expanded body whorl. Examples are the shells of the cephalaspidean *Philine*, the anaspidean *Aplysia* (Figures 1E and 1H, respectively) and most shelled members of the notaspidean family Pleurobranchidae (Willan, 1983).

The rounded shells of involute tectibranchs gave rise to the long-used common name "bubble shells" for this group in general (now restricted to Cephalaspidea). Architectibranchia, Cephalaspidea and Sacoglossa all include both involute taxa plus those with exposed spires, while an exposed spire is the rule throughout the shelled Anaspidea and Notaspidea, and in

Acteon. An exposed spire is defined here as having the protoconch still visible, regardless of spire height, although the latter can vary greatly. Many tectibranchs (e.g., *Ascobulla, Micromelo*) have "flat" or even slightly indented spires, but with the protoconch visible at the center and thus technically "exposed". All gastropod larvae also technically have exposed spires that become involute only with the growth of subsequent whorls. The shelled anaspideans and notaspideans (e.g., *Aplysia, Umbraculum*), whose shells are nearly uncoiled cap or plate shaped, thus retain the protoconch in an exposed position, albeit without a well-defined prosobranch-like spire.

Height of the spire is intra- and interspecifically variable in members of many tectibranch genera. A good example is from the family Aplustridae (Architectibranchia) that includes taxa with exposed spires (*Hydatina amplustre* (Linné, 1758), *Micromelo* spp.) and those that are involute (most *Hydatina* spp.); a variant of *H. physis* (Linné, 1758), named *H. verrilli* by Pilsbry (1949) based on a single shell, has an abnormally exposed spire.

The phrase "shell reduced" is notorious for its vagueness in its application to opisthobranchs (Mikkelsen, 1998b). Shells can be reduced in relative size and/or in thickness, two features that vary independently with one another. "Reduced" has also often been used to signify completely reduced, that is, as a synonym for absent or lost. For externally shelled taxa, relative size of the shell has often been expressed by whether or not the animal can retract fully into its shell.

Relatively robust external shells are present in adult members of Acteonidae, Ringiculidae, Bullidae, Cylichnidae, Umbraculidae and a few other tectibranch families. A thick-walled shell is considered plesiomorphic to a thin-walled shell in tectibranchs, although according to phylogenetic analysis (Mikkelsen, 1996), there are basal clades that include thin-walled species (e.g., *Hydatina*, in Architectibranchia) and derived clades that include thick-walled species (e.g., *Acteocina*, in Cephalaspidea: Philinoidea).

Poulicek *et al.* (1991) found that the microstructural and chemical composition of opisthobranch shells shifts in line with shell internalization and reduction (in size and thickness, in an analysis including one architectibranch, eight cephalaspids, four anaspids, one sacoglossan and four notaspids). Because the same changes occur in other mollusks (e.g., cephalopods, pulmonates), these changes do not appear to be phylogenetically linked throughout Mollusca. However, this does not exclude their cladistic utility within certain tectibranch lineages. In the same study, the biochemical components recorded in the thin shells of various tectibranchs have been interpreted as means to increase shell flexibility. Anaspidea and Sacoglossa have increased amounts of the amino acid

OH-Proline; Cephalaspidea and Notaspidea, along with anaspideans and convergently with other thin-shelled mollusks, have increased chitin content.

3.2. Operculum

The operculum (the chitinous "trap door" to the shell aperture) is present in *Acteon* and allied genera but is absent in nearly all other adult opisthobranchs. One member of Acteonidae has been recorded without an operculum, i.e. *Rictaxis punctocaelatus* (Carpenter, 1864); Eveline Marcus (1972: 175, 178; also Gosliner, 1981) noted that in her specimens the operculum was "lost . . ., but the pigmented opercular lobe is distinct". Unexpectedly, opercula have been found in adults of three species of *Retusa* (Cephalaspidea: Philinoidea; Minichev, 1966; Burn and Bell, 1974). Although some authors have assumed that loss of the operculum was "irreversible" in opisthobranch evolution (Thompson, 1976), this assumption is not necessarily correct. Most known tectibranch larvae have opercula that are discarded early in post-larval life (or perhaps later, in the case of *R. punctocaelatus*); in such cases, it is simple retention of a larval feature to affect a seeming "reversal" of opercular loss.

3.3. External anatomy

Most shelled opisthobranchs, especially the Cephalaspidea, are rather globular animals, with large, smoothed heads and rounded bodies. Exemplified by *Bulla* (Figure 3), prominent external features include the parapodia or lateral expansions of the foot that overlie at least the anterior part of the shell. The parapodia often extend over each other, and are in turn overlain by the cephalic tentacles, flattened posterior expansions of the shovel-shaped head or cephalic shield. Such devices serve to streamline the body and protect the entrance to the mantle cavity (containing the fragile gills), providing obvious advantages to animals that burrow through sand, mud or algae. Conversely, epifaunal tectibranchs, as are common in Sacoglossa, Anaspidea and Notaspidea, are often significantly not streamlined with respect to body form. Processes from the head (rhinophores, oral veil and/or tentacles) and dorsum (cerata) are prevalent and are themselves indicative of a lifestyle not centered on burrowing. Brace (1977: 3) considered that "the general pattern of opisthobranch evolution was one of initial assumption of infaunal life, followed, after varying intervals of time, by return to the surface". According to this interpretation, features like cerata, rhinophores and exposed gills would be secondary (novel) adaptations, apomorphic among opisthobranchs.

Although the cephalic shield classically hallmarks the Cephalaspidea, it is also present in taxa no longer included in that group (e.g., *Akera, Volvatella*). No special innervation or other evidence has been identified to demonstrate homology of this structure across tectibranch taxa (Mikkelsen, 1996) and its form varies tremendously. Wägele and Willan (2000) speculated that the oral veil of notaspideans (also some nudibranchs, possibly some sacoglossans) might be homologous with the cephalic shield. Although the expanded oral veil of the burrowing notaspidean *Euselenops* (Rudman, 1999) does effectively emulate a cephalic shield, data to support homology are not available. Similar shovel-shaped heads exist among burrowing prosobranchs (e.g., *Natica, Oliva*), more strongly suggesting several independent pathways toward this type of ecophenotypic adaptation.

Along similar lines is the notum, a dorsal shield of soft tissue in forms with internal or absent shells. While "presence of a notum" is a traditional notaspidean character, Wägele and Willan (2000) have discussed the relative size and ornamentation of the notum in nudibranchs, cephalaspideans, sacoglossans and anaspideans. But like the cephalic shield, convincing evidence for homology of the notum throughout Opisthobranchia has not been presented.

Lateral extensions of the foot, called parapodia, are present in most tectibranchs. Like the cephalic shield, parapodia serve to streamline the body and prevent sand particles from damaging delicate organs in the mantle cavity. The presence of parapodia could also be considered ecophenotypic, although their morphology is more uniform than the cephalic shield. The ability to swim is correlated with very large parapodia in *Gastropteron*, *Aplysia* and *Akera* (see Thompson, 1976: plates 16–17; Rudman and Willan, 1998: figure 16.9).

When the shell is present in opisthobranchs, it is generally carried horizontally, effectively being dragged across (or through) the substratum. Prosobranchs that do this usually rest the posterior part of the shell on the operculum at the posterodorsal end of a long foot. Perhaps correlated with the loss of the operculum, the foot of most tectibranchs has become shortened, extending only under the anterior part of the crawling animal. In some short-footed tectibranchs whose aperture is elongated laterally (e.g., *Bulla, Acteocina, Retusa*), the mantle edge at the posterior end of the shell aperture has become secondarily expanded and thickened, and augments the foot by supporting the rear of the shell. This structure has been variously called the posterior foot, posterior pallial lobe or infrapallial lobe, and appears to be unique to tectibranchs. Prosobranchs with similarly elongated shell apertures (e.g., *Conus*) have retained a slender foot without such an auxiliary structure.

The external chemosensory organs of shelled opisthobranchs fall into two types. Hancock's organs are a series of simple to elaborate folds in the anterior cerebropedal groove, i.e., on either side of the mouth. Once considered a synapomorphy for the entire Opisthobranchia (Gosliner, 1980), it is now recognized as present in many tectibranchs, including cephalaspideans, *Acteon*, *Akera* and shelled sacoglossans, innervated by nerves emanating from the cerebral ganglia. In taxa with Hancock's organs, the body is bereft of exposed cephalic tentacles (not to be confused with the posterior cephalic flaps of the cephalic shield). Similarly, the tentacle-like processes called rhinophores of some anaspideans, some sacoglossans, notaspideans and especially nudibranchs, also often include folds or ridges (called rhinophoral organs; rhinophores are technically the "carrier", analogous to the cerebropedal groove for Hancock's organs). Some authors (e.g., Odhner, 1914) have noted the morphological similarity of Hancock's organs and rhinophoral organs, thus implying their homology, while others (Rudman and Willan, 1998) have maintained that the two are independently derived. Definitive data for either interpretation, including studies of common innervation, are lacking and the question remains unresolved. Even the various types of Hancock's organs (irregular folds, transverse folds, bipinnate; Edlinger, 1980b) and rhinophoral organs (solid, rolled, perfoliate, annulate, papillate; Gosliner, 1994; but see Wägele and Willan, 2000) are of questionable phylogenetic homology within each organ group. Odhner (1914) also noted subtle differences in position or form of Hancock's organs within the cerebropedal groove, e.g., laterally on the head in *Haminoea* or on the underside of the cephalic lobe in *Bulla*. Homologous or not, the locations of these chemosensory organs (in a groove or on rhinophores) are correlated with lifestyle; burrowing forms generally have Hancock's organs, protected beneath tissue folds, while epifaunal species carry rhinophoral organs on more exposed cephalic tentacles.

Oral tentacles (tentacle-like structures flanking the mouth) and propodial tentacles (extended anterior foot corners) are each present in scattered taxa of Architectibranchia, Sacoglossa, Anaspidea and Notaspidea. These are presumably mechanosensory in function (Gosliner, 1994) but like other features of body shape discussed above, their homologies have not been studied in detail.

Opisthobranch eyes are always sessile, never stalked or extensible as in some prosobranchs or pulmonates. Structurally each eye is simple, with a solid usually spherical lens, a dark-staining reflective layer over sensory cells, and an optic nerve originating in the cerebral ganglia; the eyes are plesiomorphically superficial, lying just below the translucent integument on the dorsal surface of the headshield. In some tectibranchs, the eyes are structurally reduced (e.g., *Cylichna cylindracea*, Pennant, 1777) or absent (e.g., *Retusa obtusa*, Montagu, 1803). Some shelled sacoglossans have a strange, irregularly shaped lens of uncertain function (Mikkelsen, 1996: figure 40). In some cases (e.g., *Cylindrobulla beauii*, P. Fischer, 1856), the eyes are deeply buried in the headshield, rendering them undetectable externally, a condition sometimes misinterpreted as absent. "Direction of sight" has been inferred from position of the lens or optic nerve relative to the head surface; deeply situated, laterally directed eyes have been proposed as associated (although not strongly correlated) with a burrowing lifestyle (Mikkelsen, 1996). Eye characters are rarely discernible through gross dissection, requiring histological sectioning for best resolution.

3.4. Mantle cavity

Opisthobranchs are classically regarded as "detorted", with their bodies and internal organs moving away from the classic "twisted snail" appearance toward bilateral symmetry. Brace (1977) provided a convincing argument for this process resulting from differential growth, producing the effect of detorsion. Thus, the opening of the mantle (=pallial) cavity effectively rotated from the primitive anterior position (as in *Acteon*) to a more derived lateral or posterior one. Mikkelsen (1996: figures 1–17) developed a convenient system of using certain anatomical landmarks (e.g., osphradium, anus) on a generalized quadrant map of the body to determine orientation of the mantle opening. The opening is also repeated in the posture of the gill, e.g., an "anteriorly directed" gill (Gosliner, 1981: 220). Like so many other tectibranch attributes, the posterolateral shift of the mantle opening is associated with the burrowing habit, moving the entranceway of the gill chamber away from potentially damaging, oncoming particles.

Opisthobranchs have a variety of different types of gills, from compound folds suspended from the roof of the mantle cavity in cephalaspideans, to bipectinate leaves attached at a common base in notaspideans, to the retractable, dorsal, circumanal plumes in nudibranchs. Not all of these are homologous with any degree of certainty to each other or to the prosobranch ctenidium, the latter characterized by parallel triangular leaflets, characteristic ciliation and supporting rods (Morton, 1972). Gosliner (1978: 116; 1981: 220) called the gill of the HAO (Hypothetical Ancestral Opisthobranch) a "plicate ctenidium" but from the wording and figure, this is intended to suggest a cephalaspid-like plicatidium rather than the classic prosobranch ctenidium. Thompson (1976) believed that the opisthobranch gill (exemplified by the bipectinate gill of notaspideans) was derived by simplification of the prosobranch ctenidium (thus homologous). The two-sided plicatidium of Cephalaspidea and one-sided lamellate gill of some shelled Sacoglossa (these two believed homologous; Mikkelsen, 1996) might, however, be a novel structure (Brace, 1977). Wägele and Willan (2000) postulated that gills near the anus was plesiomorphic for Nudibranchia, although the homology of this gill with those of other opisthobranchs was contentious; the anthobranchian configuration of a circle around the anus is a derived form. The gill is absent (lost) in some sacoglossans.

The so-called ciliated strips or raphae in the mantle cavity of externally shelled heterobranchs serve to increase water circulation within the cavity (Robertson, 1985). Their development is presumably associated with reduction or modification (or in some cases, absence) of the gill. The strips generally lie parallel to the gill (when present), on both the roof and floor of the mantle cavity, leading inward from the mantle opening. In a few taxa (e.g., *Acteon, Scaphander, Akera*), the strips continue into a winding blind caecum that parallels the visceral mass into the coil of the shell (Figures 5 and 8). Ciliated strips are believed to be homologous among heterobranchs (Robertson, 1985), and their presence is probably a synapomorphy of Heterobranchia. Cephalaspids (e.g., *Scaphander, Philine*) show flexure of the end of the strip near the mantle opening, in a presumed synapomorphy for the clade; bulloidean cephalaspids have exogyrous strips (also synapomorphic), curved into a C-shape.

The osphradium is a well-studied chemosensory organ in gastropods (Haszprunar, 1986). It lies on the roof of the mantle cavity and is innervated from its own osphradial ganglion and nerve emanating from the supraesophageal ganglion. In lower heterobranchs, the osphradium is on the left side of the snail. In most (detorted) shelled opisthobranchs, it has migrated (along with the supraesophageal and osphradial ganglia) to the midline or right side. Unlike that of prosobranchs, which is complex in structure, the osphradium of tectibranchs is a simple knob. Its presence has been linked with that of the gill in pulmonates (Voltzow, 1994), therefore its reduction in opisthobranchs could be linked to loss or modification of the gill. This could also be associated with the elaboration of other chemosensory organs (e.g., Hancock's Organs). Edlinger (1980a, b) surveyed cephalaspid osphradia and suggested a pattern of general reduction among key genera. Little is known of the osphradium of other tectibranchs.

An adductor muscle, serving to compress the sides of a thin external shell from within, is present in *Akera* (Anaspidea) and a number of shelled sacoglossans. The bivalve-like *Julia,* like *Akera,* has two adductor muscles, while *Cylindrobulla* and *Ascobulla* have a single muscle. These are most likely derived from the columellar muscle (Morton, 1972). The mention of an adductor muscle in the shelled notaspidean *Bathyberthella* by Wägele and Willan (1994) refers to a pharyngeal retractor muscle and, as such, is not homologous.

3.5. Digestive system

The digestive system has been considered one of the most important character systems in both traditional and phylogenetic studies of tectibranchs. Thompson (1976) reflected this emphasis in his claim that diet was the driving force behind opisthobranch evolution, an opinion corroborated by phylogenetic hypotheses for Cephalaspidea, Anaspidea and Sacoglossa (Mikkelsen, 1996). Digestive system characters provide many clade-defining synapomorphies, e.g., a stomach caecum and a filter chamber in Anaspidea, a gizzard with three plates in Cephalaspidea and an ascus in Sacoglossa.

Jaws, or plates of parallel cuticularized rods in the oral cavity, are present in most gastropods. Mikkelsen (1996) restricted the definition of jaws to those structures originating in a generative groove, as identifiable in histological cross-sections. Jaws are repeatedly cited as absent in opisthobranchs and, indeed, Mikkelsen's (1996) phylogenetic analysis indicated that jaws were lost several times across the cladogram. Similarly in Notaspidea, jaws are present plesiomorphically, are retained in the Pleurobranchoidea, and are absent (lost) in the Umbraculoidea (Willan, 1987). The individual elements of the jaws range from simple to denticulate; these differences are useful at lower taxonomic levels.

The rodless cuticularized lining of the oral cavity is independently present or absent and is not homologous with jaws. The cuticle most often exists as a thin lining, but may be elaborately denticled (*Acteon*; Gosliner, 1994: figure 11, labelled as "jaws") or thickened (synapomorphy of Sacoglossa – Mikkelsen, 1996; also in Umbraculoidea – Willan, 1987; homology of these two undetermined).

The radula, a ribbon of chitinous denticles housed within the buccal mass (=pharynx), has long been considered a molluscan synapomorphy. It is well accepted that its presence is plesiomorphic in gastropods. The radula can be readily extracted from a fresh, preserved or dried soft body by dissolving away the organic tissues with potassium or sodium hydroxide. As such, the radula is the easiest "anatomical" feature to investigate and has been described in many, many tectibranchs, often as the sole non-conchological feature. The radula is absent in several tectibranchs that feed by ingesting their prey items whole, followed by crushing them in a muscularized "armed" gizzard (e.g., Retusidae) or by dissolving them with digestive chemicals (e.g., Aglajidae). Aglajid cephalaspids are mostly without a radula, but one species, *Odontoglaja guamensis* Rudman, 1978, has been described with a radula in the expected philinoidean form.

The radula is almost always a bilaterally symmetrical structure, composed of a central (rachidian) row flanked by rows of teeth (called laterals and marginals, based on differing size and/or denticulation) on either side. The relative width of the radula, expressed as the number of teeth per row, has been frequently used in tectibranch classification. Here again, the derived state is reduction, e.g., plesiomorphic *Haminoea* with a radular formula of (5–63).1.1.1.(5–63) [denoting 5–63 marginals and a single lateral on each side of the rachidian] versus *Acteocina* with 1.1.1, *Philine* with 1.0.1 and *Ascobulla* with 0.1.0. In these cases, the radular morphology is correlated with dietary specialization. The generalized grazer *Haminoea* has a broad unspecialized prosobranch-like (taenioglossate) radula, with quadrate-cuspidate rachidians, and varying numbers of lateral and marginal teeth. *Acteocina* and *Philine* have sets of pincer-like teeth for selecting individual prey items, while *Ascobulla*'s single dagger-tooth is used to pierce individual algal cells. Broad radulae also characterize *Acteon* and notaspideans in general, however, their many similar teeth per row are clearly derived, probably in part in response to specialized feeding on polychaetes or sponges, respectively.

Presence or absence of a rachidian row is highly variable within some genera (e.g., *Ringicula, Hydatina, Philine*), sometimes even within a single species. For example, the rachidian of *Scaphander* has been called "caducous", or weakly attached to the radular ribbon, so as to frequently fall off in the functional part of the radula (Marcus and Marcus, 1967b). As such, the mere presence or absence of a rachidian is useful mainly at the species level. In some taxa, the shape of the rachidian tooth has been shown to be of systematic value; among cephalaspideans, members of Bullidae have an elongated denticulate plate, while philinoideans usually have a bilobed (centrally indented) rachidian. One other tooth shape has been found useful; the presence of enlarged sickle-shaped, often-denticulate lateral teeth is a synapomorphy of philinoid cephalaspideans, although it is also known in other tectibranchs and non-tectibranchs (e.g., *Ringicula* – Thompson *et al.*, 1985; Aplacophora – Scheltema, 1985).

The fate of used radular teeth in sacoglossans is unique among gastropods. While other gastropods shed used teeth from the front (oldest) end of the radular ribbon, sacoglossans retain them, funneling them into the ascus (= saccus), a blind sack within the buccal mass. Here, used teeth are stored and presumably resorbed. This physiology leads to a unique orientation of the radula, with an added "descending limb" between the forward-placed functional tooth area and the ascus underneath it (other opisthobranchs have an "ascending limb" between a broader functional area and the radular sac, where new teeth are formed, behind and beneath it; Mikkelsen, 1996: figures 28–29). The presence of a descending limb and an ascus are clear synapomorphies of Sacoglossa (Mikkelsen, 1996).

Pharyngeal pouches (ventrolateral out-pockets of the buccal mass) serve as part of the elaborate suctorial feeding mechanism of sacoglossans, and have been interpreted as synapomorphic for the clade (Jensen, 1991; Mikkelsen, 1996). A single dorsal pouch is present in a few taxa of Nudibranchia that also feed suctorially (Wägele and Willan, 2000), but is probably analogous rather than homologous to those of Sacoglossa.

The crop is an expanded section of the esophagus behind the buccal mass but preceding the stomach. Its function is presumably to accommodate ingested food prior to processing by the gizzard or stomach when the rate of processing might be slower than the rate of ingestion. Its appearance in preserved material is variable depending upon the degree of distension at the time of death; an empty crop may appear no wider in diameter than the surrounding esophagus, and no histological differences in this region have been reported. The occurrence of a crop is independent of a more muscular gizzard (see below). The crop alone is present in many taxa (e.g., *Acteon, Hydatina, Cylindrobulla*) and is present just anterior to the gizzard in others (e.g., *Scaphander, Aplysia, Haminoea*). Their co-occurrence (although recognizing the inexact anatomical distinction of the crop) contradicts Gosliner's (1981, 1994) suggestion that the crop is the plesiomorphic form of the more derived gizzard.

A muscular esophageal gizzard, containing hardened gizzard plates and, in some cases, additional smaller gizzard spines for processing food items, is present in members of Cephalaspidea and Anaspidea. Mikkelsen (1996) suggested a sequence for its evolutionary development: (a) first appearance in members of Anaspidea, with multiple corneous plates and spines, (b) consolidation of the plates in Cephalaspidea, (c) the loss of gizzard spines in carnivorous Philinoidea (retained in herbivorous Bulloidea) and (d) elaboration of the plates (by tubercles, ridges, or calcification) in various members of Cephalaspidea. The Runcinoidea, a traditional group of Cephalaspidea now of uncertain status, is characterized in part by four gizzard plates that bear superficial resemblance to those of *Haminoea* spp. (Cephalaspidea: Bulloidea) (see Gosliner, 1994: figure 40C–D). Their homology with the three gizzard plates typical of cephalaspideans and their place in the above-mentioned scenario has not been tested. Thecosome pteropods (pelagic shelled opisthobranchs) also have a gizzard with four or more plates that probably evolved independently of cephalaspideans and runcinoideans (Lalli and Gilmer, 1989; Gosliner, 1994). The "gizzard" described in *Tylodina,* containing hardened ridges or "plates", is actually the stomach and not homologous to the esophageal gizzard.

The gastropod stomach is renowned for its elaborate internal structure. It usually includes an extensive blind pouch equipped with (depending on the group) a rotating enzymatic crystalline style and/or ciliated folds that act, respectively, in grinding and sorting food particles (summarized by Voltzow, 1994). The esophagus and intestine often enter and exit, respectively, in relatively close proximity at one end of such a stomach. *Acteon* and architectibranchs (e.g., *Ringicula, Hydatina*) have this kind of stomach, but generally lack the elaborate food-processing mechanisms, although *Ringicula* has thickened musculature in the posterior pouch for crushing particles (Fretter, 1960). Most other opisthobranchs have lost the pouch-like chamber, reducing the stomach to a flow-through configuration, with the esophagus entering anteriorly and intestine exiting posteriorly. In dissection or histological sections, the opisthobranch stomach is often identifiable only by its expanded diameter (relative to the esophagus and intestine) and by the openings to the digestive gland within which it is generally surrounded. In *Tylodina*, the stomach is expanded into a food-processing organ, analogous to the gizzard, with thickened ridges in the anterior portion. Anaspideans, some nudibranchs, and pteropods have a blind caecum associated with the stomach. In anaspideans the caecum contains two long typhlosoles, while in nudibranchs it is a lobe of the digestive gland (Wägele and Willan, 2000); the two types do not appear to be homologous. The caecae of Anaspidea and Thecosomata have been called identical in older literature (see Dayrat *et al.,* 2001), however, this supposition should be re-evaluated under modern standards of character homology.

The presence and/or extent of a typhlosole in the intestine have been considered of potential systematic value in tectibranchs and nudibranchs (Mikkelsen, 1996; Wägele and Willan, 2000). The function of this thickened longitudinal fold is uncertain, but has been given either as a mechanism for fecal compaction or for alignment of potentially damaging hard structures (e.g., sponge spicules) in the process of elimination. A short typhlosole has been postulated as plesiomorphic for Cephalaspidea (Mikkelsen, 1996); absence was considered apomorphic in both tectibranchs and nudibranchs (Wägele and Willan, 2000). Its homology with the more extensive structure in prosobranchs is assumed, and it has shown a minor phylogenetic signal for Anaspidea (as absent; Mikkelsen, 1996).

The plesiomorphic position of the anus in prosobranchs and lower heterobranchs is anterodorsal, near the forward entrance of the mantle cavity. In opisthobranchs, the anal opening has shifted in position posteriolaterally along with the mantle cavity. This shift may have occurred independently as many as three times in Opisthobranchia

(Gosliner, 1994). Although position of the anus has been utilized in three different opisthobranch phylogenetic analyses (Willan, 1987; Jensen, 1996; Wägele and Willan, 2000), description of this position differed significantly making comparisons difficult. The notaspidean anus is consistently lateral, adjacent to the gill (Willan, 1987). In Sacoglossa, Jensen (1991) considered the anus posterior, anterior, or lateral, or in her analysis (Jensen, 1996) dorsal or lateral. Wägele and Willan (2000) described the anal position in various opisthobranchs as plesiomorphically anterolateral, and apomorphically either posterolateral or dorsal. The absolute position of the anus can also be affected by contraction during fixation, leading to misinterpretation in preserved specimens. On a generalized quadrant map of the body (as utilized by Mikkelsen, 1996), the anal position can be somewhat reliably assessed as antero- or posterolateral relative to a hypothetical midline, but the co-dependence of this character relative to mantle cavity opening, position of the osphradium, etc., must be acknowledged.

3.6. Reproductive system

The opisthobranch reproductive system is simultaneously hermaphroditic in all recorded cases, although egg and sperm production can commence asynchronously in the ovotestis of some forms (de Maintenon and Mikkelsen, 2001). Fertilization is internal, and copulation is usually reciprocal, although the relative positions of mating partners (i.e., head-to-tail, head-to-head) are variable across taxa (Jensen, 1999). Both male and female systems generally open on the right side of the animal, usually with the male opening more anterior and the female opening more mid-lateral. The female portion includes a series of glands for secreting protective layers around the ova, and one or more sperm sacks for storing (and sometimes gametolysing) allosperm (i.e., sperm from a mating partner). The male system usually includes some sort of nutritive gland ("prostate"), one or more storage areas for autosperm (i.e., sperm produced by the animal itself for transfer to a mating partner) and a copulatory organ with a usually retractable penis. The reader is directed to the review by Hadfield and Switzer-Dunlap (1984) for more information on opisthobranch reproductive systems.

The copulatory organ provides a number of taxonomic characters for opisthobranchs. The penis is permanently extended in prosobranchs, basal heterobranchs, and umbraculoidean notaspideans, while in most tectibranchs, it is retracted into a penial sheath from which it protrudes for copulation. Retractability has been corroborated as the apomorphic state by several phylogenetic hypotheses (Mikkelsen, 1996; Willan reanalysis, herein). Penial stylets or spines occur in many tectibranchs (e.g., Anaspidea, Sacoglossa, Notaspidea) as well as in nudibranchs and lower heterobranchs; in such cases, the penis is termed "armed" (or "unarmed" in their absence). Homology of these various armaments has not been established.

Transport of autosperm to the copulatory organ occurs via one of two possible routes in tectibranchs: an internal duct (vas deferens) or an external ciliated groove along the right side of the animal leading from a common ("female") genital opening to the copulatory organ. These two characters are not, however, homologous; both features co-occur in tectibranch sacoglossans (Schmekel, 1985; Mikkelsen, 1996). Here the external ciliated groove is believed to be used for egg, rather than sperm, transport. The presence of an internal duct has generally been considered plesiomorphic (relative to its absence) owing to the condition in prosobranchs, lower heterobranchs, architectibranchs and sacoglossans, a hypothesis that has been recently phylogenetically corroborated (Mikkelsen, 1998a).

The so-called prostatic gland or prostate of opisthobranchs occurs in several forms, each presumed to secrete fluids to nourish stored autosperm. In taxa with an internal sperm duct (vas deferens), prostatic glandular tissue surrounds the duct. In those forms with only an external sperm groove, the copulatory organ has associated glandular tissue (usually in the form of a discrete bulb) that has also been called prostatic. Gosliner (1981) proposed that the HAO produced prostatic secretions from glandular cells lining the external sperm groove, a condition that has not been confirmed among living tectibranchs. Although the secretions produced are almost exclusively proteinaceous (Gosliner, 1994), the homology of two categories of tissues has not been shown and, in fact, their histological structure differs considerably both between and within each type (Mikkelsen, 1996). Investigation at the biochemical level will probably be necessary to determine the origin and true nature of these various glands. Anaspideans and some notaspideans apparently lack prostatic tissues (Gosliner, 1994).

Much has been made of the complex pathways for gamete transport within the opisthobranch reproductive system. Taxa are widely described as monaulic, diaulic (either andro- or oodiaulic) or triaulic, possessing respectively, one, two or three channels to accommodate autosperm (endogenous), allosperm (exogenous) and ova (Ghiselin, 1966). For example, *Acteon* is androdiaulic, that is, with a separate passageway for autosperm plus an otherwise undivided gonoduct (for allosperm and ova). Anaspideans and some sacoglossans are oodiaulic, having a separate passageway for ova and a common gonoduct for both auto- and allosperm.

Most heterobranchs are monaulic, with an undivided (or seemingly so) gonoduct (Ghiselin, 1966). Rudman (1978: 104) interpreted monaulism as secondary simplification in Opisthobranchia, achieved through shortening of the gonoducts associated with reduction of the mantle cavity. Conversely, Schmekel (1985) considered monaulism as ancestral. Such categories, while useful in a functional context, are generalized conditions composed of a suite of morphological characters and, as such, are problematic in a phylogenetic context (Mikkelsen, 1993).

The so-called female gland mass comprises the various ducts for sperm and egg transport plus three nidamental glands (mucus, albumen, membrane) that secrete protective layers around fertilized eggs prior to spawning. The intricacy of this complex has so far prevented its successful coding as phylogenetic characters. In their broad gastropod phylogeny, Ponder and Lindberg (1997: 195) expressly ignored "accessory reproductive structures such as glands because . . . of their doubtful homology across major groups". Resolution of structural homologies within this system is requiring ontological analyses (de Maintenon and Mikkelsen, 2001). Fortunately such studies can be conducted on post-larval individuals, because differentiation of the reproductive system is largely a post-metamorphic process.

Female gastropods classically have two sacks for storing sperm received from a male at mating. The thin-walled bursa copulatrix for short-term storage differs histologically from the receptaculum seminis that stores sperm for longer periods with their heads embedded in the sac's thicker walls (Voltzow, 1994). In hermaphroditic heterobranchs, the bursa is distal and the receptaculum is proximal relative to the gonad. The bursa of tectibranchs is more appropriately called a gametolytic gland (=spermatheca), reflecting its apparent role in dissolving waste reproductive products. The homology of the prosobranch bursa and the tectibranch gametolytic gland across various groups, once called into question by virtue of their cellular structure, has been recently corroborated (Mikkelsen, 1998a). *Acteon* and the lower heterobranch *Gegania* share a unique configuration of a proximal gametolytic gland and receptaculum seminis in tandem on a common duct (Haszprunar, 1985c; Mikkelsen, 1996). In sacoglossans, the bursa is absent and the receptaculum seminis has migrated distally taking the former position of the bursa; these taxa appear (from histological evidence) to have a novel bursa copulatrix located distally on a bifurcated common duct with the receptaculum seminis (Mikkelsen, 1996).

Many opisthobranchs (e.g., in Sacoglossa, Nudibranchia) have a protrusion or tissue flap at the opening of the female reproductive tract that has usually been called a female genital papilla. Although its presence has been cited as a synapomorphy for shelled sacoglossans (Oxynoidea), its homology with the like-named structure in nudibranchs has not been investigated.

3.7. Nervous system

The gastropod nervous system consists of two loops of ganglia and cords, the latter called commissures (between like ganglia pairs) or connectives (between unlike ganglia). The first loop is the circumesophageal nerve ring, comprising plesiomorphically two cerebral, two pedal and two pleural ganglia, surrounding the anterior esophagus as the name implies. The position of this ring, either in front of or behind the pharynx (termed pre- or post-pharyngeal, respectively), has long been considered taxonomically useful in gastropods. A pre-pharyngeal location is considered most plesiomorphic in opisthobranchs, although this state is apomorphic for gastropods in general (Gosliner, 1994). The relative length of the ganglionic commissures is variable and can also be phylogenetically informative; for example, an apomorphically short cerebral commissure was a moderately supportive character for Anaspidea in Mikkelsen's (1996) analysis.

The second loop is the long visceral nerve loop, emanating from the pleural ganglia of the nerve ring, and including five ganglia in heterobranchs along its length: two pallial, two esophageal (sub- and supra-; also called intestinal) and a single visceral ganglion (Williams, 1975: figure 1). Gosliner's (1994: figure 2) HAO is incomplete in this regard, with only three ganglia on the visceral loop, excluding the pallial ganglia. Haszprunar (1985b) used this plesiomorphic configuration of five ganglia as the basis for his new taxon Pentaganglionata, a group similar in content to the Euthyneura. Other ganglia, including two buccal ganglia (from the cerebrals), a single genital ganglion (from the visceral) and a single osphradial ganglion (from the supraesophageal), are also prominent and often cited as configurationally important. The genital ganglion occurs in two different locations among tectibranchs, either directly on the visceral nerve loop (e.g., in architectibranchs) or on a separate genital nerve emanating from the visceral ganglion (Mikkelsen, 1996).

Within the circumesophageal nerve ring, certain tectibranchs (e.g., *Acteon, Cylindrobulla*) show fusion of the cerebral and pleural ganglia. This fusion occurs throughout Opisthobranchia in general, including in some sacoglossans and nudibranchs (Wägele and Willan, 2000), although its occurrence is not uniform within any clade. In notaspideans, the pleural, parietal and visceral ganglia are fused, a condition called notoneury by Thiele (1935 in 1929–1935) and considered a synapomorphy for Umbraculoidea by Schmekel (1985). Complete fusion of these and any other ganglia, evidenced by a double connective between the fused ganglion and its adjacent ganglion, should be confirmed at the histological level; closely adjacent ganglia can appear fused in gross dissection while still maintaining separate connectives.

The most obvious hallmark of torsion in gastropod mollusks is the crossed nerve cords of the visceral loop, in a figure-eight configuration called streptoneury (=chiastoneury). This is the condition in prosobranchs, lower heterobranchs and some architectibranchs (Haszprunar, 1985a; Mikkelsen, 1996). As part of the process of "detorsion", opisthobranchs become euthyneurous, with the visceral loop uncrossed. However, like posterior migration of the mantle cavity, the transition from streptoneurous and euthyneurous is a gradual continuum in tectibranchs. Williams (1975) coined the term "slight streptoneury" to reflect an incomplete state where the nerve cords are uncrossed in dorsal view, but where the right cord (with supraesophageal ganglion) is higher (more dorsal) than the left cord (with subesophageal ganglion). However, there are variable degrees in this category as well; for instance, in *Philine* sp., the nerve loop is uncrossed when judged at the level of the esophageal ganglia, although it remains crossed posteriorly, just in front of the visceral ganglion. Similarly, in the basal pulmonate *Chilina*, the nerve loop is uncrossed and is probably slightly streptoneurous (Harry, 1964; Brace, 1983); Harry (1964: 367), however, noted a peculiar twisting of the loop in the vicinity of the supraesophageal ganglion that "give the distinct impression of chiastoneury".

In groups such as tectibranchs where many morphological features are questionably ecophenotypic, the nervous system has been regarded as less affected by external influences than features, for example, of external anatomy (Boettger, 1955; Mikkelsen, 1993). Yet the euthyneurous state is itself homoplastic, brought about either by "detorsion" or by concentration of the nervous system, which effectively untwists the visceral loop in the process of shortening the cords. Extreme concentration results in a tightly packed and therefore uncrossed ring of ganglia immediately behind the circumesophageal nerve ring. Such a condition exists in unshelled sacoglossans, most notaspideans, *Notarchus* (Anaspidea) and nudibranchs (Schmekel, 1985).

Heterobranch descriptions (e.g., Schmekel, 1985) frequently refer to the number of ganglia on the visceral loop as five, three, or fewer than three. In opisthobranchs, fewer than five ganglia is related not to loss, but to migration of the ganglia along the visceral loop followed in some cases by fusion. Therefore, the mere counting of ganglia on the visceral loop can be potentially misleading (Dayrat and Tillier, 2000). Mikkelsen (1966) developed a notation for expressing ganglionic positions along the visceral nerve loop. The formula for a basal opisthobranch visceral loop (e.g.,

Williams, 1975: figure 1) is L-A-B-V-P-A-L (where L = pleural ganglion (paired, part of circumesophageal nerve ring), A = pallial ganglion (paired), B = subesophageal ganglion, V = visceral ganglion, P = supra-esophageal ganglion, dash = connective). The nervous system of *Acteon* reflects this unmodified formula. A cephalaspid (*Bulla*) diverging from this formula is expressed as L--A-BV-P--AL, indicating that the left pallial and subesophageal ganglia have migrated posteriorly (the latter close to the visceral; the double-dash representing a lengthened connective), and that the right pallial has migrated anteriorly, to lie close to the right pleural. (Adjacency and fusion of ganglia are expressed identically in this formula.) The formula for a species with complete concentration of the nerve loop (e.g., *Tylodina*) would be LABVPAL, the lack of dashes indicating the lack of connectives. According to the phylogenetic analysis, the plesiomorphic formula for Cephalaspidea is LA--B-V-P--AL. Migration of the supraesophageal ganglion toward the visceral ganglion (e.g., -VP-) was found apomorphic (although homoplastic) for Anaspidea (Mikkelsen, 1996). Similar consistent configurations have not been determined for other tectibranch groups.

Wägele and Willan (2000) postulated that the configuration of the radular nerve(s), emanating from the buccal ganglia to the pharynx, might be clade-specific in Opisthobranchia; unfortunately, they lacked sufficient data to include this in their phylogenetic analysis. The presence of separate radular nerves from the ganglia was suggested as the plesiomorphic state, with a single nerve from the buccal commissure judged as apomorphic. The basal pulmonate *Chilina* also has a single radular nerve (Harry, 1964).

3.8. Other characters

Chromosome number, now known for a wide variety of opisthobranchs, has been summarized by Schmekel (1985) and Curini-Galletti (1988). Fifteen pairs are known in *Hydatina* (Architectibranchia). Seventeen pairs are usual for Cephalaspidea, although members of *Smaragdinella* have 18 and *Philinoglossa* have 13. Seventeen is also recorded for Anaspidea and Sacoglossa, 12 for Pleurobranchoidea, 13 for Nudibranchia. This character has not yet shown a strong phylogenetic pattern.

Hickman (1996) advocated coding complex behavioral traits such as those seen in locomotory patterns of trochoid gastropods, by virtue of their underlying structural basis. Several attributes of opisthobranchs lend themselves to such treatment, but present understanding does not indicate homologous causal morphologies. The first is swimming, known in anaspids (*Aplysia, Akera*), cephalaspids (*Gastropteron*) and nudibranchs

(*Hexabranchus*). In the first two groups, swimming is accomplished by flapping motions of the enlarged parapodia, while in *Hexabranchus*, familiarly known as the "Spanish Dancer", the same effect is caused by flexure of an expansive mantle skirt (Thompson, 1972).

Like the well-known pigment Tyrian Purple produced from Mediterranean Muricidae during the Greco-Roman era, a purple secretion is discharged upon handling in members of *Acteon, Akera, Aplysia,* Philinidae and *Haminoea* (Ansell, 1968; Winner and Mikkelsen, 1986). The most likely source of this substance is the hypobranchial gland, a mucus-secreting gland adjacent to the gill in most gastropods. However, a specialized purple or Blochman's gland has been identified in anaspideans that differs histologically from a hypobranchial gland. Most authors have maintained that these secretions function as anti-predator devices, although others claim that they are mere excretory products. Comparative biochemistry and the homologies of these secretions or the glands that produce them have not been elucidated.

Diet is known for most opisthobranchs, and dietary specialization is very common. Guiart (1901) used diet to classify cephalaspids into carnivorous Philinoidea and herbivorous Bulloidea. Gosliner (1994) assumed his HAO to be a grazing omnivore. Although diet is a well accepted force in opisthobranch evolution, the use of carnivore/herbivore/omnivore as character states is problematic in phylogenetics for the same reasons as cited for monaulism and swimming behavior (although it has been so used; e.g., Haszprunar, 1986). Authors have frequently discussed whether carnivory or herbivory is most plesiomorphic in opisthobranchs, a question related to the origin of the Opisthobranchia. Haszprunar (1985b) maintained that the most primitive condition was carnivory, based largely on the Architectibranchia, but Mikkelsen (1996), in agreement with Purchon, (1977) found herbivory most primitive in accordance with radular and gizzard plate morphologies. The matter is not yet convincingly settled. Regardless, different types of carnivory and herbivory must also be acknowledged. For example, sacoglossans and *Haminoea* spp. are all herbivores, but the former are suctorial feeders on green algae using knife-like puncturing teeth while the latter handle filamentous algae with rather generalized radulae and crushing gizzard plates. Mikkelsen (1996) handled this inconsistent "character" within a phylogenetic context by noting patterns produced by mapping diet subsequently on a parsimony-generated cladogram.

Features of spermatozoa are among the most intriguing characters with potential for taxonomic and phylogenetic significance. Mikkelsen (1996) summarized the available data for Heterobranchs (especially the works of Healy, e.g., Healy and Willan, 1984, 1991), including length of the nucleus, midpiece membranes, acrosomal morphology, glycogen piece and midpiece helices. Many of these characters can only be assessed at the ultrastructural level. Taxonomic coverage of these data is unfortunately still too sparse for phylogenetic conclusions to be drawn.

Because of their bright coloration, "defenseless" (shell-less) bodies and concomitant potential for harboring defensive chemicals, opisthobranchs have been prime targets in recent studies in chemical ecology and natural products chemistry. *Aplysia* spp. were the first opisthobranchs to be so tested in the 1960s. In a 1995 review of taxa and chemicals known to that date, Avila (1995) listed 67 species of tectibranchs and over 350 extracts ranging from generalized "acid secretions" to newly characterized compounds named for their sources (e.g., haminol-A, aplysianin-E). Many additional unshelled sacoglossans, notaspids, and nudibranchs were also listed. Cimino *et al.* (1999) made one of the first attempts to place these data into a systematic context, suggesting evolutionary pathways from sequestering biochemicals from algal foods to *de novo* biosynthesis. To date, no phylogenetic analysis using these data has yet been attempted.

4. HISTORY AND TECTIBRANCH PHYLOGENETICS (Table 2)

The strength of phylogenetic analyses lies not only in revising taxonomic classification based on implied relationships, but also in studying characters and the presumed evolutionary changes that they have undergone. Cladistics has become a routine method in malacology only during the past two decades; however, it is also useful to analyze the evolutionary relationships discussed by early authors in a phylogenetic light. In many ways, forcing parsimony upon these earlier authors seems unjustified, yet such a test can highlight the degree of consistency between the characters that an author used and the classification that was deduced from them. Modern phylogenetic studies can then be better interpreted within the context of traditional characters and dogmatic classifications. To end this section, a synthesis will be attempted, incorporating exemplar taxa from each tectibranch clade along with most of their currently accepted synapomorphies plus the common characters that can be identified across taxa.

4.1. Henry Pilsbry

From 1894–1896, in the *Manual of Conchology*, Henry Pilsbry monographed the Tectibranchia, comprising those opisthobranch snails with (at least primitively) "one branchial plume [gill] situated on the right side; mantle and shell developed'' (Pilsbry, 1894 in 1894–1896: 134). Although the configuration of the gill was all-important in this definition, it was the presence of a shell – usually a well-calcified external shell – which was most critical to its application. Other diagnostic features of tectibranchs at the species level included soft-tissue anatomical characters (e.g., shape of the head and body). Despite this, families were diagnosed almost exclusively by ''hard-part'' features, e.g., of the shell, operculum, radular teeth and (in the case of cephalaspideans and notaspideans) gizzard plates.

Pilsbry's Tectibranchia (Table 2) comprised nine families of archetypal bubble shells (Cephalaspidea), plus two families of sea hares (Anaspidea) and three of side-gilled slugs (Notaspidea). Because of the presence of a shell, Pilsbry placed the family Oxynoidae in Anaspidea, although he acknowledged that Oxynoidae had earlier been associated with the unshelled leaf slugs (Sacoglossa), even though Oxynoidae and sacoglossans shared the feature of dagger-shaped teeth (history summarized by Pilsbry, 1896 in 1894–1896: 161). Sacoglossa (=Ascoglossa) was thus the first opisthobranch taxon to be acknowledged as having anatomical characters equal in systematic value to those of the shell (summarized by Wägele and Willan, 2000). Pilsbry nevertheless insisted on a tectibranch arrangement, calling Oxynoidae ''intermediate between Tectibranchiata and Ascoglossa, but nearer the latter'' (Pilsbry, 1896 in 1894–1896: 162). Tectibranchia stood in juxtaposition with the unshelled sea slugs (Nudibranchia), which in Pilsbry's view had evolved directly from tectibranchs; it is unclear where Pilsbry placed Sacoglossa in this scheme, although it seems to have been equal in rank to Tectibranchia and Nudibranchia.

In his taxonomic descriptions, Pilsbry presented characters and character states that, in his view, separated or united the various groups of tectibranch families. An evolutionary tree that he might have presented, constructed from partial trees he included in his monograph plus descriptive statements about Cephalaspidea, Anaspidea and Notaspidea, appears in Figure 9A. In this tree (when discussed phylogenetically), the cephalic shield is the single synapomorphy for Cephalaspidea, as is the notum for Notaspidea. Anaspidea was based on the absence of both of these and thus has no synapomorphies. In this evolutionary scenario, loss of the operculum occurs twice, internalization/loss of the shell three times, and reduction/loss of the radula four times.

Pilsbry presented his diagnostic characters in comparable sets, a manner that is amenable to cladistic coding and analysis (see Appendix I). Perhaps not surprisingly, with the few data provided, the analysis reconstructed none of Pilsbry's three presumed clades exactly and most of the clades were weakly supported. Pilsbry's Notaspidea was consistently supported, based on two synapomorphies (notum, genital orifices close together). No

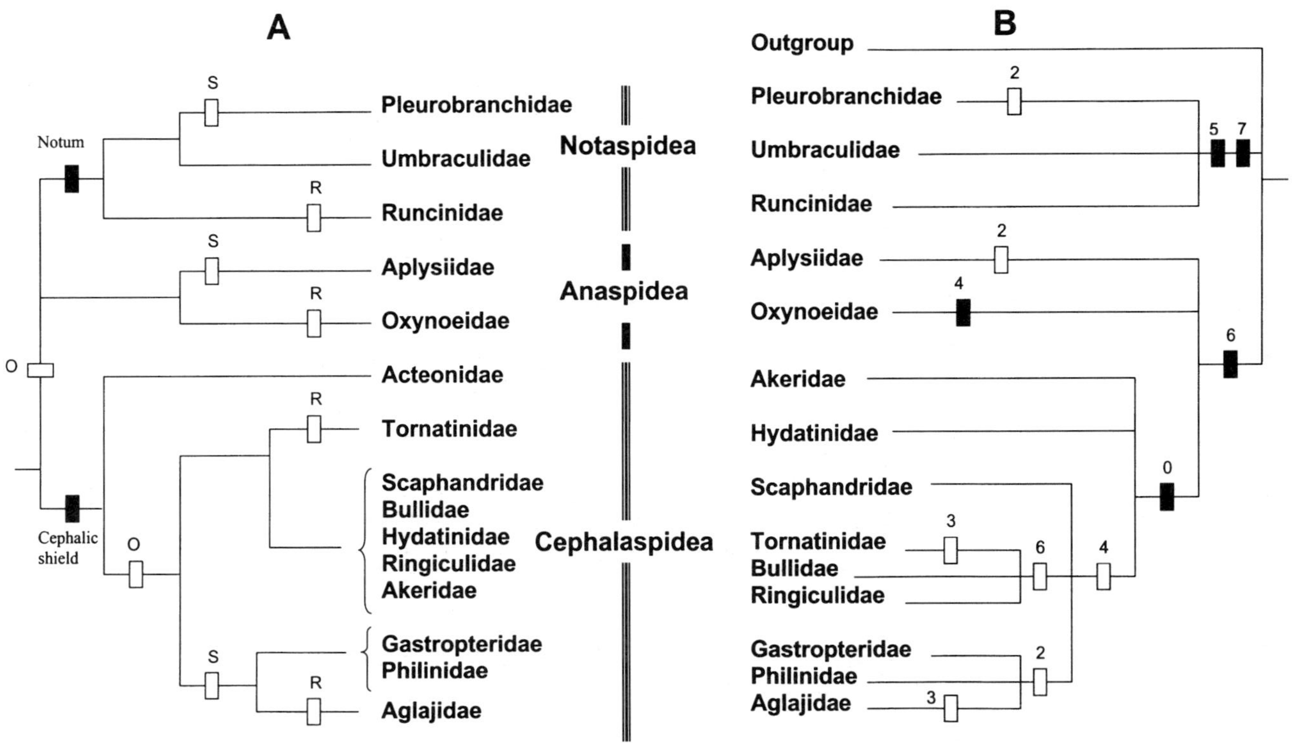

Figure 9 Pilsbry's (1894–1896) view of tectibranch evolution. Solid rectangles indicate synapomorphies; open rectangles indicate homoplastic character state changes. A. Evolutionary tree drawn from his character descriptions. O, opercular loss; R, radular loss or reduction; S, shell internalization or loss. B. Cladogram constructed from his diagnostic characters (excluding Acteonidae, 15 steps, CI 0.66, RI 0.81). See Appendix I for explanation of character numbers.

other clades were this strongly supported, however, the internally shelled cephalaspids (Aglajidae–Gastropteridae–Philinidae) were consistently reconstructed, as was the clade of Tornatinidae–Bullidae–Ringiculidae (TBR) with parapodial lobes. Acteonidae was the most widely relocating taxon, but always appeared in one of two locations: with the TBR clade or in a basal polytomy with the outgroup. With Acteonidae removed from the analysis, one tree resulted (Figure 9B) that restored Notaspidea and Cephalaspidea although neither in the topology that Pilsbry suggested; Anaspidea is not monophyletic but its members are united with Cephalaspidea as a monophyletic clade based on parapodial lobes. Pilsbry apparently recognized subtle differences in what he called epi-, para- and pleuropodial lobes. These differences were unclear and thus combined for this analysis, a decision that led directly to the final tree topology. This tree shows similar character-state changes to Pilsbry's original: operculum lost twice, shell internalized three times, and radula reduced three times.

4.2. Johannes Thiele

The molluscan classification by Johannes Thiele (1929–1935) in his classic *Handbuch der systematischen Weichtierkunde* still forms the basis of collection organization in many museums today. The second part (of four) of this work covered Opisthobranchia, and provided taxonomic characters down to genus level. By this point in malacological history, many anatomical characters were included in family diagnoses, from external anatomy and digestive, reproductive and nervous systems. Thiele initiated the break up of Tectibranchia, renaming it Pleurocoela (resurrecting a name used by Guiart, 1901), and restricting it to the shelled Cephalaspidea and Anaspidea (Table 2). Cephalaspidea received several new families (some by division of other families, some based on new forms), and one family (Runcinidae) transferred into Cephalaspidea from Notaspidea (on the basis of an external reproductive feature). The shelled Oxynoidae was reunited with the otherwise unshelled Sacoglossa, in a group of equal rank with Pleurocoela, based on the dagger-shaped teeth in a suctorial feeding apparatus. Notaspidea, with both shelled and unshelled forms, was united with unshelled Nudibranchia in the group Acoela, by virtue of a reduced mantle cavity.

Thiele did not list his diagnostic characters as dichotomously as Pilsbry, nor did he present any of his taxonomic relationships in evolutionary tree format. Therefore, it is difficult to attempt a phylogenetic test of his classification. However, in Part 4 of his handbook ("Outline of the

Phylogeny of Mollusks", pp. 1073–1121), he commented on the relationships among tectibranchs through the characters that he obviously considered important. Featured heavily were shell form, radula and configuration of the nervous and reproductive systems. Included was an extensive discussion on the relationship between prosobranchs and opisthobranchs, especially regarding the transitional nature and placement of Pyramidellidae and Acteonidae.

4.3. Nils Odhner

The opisthobranch classification developed by Odhner in the 1930s and 1940s (especially Odhner, 1939; classification detailed by Franc, 1968, from notes furnished by Odhner in 1957–1959) has been used largely unchanged since its publication (Wägele and Willan, 2000). By this time, Tectibranchia had disappeared in name and in concept. Odhner proposed 12 groups of opisthobranchs of equal (ordinal) rank: Cephalaspidea, Aplysiacea (= Anaspidea), Sacoglossa, Pleurobranchacea (= Notaspidea), two minor groups (Runcinacea, Acochlidiacea), two holoplanktonic pteropod groups (Thecosomata, Gymnosomata) and four groups of Nudibranchia (Table 2). Many sources of characters were included, ranging from shell and external anatomy to configurations of the nervous and reproductive systems.

Odhner was one of the first malacologists to consider characters as existing in two or more character states. In his 1939 paper on Norwegian opisthobranchs, Odhner provided dichotomous keys to the various higher groups, to Cephalaspidea and to several groups of Nudibranchia. His binary characters included shell external or internal, nerve ring prepharyngeal or post-pharyngeal, parapodia absent or present, to name only a few. Unfortunately, he did not provide statements that reveal his thoughts on tectibranch evolution, nor did his characters extend consistently across a sufficient number of tectibranch taxa to make construction of a data matrix possible. Nevertheless, his anatomical work and taxonomic revisions in tectibranch systematics helped to extend the transfer, begun earlier by Thiele, away from the shell and toward an emphasis on soft anatomy.

4.4. Modern cladists

Tectibranchia disappeared as an accepted taxon by the 1940s. Most of its former members have remained in its "effective replacement", Cephalaspidea, while others have been redistributed among the various heterobranch groups, most of which now include both shelled and unshelled taxa (Table 2). This rearrangement has eliminated much of the character state variability for which Tectibranchia, and later Cephalaspidea, were infamous (Mikkelsen, 1993). Several groups, notably Diaphanidae (plus Notodiaphanidae) and Runcinidae, have remained problematic and of uncertain affinities; because of this, they are sometimes allocated to their own higher-level taxa (Diaphanoidea and Runcinoidea, respectively), but usually within Cephalaspidea. A new basal taxon, Architectibranchia, now holds the less-derived former-cephalaspid families.

Only four morphological phylogenetic analyses involving tectibranchs have been published in a testable cladistic format; each of these has already been alluded to in the previous character discussions.

Mikkelsen (1996) evaluated Cephalaspidea, with exemplars from Architectibranchia, Anaspidea and shelled Sacoglossa. An analysis of 47 characters and 20 taxa (including a hypothetical all-zero and a lower heterobranch (*Gegania*) as outgroups) produced monophyletic clades representing Sacoglossa, Anaspidea and Cephalaspidea, the last in two monophyletic superfamilies. Three traditional cephalaspids (*Ringicula, Hydatina, Acteon*) were removed to the Architectibranchia or the "lower heterobranchs". Most informative in the result were characters related to the digestive system (especially the gizzard) and reproductive system (sperm-storage pouches and sperm grooves/ducts).

Jensen (1996, 1997; reanalyzed by Mikkelsen, 1998a) analyzed the Sacoglossa, finding both the shelled (Oxynoidea) and unshelled (Plakobranchoidea) sacoglossans to form monophyletic clades. Synapomorphies for the shelled taxa were the presence of a shell adductor muscle, a lamellate gill, and a female genital papilla. Informative characters of the unshelled taxa included absence (loss) of the shell and cephalic shield, the presence of rhinophores, and details of the nervous and reproductive systems.

Willan (1987) reappraised Notaspidea both cladistically and phenetically with similar results. His phylogenetic analysis was "hand-drawn", indicated no clear outgroup, and included many characters (e.g., autapomorphies, polarity conflicts) that are considered problematic by today's standards. For these reasons, his dataset has been reanalyzed here (see Appendix II). The final result, using 41 characters and 12 taxa (including a hypothetical all-zero outgroup), is very similar to Willan's (1987: figure 20) original hypothesis and is well supported by numerous synapomorphies. The subgroup Pleurobranchoidea, as well as its three unshelled members (*Euselenops*, *Pleurobranchaea* and *Pleurobranchella*), were better resolved than in Willan's original analysis, however, the three externally shelled notaspideans (i.e., the "tectibranchs" *Umbraculum*, *Tylodina* and

Anidolyta) remained unresolved. Characters of the gill, oral tube, pedal gland and copulatory organ provided synapomorphies for the two major clades.

In 1991, Wägele and Hain described a new genus and species of notaspid, *Tomthompsonia spiroconchalis*, with a suite of unique attributes for Pleurobranchoidea in which it was placed. Importantly, this taxon has an internal spirally coiled shell, with a definitely heterostrophic protoconch (discussed earlier). The excellent original description easily allowed this genus to be added to the notaspid analysis and rerun. The result was four most-parsimonious trees of 75 steps (CI 0.74, RI 0.76), identical to the previous result but with *Tomthompsonia* added in one of two locations: (1) basally in the Pleurobranchoidea, or (2) unexpectedly, basally within the clade of unshelled pleurobranchoideans (*Euselenops–Pleurobranchella–Pleurobranchaea*). The latter placement was supported by the location of the anus (synapomorphy; in front of the hind end of the gill basement membrane) and the prostate gland (homoplastic; as a discrete organ). Nevertheless, *Tomthompsonia* shares four character states with all or some of the shelled pleurobranchoideans (11, 29, 36, 38); clearly the relationships of this interesting taxon need to be carefully re-evaluated. The consensus tree naturally leaves *Tomthompsonia* in an unresolved polytomy with the two clades of pleurobranchoideans (Figure 10).

Willan (1987) provided no synapomorphies for the shelled pleurobranchoideans, or family Pleurobranchidae. From this reanalysis, there are three: rectangular shell outline (as opposed to circular), auriculate-oval shell shape and a narrow oral veil relative to the body.

Although a full discussion is beyond the scope of this review, Wägele and Willan's (2000) phylogeny of Nudibranchia considered many characters and character states throughout Opisthobranchia (discussed above). One conclusion from their study is of particular interest to this discussion, namely the description of a new taxon, Nudipleura, comprising the Nudibranchia plus one of the two major groups from Notaspidea, the Pleurobranchoidea. This clade was identified earlier, although not named, by Schmekel (1985). Cited synapomorphies are the presence of a blood gland, an androdiaulic reproductive system, and loss of the osphradium. Despite the many points of anatomical evidence, it must be noted that the cladograms produced by their analysis did not support monophyly of this new taxon.

4.5. An attempt at synthesis

A phylogenetic analysis of all opisthobranchs, shelled and unshelled, has yet to be presented. Lack of such a synthesis leaves the interrelationships

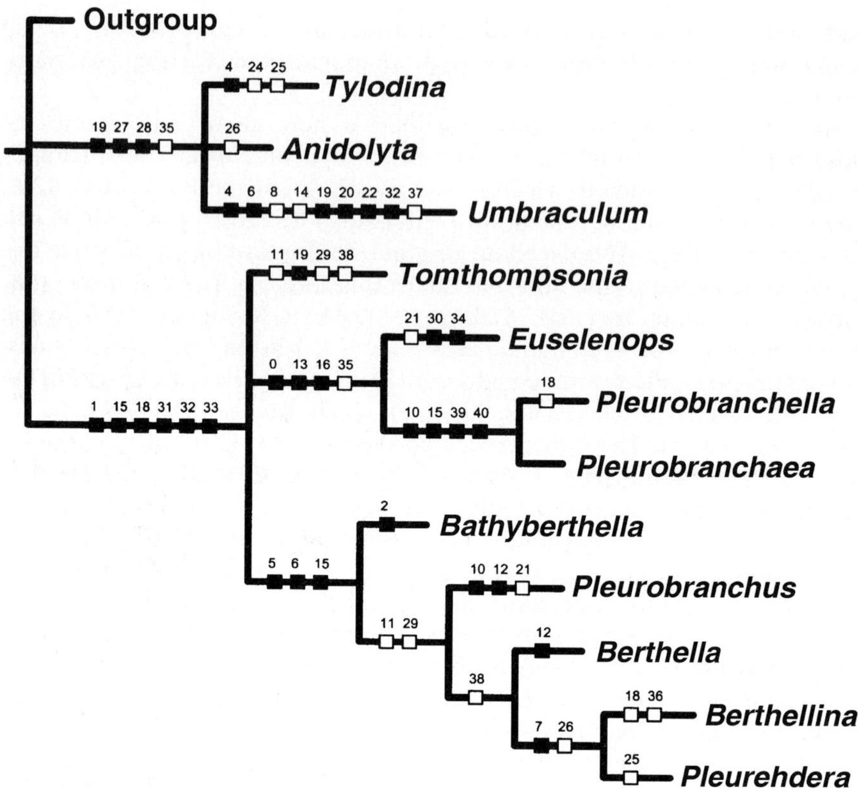

Figure 10 Cladogram (length 75 steps, CI 0.74, RI 0.76) from reanalysis of Willan's (1987) phylogenetic analysis of Notaspidea, modified as outlined in Appendix II and with the addition of *Tomthompsonia*. Symbols as in Figure 9. See Appendix II for explanation of character numbers.

of the tectibranch taxa unresolved, and questions such as the possible polyphyly of Notaspidea unanswered (Willan, 1983; Schmekel, 1985; Wägele and Willan, 2000). Combining the four existing morphological analyses of opisthobranchs (the above three, plus that for Nudibranchia by Wägele and Willan, 2000) is complicated by few overlapping characters and untested character state homologies. Nevertheless, a preliminary attempt is made here, to judge our present state of knowledge and to identify areas of needed research.

Twenty-six genus-level tectibranch taxa were selected, representing the major clades confirmed in each of the previous analyses (Willan, 1987; Jensen, 1996, 1997; Mikkelsen, 1996; Wägele and Willan, 2000). These comprised two architectibranchs, four cephalaspids, two anaspids, six sacoglossans, six notaspideans and four nudibranchs, plus *Acteon* and *Gegania* representing the lower heterobranchs. Of the 187 total characters in these four analyses, 41 overlapped in part, were useful at this level, and/or were appropriate for coding across the range of taxa (see Appendix III).

The resulting cladogram (Figure 11) suggested convincingly that *Acteon* should be placed within the lower heterobranchs, and that Opisthobranchia, Cephalaspidea and Sacoglossa are monophyletic groups. Three synapomorphies support Opisthobranchia: loss of the adult operculum, presence of parapodia (with three reversals) and loss of the posterior stomach chamber (resulting in a flow-through configuration; reversed in nudibranchs). Cephalaspidea showed two synapomorphies: three gizzard plates and flexed/exogyrous ciliated strips; a posterior foot was present in cephalaspids and in *Akera*. Sacoglossa showed three synapomorphies: loss of the lateral radular teeth, pharyngeal pouches (with two reversals) and presence of an ascus; a shortened visceral nerve loop was present, reversed in *Cylindrobulla* and also present in the Nudibranchia–Notaspidea combined clade. Cephalaspidea (without *Runcina, Acteon* and the acteonid *Pupa*) and Sacoglossa were also monophyletic in the 28S rRNA molecular phylogeny by Dayrat *et al.* (2001); Opisthobranchia was not found to be monophyletic by this study.

Other traditional groups were less consistently represented in the final trees. Anaspidea was monophyletic in only 3 of the 45 trees, supported by one synapomorphy (filter chamber) and absence of a typhlosole (highly homoplastic). Within Cephalaspidea, Philinoidea was consistently monophyletic (with three synapomorphies in the indented rachidian, absence/loss of gizzard spines and flexed ciliated strips), but Bulloidea was paraphyletic in all trees. Similarly within Sacoglossa, the unshelled Plakobranchoidea was usually monophyletic (in 44 of the 45 trees; weakly supported by the absence of a shell), but the shelled Oxynoidea was usually paraphyletic (in 43 of 45 trees). Only 3 of the 45 trees found Notaspidea monophyletic (always as the "crown group" of a paraphyletic Nudibranchia), weakly supported by enrolled rhinophores (homoplastic) and a protrusible oral tube (homoplastic and reversed in *Umbraculum*). The shelled notaspideans (Pleurobranchidae, represented by *Bathyberthella* and *Pleurobranchus*) were monophyletic in only 12 of the 45 trees, weakly supported by an internalized rectangular shell. Non-monophyly of the Anaspidea and Notaspidea was also found in the 28S rRNA molecular phylogeny by Dayrat *et al.* (2001).

Two non-traditional clades showed relatively strong support. In the 6 trees where Anaspidea was monophyletic and in 28 of the remaining 39 cases (excluding *Aplysia*), Anaspidea formed a monophyletic clade with Cephalaspidea (a result suggested earlier by Mikkelsen, 1996, and Dayrat

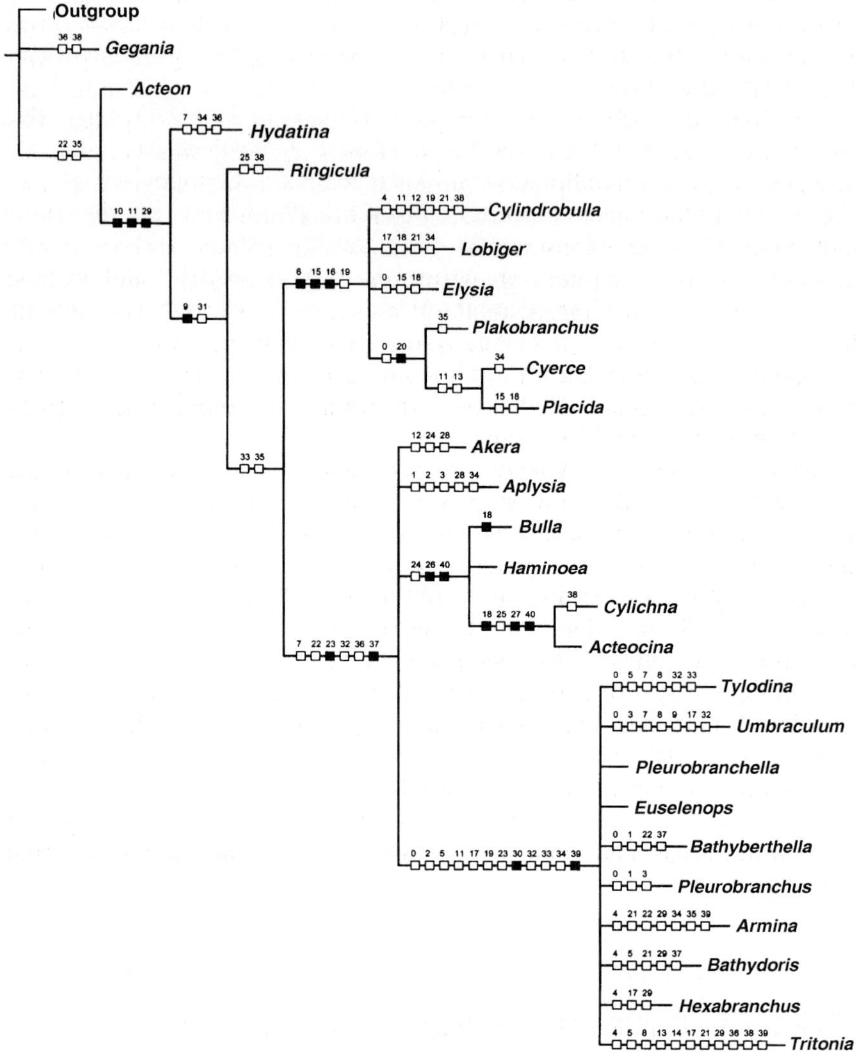

Figure 11 Strict consensus tree (length 140 steps, CI 0.32, RI 0.55) from the combined analysis of representative lower heterobranchs, architectibranchs, cephalaspids, anaspids, sacoglossans, notaspids and nudibranchs, using characters from the four published phylogenies (see Appendix III). Symbols as in Figure 9. See Appendix III for explanation of character numbers.

and Tillier, in press). Four characters lent support to this clade (depending on configuration of the individual tree): an esophageal gizzard with gizzard plates, a posterior foot (reversed in *Aplysia*), absence of the internal sperm duct (homoplastic) and presence of the external sperm groove (homoplastic). Finally, Wägele and Willan's (2000) new taxon Nudipleura (comprising the Nudibranchia plus pleurobranchoidean Notaspidea) was monophyletic in 42 of the 45 trees. However, only two homoplastic characters unite the Nudipleura: presence of an internal sperm duct and presence of a blood gland (reversed in *Armina-Tritonia*). A monophyletic clade comprising the Nudibranchia plus all of Notaspidea was consistently supported by two synapomorphies (presence of an oral veil and a blood gland (ambiguous in Umbraculoidea)) and a suite of other homoplastic characters (see Figure 11). A clade comprising Anaspidea, Cephalaspidea, Notaspidea and Nudibranchia was supported by the presence of a gizzard (reversed in Notaspidea–Nudibranchia) and a fused radular nerve (with several ambiguities and reversals).

Most of these series of synapomorphies are from non-shell, soft-anatomy characters. This result is not surprising given the fact mentioned earlier that tectibranch shell characters are useful mainly at the species level; these analyses have used genera as taxonomic units. Most loss-type characters (shell, rachidian, jaws, typhlosole) showed high degrees of convergence. Some derived character state convergences suggest homology problems and the need for further study. According to this tree, rhinophores have evolved three times, thickened labial cuticle four times, oral tentacles five times and propodial tentacles three times, recommending re-interpretation of these structures in the various opisthobranch groups. Likewise, development of a lateral external groove has occurred three times on this tree (and reversed twice), calling into question the origin and homology of what ultimately appears as a simple ciliated groove.

5. WHENCE *ACTEON*, THE "TYPICAL" OPISTHOBRANCH?

Acteon, typified by the European genotype, *A. tornatilis* (Linné, 1758), has repeatedly served as the basal representative of all Opisthobranchia, i.e., as the typical and/or most primitive opisthobranch. Although it is usually portrayed with a "transitional" combination of prosobranch and opisthobranch characters (Fretter and Graham, 1954: 582; Boss, 1982), "as a matter of convenience", *Acteon* has been placed as a basal opisthobranch (i.e., in Cephalaspidea) rather than among the derived prosobranchs or in its own group. As discussed previously, it continues today to serve as an opisthobranch "token" in larger gastropod studies.

Yet, anatomical evidence shows *Acteon* to be highly derived in morphology. A prime example is its radular teeth, which are minute, uniform, denticulate hooks multiplied greatly in number (ca. 100–200 teeth each side, but often expressed as "∞") for feeding on polychaete worms. When taken as part of the most basal opisthobranch, this radular structure ("broad with many similar teeth") has been considered plesiomorphic, although the unlikelihood of this interpretation has not gone unnoticed. Hurst (1965) remarked that this type of radula was unlikely to have given rise to more derived forms, e.g., the pincer-type in Philinoidea. Rudman (1972d) called *Acteon*'s radula "anomalous", while Beeman and Williams (1980) considered it "degenerate". More meaningfully, Fretter (1939; also Fretter and Graham, 1954: 582) interpreted its "apparent simplicity" as secondary specialization associated with diet.

There is a mounting body of evidence, historic and modern, for transferring *Acteon* to the "lower heterobranchs". Guiart (1901) separated it from other existing tectibranch taxa, as *Pleurocoeles: Diaules*, or Pleurocoela (Cephalaspidea plus Anaspidea) with a diaulic (rather than monaulic) reproductive system. Thiele (1935 in 1929–1935: 1096) supported a close relationship of *Acteon* with Pyramidellidae (currently among the "lower heterobranchs"), citing similarities in the shell, operculum, radula, nervous system, ciliated strips and eye position. Thiele considered Pyramidellidae a prosobranch mesogastropod group and used the similarity between it and Acteonidae as evidence for a mesogastropod origin of opisthobranchs. In his study of opisthobranch reproductive morphology, Ghiselin (1966: figure 7) configured the two families Acteonidae and Hydatinidae as a clade quite separate from Cephalaspidea s. s., actually as in a sister-relationship to the nudibranchs-plus-notaspids clade. Beeman (1973) wrote, "I agree with Dr. [Hennig] Lemche [famous Copenhagen anatomist in the 1950s through 1970s] that neither the Pyramidellidae nor the Acteonidae should be included in the Cephalaspidea . . .", although no publication by Lemche can be located that expresses this viewpoint. Yonow and Renwrantz (1986) found granular cells in the hemolymph of *A. tornatilis* with large globular inclusions; to date, globule-containing hemocytes have only been found in Pyramidellidae (although very few other heterobranchs have been examined).

Mikkelsen's (1996) phylogenetic analysis of cephalaspids and selected other tectibranchs suggested that *Acteon* more logically belongs in the "lower heterobranchs," outside of the true opisthobranchs. This was based on a single synapomorphy – the gametolytic gland and receptaculum seminis on a single stalk – a character state that has been recorded in three other lower heterobranchs (two in Mathildidae; one in Architectonicidae).

The extent of this character throughout lower heterobranchs, however, has not yet been determined, and thus its taxonomic value is untested. Other recent phylogenies seem to support the move of *Acteon* out of Opisthobranchia. In their gastropod phylogeny, Ponder and Lindberg (1997: figure 5) found *Acteon* basal to other euthyneurans (i.e., among the "lower heterobranchs"), but nevertheless retained it within Euthyneura. Three molecular studies have further supported its transfer to the "lower heterobranchs": (1) a 28S rRNA phylogeny of Gastropoda by Tillier *et al.* (1992); (2) a 16S rRNA phylogeny of Euthyneura by Thollesson (1999); and a 28S rRNA phylogeny of Euthyneura by Dayrat *et al.* (2001). Finally, this analysis, as preliminary as it is, corroborates *Acteon*'s placement outside of Opisthobranchia to (at least for the moment) the "lower heterobranchs".

It should be noted that this discussion is centered upon the genus *Acteon*, and not on other members of the family Acteonidae as presently understood or on the family as a whole. For instance, other traditional acteonids (e.g., *Pupa*) have a radular formula of (20–100).0.(20–100), with relatively large hook-shaped laterals increasing in size and decreasing in denticulation peripherally (see e.g., Gosliner, 1994: figure 19B). Aspects of their anatomy relative to other characters discussed above have not been extensively investigated. It is conceivable that the traditional members of Acteonidae are not monophyletic; in view of the great historical and phylogenetic importance of this taxon, this is a hypothesis in need of further investigation.

ACKNOWLEDGEMENTS

I thank my associate Rüdiger Bieler (Field Museum of Natural History, Chicago) for use of the *Aplysia* shell photograph (Figure 1H) and, as always, for improving my coverage of the gastropod literature and my writing style. Heike Wägele (Ruhr-Universität, Bochum, Germany) and Richard Willan (Museum and Art Gallery of the Northern Territory Museum, Darwin, Australia) allowed my use of their nudibranch phylogeny paper prior to its publication. Arthur E. Bogan (North Carolina State Museum of Natural Sciences, Raleigh) and an anonymous reviewer brought important pieces of literature to my attention. Steve Thurston (AMNH) assisted with other shell photography and completed the artistic renderings of Figures 2–8 and the cover. Part of this paper was presented at a symposium on "The Phylogeny and Biogeography of Gastropods" hosted by the Malacological Society of Japan, Yamaguchi, in January 2000. The generous hospitality of my hosts at this conference, Drs Takashi Okutani, Takahiro Asami and Hiroshi Fukuda, is gratefully acknowledged.

REFERENCES

Ansell, A. D. (1968). Defensive adaptations to predation in the Mollusca. Proceedings of the Symposium on Mollusca: held at Cochin, 12–16 January, 1968, 3 volumes, *Marine Biological Association of India* **2**, 487–512.

Avila, C. (1995). Natural products of opisthobranch molluscs: a biological review. *Oceanography and Marine Biology: an Annual Review* **33**, 487–559.

Bandel, K. (1994). Triassic Euthyneura (Gastropoda) from St. Cassian Formation (Italian Alps) with a discussion on the evolution of the Heterostropha. *Freiberger Forschungsheft* **452**, 79–100.

Beeman, R. D. (1973). Reader forum. *Opisthobranch Newsletter* **5**, 1–2.

Beeman, R. D. and Williams, G. C. (1980). Opisthobranchia and Pulmonata: the sea slugs and allies. *In* "Intertidal Invertebrates of California" (R. H. Morris, D. P. Abbott, and E. C. Haderlie, eds), pp. 308–354. Stanford University Press, Stanford, California.

Bergh, R. S. (1901). Malacologische Untersuchungen. 4. Abtheilung. 3. Abschnitt. Bullacea. *In* "Reisen im Archipel der Philippinen" (C. Semper, ed.), pp. 209–312, plates 17–24. Wissenschaftliche Resultate, 7. Band, C. W. Kreidel, Wiesbaden.

Bieler, R. (1992). Gastropod phylogeny and systematics. *Annual Reviews of Ecology and Systematics* **23**, 311–338.

Bieler, R., Ball, A. D. and Mikkelsen, P. M. (1998). Marine Valvatoidea – comments on anatomy and systematics with description of a new species from Florida (Heterobranchia: Cornirostridae). *Malacologia* **40**, 305–320.

Boettger, C. R. (1955 [1954]). Die Systematik der euthyneuren Schnecken. *Verhandlungen der Deutschen Zoologischen Gesellschaft (Zoologischer Anzeiger: Supplementband)* **18**, 253–280.

Boss, K. J. (1982). Mollusca. *In* "Synopsis and Classification of Living Organisms, Volume 1" (S. P. Parker, ed.), pp. 945–1166. McGraw-Hill, New York.

Brace, R. C. (1977). The functional anatomy of the mantle complex and columellar muscle of tectibranch molluscs (Gastropoda: Opisthobranchia), and its bearing on the evolution of opisthobranch organization. *Philosophical Transactions of the Royal Society of London* B **277**, 1–56.

Brace, R. C. (1983). Observations on the morphology and behaviour of *Chilina fluctuosa* Gray (Chilinidae) with a discussion on the early evolution of pulmonate gastropods. *Philosophical Transactions of the Royal Society of London* B **300**, 463–491.

Brezina, V., Orekhova, I. V. and Weiss, K. R. (1996). Functional uncoupling of linked neurotransmitter effects by combinatorial convergence. *Science, NY* **273**, 806–810.

Burn, R. and Bell, K. N. (1974). Description of *Retusa pelyx* Burn sp. nov. (Opisthobranchia) and its food resources from Swan Bay, Victoria. *Journal of the Malacological Society of Australia* **3**, 37–42.

Cimino, G., Fontana, A. and Gavagnin, M. (1999). Marine opisthobranch mollusks: chemistry and ecology in sacoglossans and dorids. *Current Organic Chemistry* **3**, 327–372.

Clark, K. B., Jensen, K. R. and Stirts, H. M. (1990). Survey for functional kleptoplasty among West Atlantic Ascoglossa (= Sacoglossa) (Mollusca: Opisthobranchia). *The Veliger* **33**, 339–345.

Curini-Galletti, M. C. (1988). Analyse du caryotype de *Runcina coronata* (Gastropoda Cephalaspidea). *Cahiers de Biologie Marine* **29**, 313–318.

Dayrat, B. and Tillier, S. (2000). Taxon sampling, character sampling and systematics: how gradist presuppositions created additional ganglia in euthyneuran taxa. *Zoological Journal of the Linnean Society* **129**, 403–418.

Dayrat, B. and Tillier, S. (in press). Evolutionary relationships of euthyneuran gastropods (Mollusca): a cladistic re-evaluation of morphological characters. *Zoological Journal of the Linnean Society* [cited by Dayrat *et al.*, 2001].

Dayrat, B., Tillier, A., Lecointre, G. and Tillier, S. (2001). New clades of euthyneuran gastropods (Mollusca) from 28S rRNA sequences. *Molecular Phylogenetics and Evolution* **19**, 225–235.

de Maintenon, M. and Mikkelsen, P. M. (2001). Late reproductive system development in two cephalaspideans (Gastropoda, Opisthobranchia): *Bulla striata* Bruguière, 1792, and *Acteocina atrata* Mikkelsen & Mikkelsen, 1984. *The Veliger* **44**, 237–260.

Eales, N. B. (1938). A systematic and anatomical account of the Opisthobranchia. *Scientific Reports of the John Murray Expedition* **5**, 77–122, plate 1.

Eales, N. B. (1960). Revision of the world species of *Aplysia* (Gastropoda, Opisthobranchia). *Bulletin of the British Museum (Natural History), Zoology* **5**, 269–404.

Edlinger, K. (1980a). Beitrage zur Anatomie, Histologie, Ultrastruktur und Physiologie der chemischen Sinnesorgane einiger Cephalaspidea (Mollusca, Opisthobranchia). *Zoologischer Anzeiger* **205**, 90–112.

Edlinger, K. (1980b). Zur Phylogenie der chemischen Sinnesorgane einiger Cephalaspidea (Mollusca-Opisthobranchia). *Zeitschrift für Zoologische Systematik und Evolutionsforschung* **18**, 241–256.

Eliot, C. (1906). Nudibranchs and tectibranchs from the Indo-Pacific. II. Notes on *Lophocercus*, *Lobiger*, *Haminaea*, and *Newnesia*. *Journal of Conchology* **11**, 298–315, plate 5.

Emberton, K. C. (1991). The genitalic, allozymic and conchological evolution of the tribe Mesodontini (Pulmonata: Stylommatophora: Polygyridae). *Malacologia* **33**, 71–178.

Emberton, K. C., Kuncio, G. S., Davis, G. M., Phillips, S. M., Monderewicz, K. M. and Guo, Y. H. (1990). Comparison of recent classifications of stylommatophoran land-snail families, and evaluation of large-ribosomal-RNA sequencing for their phylogenetics. *Malacologia* **31**, 327–352.

Franc, A. (1968). Sous-classe des opisthobranches. *In* "Traité de Zoologie: Anatomie, Systématique, Biologie, Tome V Mollusques Gastéropodes et Scaphopodes (Fascicule III)" (P.-P. Grasse, ed.), pp. 608–893. Masson et Cie Editeurs, Paris.

Fretter, V. (1939). The structure and function of the alimentary canal of some tectibranch molluscs, with a note on excretion. *Transactions of the Royal Society of Edinburgh* **59**, 599–646.

Fretter, V. (1960). Observations on the tectibranch *Ringicula buccinea* (Brocchi). *Proceedings of the Zoological Society of London* **135**, 537–549.

Fretter, V. and Graham, A. (1954). Observations on the opisthobranch mollusc *Acteon tornatilis* (L). *Journal of the Marine Biological Association of the United Kingdom* **33**, 565–585.

Gascoigne, T. and Sordi, M. (1980). A redescription of *Placida viridis* Trinchese, 1873 (Gastropoda: Ascoglossa). *Journal of Conchology* **30**, 167–179.

Geiger, D. L. and Jung, P. (1996). A shell of *Floribella aldrichi* (Dall, 1890), a large seahare (Mollusca: Opisthobranchia: Aplysiidae) from the Neogene of the northern Dominican Republic. *Journal of Conchology* **35**, 437–444.

Ghiselin, M. T. (1966 [1965]). Reproductive function and the phylogeny of opisthobranch gastropods. *Malacologia* **3**, 327–378.

Gohar, H. A. F. and Soliman, G. N. (1963). The biology and development of *Hexabranchus sanguineus* (Rüpp. and Leuck.) (Gastropoda, Nudibranchiata). *Publications of the Marine Biological Station, Al-Ghardaqa* **12**, 219–247, 2 plates.

Gosliner, T. M. (1978). "The Evolution of the Cephalaspidea (Mollusca: Gastropoda) and its Implications to the Origins and Phylogeny of the Opisthobranchia". PhD thesis, University of New Hampshire, Durham, New Hampshire.

Gosliner, T. M. (1980). Systematics and phylogeny of the Aglajidae (Opisthobranchia: Mollusca). *Zoological Journal of the Linnean Society* **68**, 325–360.

Gosliner, T. M. (1981). Origins and relationships of primitive members of the Opisthobranchia (Mollusca: Gastropoda). *Biological Journal of the Linnean Society* **16,** 197–225.

Gosliner, T. M. (1994). Gastropoda: Opisthobranchia. *In* "Microscopic Anatomy of Invertebrates, Volume 5, Mollusca I" (F. W. Harrison and A. J. Kohn, eds), pp. 253–355. Wiley Liss, New York.

Gosliner, T. M. and Ghiselin, M. T. (1984). Parallel evolution in opisthobranch gastropods and its implications for phylogenetic methodology. *Systematic Zoology* **33**, 255–274.

Guiart, J. (1901). "Contribution à l'Étude des Gastéropodes Opisthobranches et en Particulier des Céphalaspides". PhD Dissertation, Université de Paris.

Hadfield, M. G. and Switzer-Dunlap, M. (1984). Opisthobranchs. *In* "The Mollusca, Volume 7, Reproduction" (A. S. Tompa, N. H. Verdonk and J. A. M. van den Biggelaar, eds), pp. 209–350. Academic Press, Orlando, Florida.

Harry, H. W. (1964). The anatomy of *Chilina fluctuosa* Gray reexamined, with prolegomena on the phylogeny of the higher limnic Basommatophora (Gastropoda: Pulmonata). *Malacologia* **1**, 355–385.

Hartley, T. W. (1964). Egg laying and early development of *Umbraculum sinicum* Gmelin. *Journal of the Malacological Society of Australia* **8**, 33–35.

Haszprunar, G. (1985a). Zur Anatomie und systematischen Stellung der Architectonicidae (Mollusca, Allogastropoda). *Zoologica Scripta* **14**, 25–43.

Haszprunar, G. (1985b). The Heterobranchia – a new concept of the phylogeny of the higher Gastropoda. *Zeitschrift für Zoologische Systematik und Evolutionsforschung* **23**, 15–37.

Haszprunar, G. (1985c). On the anatomy and systematic position of the Mathildidae (Mollusca, Allogastropoda). *Zoologica Scripta* **14**, 201–213.

Haszprunar, G. (1986). Fine morphology of gastropod osphradia. *Proceedings of the Eighth International Malacological Congress, Budapest, 1983*, pp. 101–104. Hungarian Natural History Museum, Budapest.

Healy, J. M. and Willan, R. C. (1984). Ultrastructure and phylogenetic significance of notaspidean spermatozoa (Mollusca, Gastropoda, Opisthobranchia). *Zoologica Scripta* **13**, 107–120.

Healy, J. M. and Willan, R. C. (1991). Nudibranch spermatozoa: comparative ultrastructure and systematic importance. *The Veliger* **34**, 134–165.

Hickman, C. S. (1996). Phylogeny and patterns of evolutionary radiation in trochoidean gastropods. *In* "Origin and Evolutionary Radiation of the Mollusca" (J. Taylor, ed.), pp. 177–198. Oxford University Press, New York.

Hoffmann, H. (1932–1939). "Dr. H. G. Bronns Klassen und Ordnungen des Tierrechs. 3. Band: Mollusca. II. Abteilung: Gastropoda. 3. Buch: Opisthobranchia". Akademische Verlagsgesellschaft M. B. H., Leipzig.

Hurst, A. (1965). Studies on the structure and function of the feeding apparatus of *Philine aperta* with a comparative consideration of some other opisthobranchs. *Malacologia* **2**, 281–347.

Jensen, K. R. (1980). *Oxynoe azuropunctata*, n. sp., a new sacoglossan from the Florida Keys (Mollusca: Opisthobranchia). *Journal of Molluscan Studies* **46**, 282–292.

Jensen, K. R. (1991). Comparison of alimentary systems in shelled and non-shelled Sacoglossa (Mollusca, Opisthobranchia). *Acta Zoologica* **72,** 143–150.

Jensen, K. R. (1992). Anatomy of some Indo-Pacific Elysiidae (Opisthobranchia: Sacoglossa (= Ascoglossa)), with a discussion of the generic division and phylogeny. *Journal of Molluscan Studies* **58**, 257–296.

Jensen, K. R. (1996). The Diaphanidae as a possible sister group of the Sacoglossa (Gastropoda, Opisthobranchia). *In* "Origin and Evolutionary Radiation of the Mollusca" (J. Taylor, ed.), pp. 231–247. Oxford University Press, New York.

Jensen, K. R. (1997). "Sacoglossernes systematik, fylogeni og evolution (Mollusca, Opisthobranchia)". Vestjydsk Forlag, Copenhagen.

Jensen, K. R. (1999). Copulatory behaviour in three shelled and five non-shelled sacoglossans (Mollusca, Opisthobranchia), with a discussion of the phylogenetic significance of copulatory behaviour. *Ophelia* **51**, 93–106.

Kandel, E. R. (1979). "Behavioral Biology of *Aplysia*: a Contribution to the Comparative Study of Opisthobranch Molluscs". W. H. Freeman, San Francisco.

Kay, E. A. (1968). A review of the bivalved gastropods and a discussion of evolution within the Sacoglossa. *In* "Studies in the Structure, Physiology and Ecology of Molluscs" (V. Fretter, ed.), pp. 109–134. Academic Press, New York.

Kolb, A. (1998). Morphology, anatomy and histology of four species of *Armina* Rafinesque, 1815 (Nudibranchia, Arminoidea, Arminidae) from the Mediterranean Sea and the Atlantic Ocean. *Journal of Molluscan Studies* **64**, 355–386.

Lalli, C. M. and Gilmer, R. W. (1989). "Pelagic Snails: the Biology of Holoplanktonic Gastropod Mollusks". Stanford University Press, Stanford, California.

Lemche, H. (1956). The anatomy and histology of *Cylichna* (Gastropoda Tectibranchia). *Skrifter Udgivet af Universitetets Zoologiske Museum, Kobenhavn*, **16**, 1–278, 46 plates.

MacFarland, F. M. (1966). Studies of opisthobranchiate mollusks of the Pacific coast of North America. *Memoires of the California Academy of Sciences* **7**, 1–546, 72 plates.

Marcus, E. d. B.-R. (1972). On some Acteonidae (Gastropoda, Opisthobranchia). *Papéis Avulsos do Departamento de Zoologia, São Paulo* **25**, 167–188, 1 plate.

Marcus, E. d. B.-R. (1977). On the genus *Tornatina* and related forms. *Journal of Molluscan Studies Supplement* **2**, 1–35.

Marcus, E. d. B.-R. (1982). Systematics of the genera of the order Ascoglossa (Gastropoda). *Journal of Molluscan Studies Supplement* **10**, 1–31.

Marcus, E. d. B.-R. (1985). Catalogue of the western Atlantic warm water Opisthobranchia. 10. The western Atlantic warm water Notaspidea (Gastropoda, Opisthobranchia), Part 3. Umbraculacea. *Boletim de Zoologia, Universidade de São Paulo* **9**, 1–15.

Marcus, E. d. B.-R. and Gosliner, T. M. (1984). Review of the family Pleurobranchaeidae (Mollusca, Opisthobranchia). *Annals of the South African Museum* **93**, 1–52.

Marcus, E. d. B.-R. and Marcus, E. (1967a). American opisthobranch mollusks. Part 1. Tropical American opisthobranchs. Part 2. Opisthobranchs from the Gulf of California. *Studies in Tropical Oceanography, Miami* **6**, 1–256, 1 plate.

Marcus, E. d. B.-R. and Marcus, E. (1967b). Opisthobranchs from the southwestern Caribbean Sea. *Bulletin of Marine Science* **17**, 597–628.

Martínez, E. and Ortea, J. (1997). *Haminaea elegans* (Gray, 1825) (Opisthobranchia: Cephalaspidea), a truly amphiatlantic species. *The Veliger* **40**, 281–291.

Medina, M. and Walsh, P. J. (2000). Molecular systematics of the order Anaspidea based on mitochondrial DNA sequence (12S, 16S, and COI). *Molecular Phylogenetics and Evolution* **15**, 41–58.

Mikkelsen, P. M. (1985). *Acteocina* (and related forms) in Texas. *Texas Conchologist* **21**, 62–72.

Mikkelsen, P. M. (1993). Monophyly versus the Cephalaspidea (Gastropoda, Opisthobranchia), with an analysis of traditional cephalaspid characters. *Bolletino Malacologico* **29**, 115–138.

Mikkelsen, P. M. (1996). The evolutionary relationships of Cephalaspidea s. l. (Gastropoda: Opisthobranchia): a phylogenetic analysis. *Malacologia* **37**, 375–442.

Mikkelsen, P. M. (1998a). *Cylindrobulla* and *Ascobulla* in the western Atlantic (Gastropoda, Opisthobranchia, Sacoglossa): systematic review, description of a new species, and phylogenetic reanalysis. *Zoologica Scripta* **27**, 49–71.

Mikkelsen, P. M. (1998b). Review of shell reduction and loss in traditional and phylogenetic molluscan systematics, with experimental manipulation of a negative gain character. *American Malacological Bulletin* **14**, 201–218.

Mikkelsen, P. S. and Mikkelsen, P. M. (1984). Comparison of *Acteocina canaliculata* (Say, 1826), *A. candei* (d'Orbigny, 1841), and *A. atrata* spec. nov. (Gastropoda: Cephalaspidea). *The Veliger* **27**, 164–192.

Millard, V. G. (1997). "Classification of Mollusca". Privately produced electronic file, Rhine Road, South Africa.

Minichev, Y. S. (1966). Morphological peculiarities of abyssal Cephalaspidea (Gastropoda, Opisthobranchia). *Zoologischeskii Zhurnal* **45**, 509–517 (in Russian with English summary).

Morton, J. E. (1963). The molluscan pattern: evolutionary trends in a modern classification. *Proceedings of the Linnean Society of London*, 174 session, **1961–62**, 53–72.

Morton, J. E. (1972). The form and functioning of the pallial organs in the opisthobranch *Akera bullata* with a discussion on the nature of the gill in Notaspidea and other tectibranchs. *The Veliger* **14**, 337–349.

Morton, J. E. and Holme, N. A. (1955). The occurrence at Plymouth of the opisthobranch *Acera bullata*, with notes on its habits and relationships. *Journal of the Marine Biological Association of the United Kingdom* **34**, 101–112.

Odhner, N. J. (1914). *Ptisanula limnaeoides*, a new Arctic opisthobranchiate mollusc, its anatomy and affinities. *Arkiv für Zoologi* **8**(25), 1–18, plate 1.

Odhner, N. J. (1939). Opisthobranchiate Mollusca from the western and northern coasts of Norway. *Kongelige Norske Videnskabers Selskabs Skrifter* **1**, 1–92.

Ortea, J. A. (1984). Molluscos opistobranquios de las Islas Canarias. Primera parte: Ascoglosos. *Boletín del Instituto Español de Oceanografía* **6**, 180–199.

Pilsbry, H. A. (1894–1896). Order Opisthobranchiata. Suborder Tectibranchiata. *Manual of Conchology* **15**, 133–436, plates 18–50, 59–61; **16**, 1–262, plates 1–74.

Pilsbry, H. A. (1949). A West Indian *Hydatina*. *The Nautilus* **63**, 15–17, plate 1.

Ponder, W. F., ed. (1988). Prosobranch Phylogeny. *Malacological Review* **4** (suppl.).

Ponder, W. F. and Lindberg, D. R. (1997). Towards a phylogeny of gastropod molluscs: an analysis using morphological characters. *Zoological Journal of the Linnean Society* **119**, 83–265.

Poulicek, M., Voss-Foucart, M.-F. and Jeuniaux, C. (1991). Regressive shell evolution among opisthobranch gastropods. *Malacologia* **32**, 223–232.

Purchon, R. D. (1977). "The Biology of the Mollusca," 2nd edn. Pergamon Press, New York.

Robertson, R. (1985). Four characters and the higher category systematics of gastropods. *American Malacological Bulletin Special Edition* **1**, 1–22.

Rudman, W. B. (1971). Structure and functioning of the gut in the Bullomorpha (Opisthobranchia). Part 1. Herbivores. *Journal of Natural History* **5**, 647–675.

Rudman, W. B. (1972a). Structure and functioning of the gut in the Bullomorpha (Opisthobranchia). Part 2. Acteonidae. *Journal of Natural History* **6**, 311–324.

Rudman, W. B. (1972b). Structure and functioning of the gut in the Bullomorpha (Opisthobranchia). Part 4. Aglajidae. *Journal of Natural History* **6**, 547–560.

Rudman, W. B. (1972c). Structure and functioning of the gut in the Bullomorpha (Opisthobranchia). Part 3. Philinidae. *Journal of Natural History* **6**, 459–474.

Rudman, W. B. (1972d). A study of the anatomy of *Pupa* and *Maxacteon* (Acteonidae, Opisthobranchia), with an account of the breeding cycle of *Pupa kirki*. *Journal of Natural History* **6**, 603–619.

Rudman, W. B. (1978). A new species and genus of the Aglajidae and the evolution of the philinacean opisthobranch molluscs. *Zoological Journal of the Linnean Society* **62**, 89–107.

Rudman, W. B. (1998, December 11). *Plakobranchus ocellatus* van Hasselt, 1824. "Sea Slug Forum", http://seaslugforum.net/placocel.htm, accessed 21 March 2001.

Rudman, W. B. (1999, December 1). *Euselenops luniceps*, page 2 – adaptations for a sandy habitat. "Sea Slug Forum", http://seaslugforum.net/euseluni2.htm, accessed 25 March 2001.

Rudman, W. B. and Willan, R. C. (1998). Opisthobranchia Introduction. *In* "Mollusca: The Southern Synthesis" (P. L. Beesley, G. J. B. Ross and A. Wells, eds), pp. 915–942. CSIRO Publishing, Melbourne.

Salvini-Plawen, L. v. (1988). The structure and function of molluscan digestive systems. *In* "The Mollusca, Volume 11, Form and Function" (E. R. Trueman and M. R. Clarke, eds), pp. 301–379. Academic Press, San Diego.

Salvini-Plawen, L. v. (1991a [1990]). Origin, phylogeny and classification of the phylum Mollusca. *Iberus* **9**, 1–33.

Salvini-Plawen, L. v. (1991b). The status of the Rhodopidae (Gastropoda: Euthyneura). *Malacologia* **32**, 301–311.

Scheltema, A. H. (1985). The aplacophoran family Prochaetodermatidae in the North American basin, including *Chevroderma* n.g. and *Spathoderma* n.g. (Mollusca: Chaetodermomorpha). *Biological Bulletin, Marine Biological Laboratory, Woods Hole* **169**, 484–529.

Schiøtte, T. (1992). The tectibranch gastropods of the Faroe Islands and their relations to temperature and depth. *In* "Årbok 1991–92: Symposium on Marine Biology and Oceanography of the Faroe Islands" (Kløvstad, ed.), pp. 96–97. Nordurlandahúsid í Føroyum, Tórshavn, Faroe Islands.

Schmekel, L. (1985). Aspects of the evolution within opisthobranchs. *In* "The Mollusca, Volume 10, Evolution" (E. R. Trueman and M. R. Clarke, eds), pp. 221–267. Academic Press, Orlando.

Smith, B. J. and Stanisic, J. (1998). Pulmonata Introduction. *In* "Mollusca: The Southern Synthesis" (P. L. Beesley, G. J. B. Ross and A. Wells, eds), pp. 1037–1061. CSIRO Publishing, Melbourne.

Sohl, N. F. and Kollmann, H. A. (1985). Cretaceous actaeonellid gastropods from the Western Hemisphere. *United States Geological Survey Professional Paper* **1304**, 1–104, 23 plates.

Tarr, R. (1989). The umbrella shell. *Hawaiian Shell News* **37**, 1, 12.

Thiele, J. (1929–1935). "Handbuch der systematischen Weichtierkunde". Teil 1 (Loricata; Gastropoda: Prosobranchia), 1929, pp. 1–376. Teil 2 (Gastropoda: Opisthobranchia, Pulmonata), 1931, pp. 377–788. Teil 3 (Scaphopoda; Bivalvia; Cephalopoda), 1934, pp. 789–1022. Teil 4 (General Part), 1935, pp. 1023–1154. Gustav Fischer Verlag, Jena. [English translation: Bieler, R. and Mikkelsen, P. M., sci. eds. (1992–98). "Handbook of Systematic Malacology, Volumes I–III." Amerind Publishing Company, India, with Smithsonian Institution Libraries, Washington, DC.]

Thollesson, M. (1998). Phylogenetic analysis of dorid nudibranchs (Gastropoda, Doridacea) using the mitochondrial 16S rRNA gene. *In* "Nudibranch Systematics and Molecular Data" (M. Thollesson, ed.), pp. 1–22. Göteborg University, Göteborg.

Thollesson, M. (1999). Phylogenetic analysis of Euthyneura (Gastropoda) by means of the 16S rRNA gene: use of a 'fast' gene for 'higher-level' phylogenies. *Proceedings of the Royal Society of London* B **266**, 75–83.

Thompson, T. E. (1972). Observations on *Hexabranchus* from the Australian Great Barrier Reef (Gastropoda: Opisthobranchia). *The Veliger* **15**, 1–5.

Thompson, T. E. (1976). "Biology of Opisthobranch Molluscs, Volume 1". The Ray Society, London.

Thompson, T. E. and Seaward, D. R. (1986). Observations on the opisthobranch mollusc *Akera bullata* in the Fleet, Dorset. *In* "Biology of the Fleet" (M. Ladle, ed.), pp. 17–22. Dorset Council, Dorchester.

Thompson, T. E. and Seaward, D. R. (1989). Ecology and taxonomic status of the aplysiomorph *Akera bullata* in the British Isles. *Journal of Molluscan Studies* **55**, 489–496.

Thompson, T. E. and Slinn, D. J. (1959). On the biology of the opisthobranch *Pleurobranchus membranaceus*. *Journal of the Marine Biological Association of the United Kingdom* **38**, 507–524.

Thompson, T. E., Jarman, G. M. and Zenetos, A. (1985). Infralittoral macrobenthos of the Patras Gulf and Ionian Sea: opisthobranch molluscs. *Journal of Conchology* **32**: 71–95.

Tillier, S., Masselot, M., Hervé, P. and Tillier, A. (1992). Phylogénie moléculaire des Gastropoda (Mollusca) fondée sur le séquençage partiel de l'ARN ribosomique 28 S. *Comptes Rendus de l'Académie de Sciences (Paris), Series 3,* **134**, 79–85.

Tracey, S., Todd, J. A. and Erwin, D. H. (1993). Mollusca: Gastropoda. *In* "The Fossil Record 2" (M. J. Benton, ed.), pp. 131–167. Chapman & Hall, London.

Vayssière, A. (1879–1880). Recherches anatomiques sur les Mollusques de la famille des Bullides. *Annales des Sciences Naturelles, Zoologie* **9** (4, art. 1), 1–123, plates 1–12.

Vayssière, A. (1885). Recherches zoologiques et anatomiques sur les Mollusques Opisthobranches du Golfe de Marseille. 1er partie. Tectibranches. *Annales du Musée d'Histoire Naturelle de Marseille, Zoologie* **2(3)**, 181 pp., 6 plates.

Voltzow, J. (1994). Gastropoda: Prosobranchia. *In* "Microscopic Anatomy of Invertebrates, Volume 5, Mollusca I" (F. W. Harrison and A. J. Kohn, eds), pp. 111–252. Wiley Liss, New York.

Wade, C. M., Mordan, P. B. and Clarke, B. (2000). A phylogeny of the land snails (Gastropoda: Pulmonata). *Proceedings of the Royal Society of London* **268**, 413–422.

Wägele, H. (1989). A revision of the Antarctic species of *Bathydoris* Bergh, 1884 and comparison with other known bathydorids (Opisthobranchia, Nudibranchia). *Journal of Molluscan Studies* **55**, 343–364.

Wägele, H. (1995). The morphology and taxonomy of the Antarctic species of *Tritonia* Cuvier, 1797 (Nudibranchia: Dendronotoidea). *Zoological Journal of the Linnean Society* **113**, 21–46.

Wägele, H. (1998). Histological investigation of some organs and specialised cellular structures in Opisthobranchia (Gastropoda) with the potential to yield phylogenetically significant characters. *Zoologischer Anzeiger* **236**, 119–131.

Wägele, H. and Hain, S. (1991). Description of a new notaspidean genus and species (Opisthobranchia: Notaspidea) from the Antarctic Ocean. *Journal of Molluscan Studies* **57**, 229–242.

Wägele, H. and Willan, R. C. (1994). The morphology and anatomy of the Antarctic gastropod *Bathyberthella antarctica* (Opisthobranchia, Notaspidea, Pleurobranchidae). *Zoologica Scripta* **23(4)**, 313–324.

Wägele, H. and Willan, R. C. (2000). On the phylogeny of the Nudibranchia. *Zoological Journal of the Linnean Society* **130**, 83–181.

Warén, A. (1994). Systematic position and validity of *Ebala* Gray, 1847 (Ebalidae fam. n., Pyramidelloidea, Heterobranchia). *Bolletino Malacologico* **30**(5–9), 203–210.

Warén, A., Gofas, S. and Schander, C. (1993). Systematic position of three European heterobranch gastropods. *The Veliger* **36**, 1–15.

Wayne, N. L and Block, G. D. (1992). Effects of photoperiod and temperature on egg-laying behavior in a marine mollusk, *Aplysia californica. Biological Bulletin, Marine Biological Laboratory, Woods Hole* **182**, 8–14.

Wells, F. E., Jr. and Bryce, C. W. (1993). "Sea Slugs of Western Australia". Western Australian Museum, Perth.

Willan, R. C. (1977). A review of *Pleurobranchella* Thiele, 1925 (Opisthobranchia: Pleurobranchaeinae). *Journal of Conchology* **29**, 151–155.

Willan, R. C. (1983). New Zealand side-gilled sea slugs (Opisthobranchia: Notaspidea: Pleurobranchidae). *Malacologia* **23**, 221–270.

Willan, R. C. (1987). Phylogenetic systematics of the Notaspidea (Opisthobranchia) with reappraisal of families and genera. *American Malacological Bulletin* **5**, 215–241.

Williams, G. C. (1975). "Phylogenetic Implications of the Degree of Concentration Within the Opisthobranch Nervous System". Masters thesis, San Francisco State University, San Francisco.

Winner, B. E. and Mikkelson [sic], P. M. (1986). The purple spill: two *Haminoea* spp. in Lake Worth, Florida. *Hawaiian Shell News* **34**, 12.

Yonow, N. and Renwrantz, L. (1986). Studies on the haemocytes of *Acteon tornatilis* (L.) (Opisthobranchia: Acteonidae). *Journal of Molluscan Studies* **52**, 150–155.

Yoon, S. H. and Kim, W. (2000). Phylogeny of some gastropod mollusks derived from 18S rDNA sequences with emphasis on the Euthyneura. *The Veliger* **114**, 84–92.

Young, D. K. (1969). The functional morphology of the feeding apparatus of some Indo-West-Pacific dorid nudibranchs. *Malacologia* **9**, 421–446.

Other cited World-Wide Web pages

Bibliomania, http://www.bibliomania.com, accessed 22 March 2001.

British Marine Life Study Society, "The Seashore Rockpooler in March, by Andy Horton, http://ourworld.compuserve.com/homepages/BMLSS/March.htm, accessed 22 March 2001.

International Council for Exploration of the Sea, "Report of the ICES/HELCOM Steering Group on Quality Assurance on Biological Measurements in the Baltic Sea, 14–16 February 2000, Copenhagen, http://www.ices.dk/reports/acme/2000/sgqab/sgqab00.pdf, accessed 22 March 2001.

MarLIN (Marine Life Network for Britain and Ireland), http://www.marlin.ac.uk, accessed 22 March 2001.

Winclada and Nona can be downloaded from http://cladistics.com, accessed 27 March 2001.

APPENDIX I

Pilsbry analysis

Characters and character states from Pilsbry's (1894–1896) taxonomic descriptions:

0. Cephalic shield: (0) absent, (1) present.
1. Operculum: (0) present, (1) absent.
2. Shell: (0) external, (1) internal or internalized.
3. Radula: (0) present, (1) absent.
4. Radula: (0) multiserial, (1) uniserial, (2) many similar rows, (3) multiserial, but few rows.
5. Dorsal shield or notum: (0) absent, (1) present.
6. Epi-/para-/pleuropodial lobes: (0) absent, (1) present.
7. Male and female orifices: (0) widely separated, (1) close together.

Data matrix:

	01234567
Outgroup	00000000
Acteonidae	10002000
Tornatinidae	1101–000
Scaphandridae	11003010
Bullidae	11003000
Akeridae	11000010
Hydatinidae	11000010
Ringiculidae	11003000
Gastropteridae	11103010
Philinidae	11103010
Aglajidae	1111–010
Aplysiidae	01100010
Oxynoeidae	01001010
Runcinidae	01003101
Pleurobranchidae	01100101
Umbraculidae	01002101

Analysis: Using 15 family-level taxa, a hypothetical all-zero outgroup, 8 unweighted characters (7 binary, 1 multistate unordered), and Nona ver. 2.0 (for Windows 95/98/NT; P. A. Goloboff, 1993) within Winclada beta ver. 0.9.99m24 (Kevin C. Nixon, 1999), specifying 3000 replicates in a heuristic search: 2 most-parsimonious trees resulted, each with 17 steps (CI 0.58, RI 0.74), but variable topologies depending on the number of replicates performed. When Acteonidae was eliminated and the analysis rerun, a single most-parsimonious tree resulted, of 15 steps (CI 0.66, RI 0.79), shown in Figure 9B.

APPENDIX II

Willan Notaspidea reanalysis

Characters and character states [with noted changes or comments relative to the original analysis by Willan, 1987]:

0. Shell (0) present, (1) absent. [coding reversed]
1. Shell (0) external, (1) internal beneath mantle.
2. Shell (0) calcified, (1) uncalcified. [miscoded in Willan, 1987: table 5]
3. Periostracum (0) smooth, adhering to shell, (1) rough, lamellate.
4. Muscle scar (0) incomplete, (1) intermediate suspensor present, (2) forming a complete ring.
5. Shell shape (0) circular, (1) rectangular.
6. Shell shape (0) conical, (1) flattened or plate-like, (2) auriculate-oval, (3) spatulate-triangular. [two original characters combined]
7. Shell location relative to body (0) central, (1) anterior, (2) posterior. [repolarized and recoded]
8. Shell size relative to body (0) large, (1) medium, (2) small.
9. Mantle size relative to shell (0) same size, (1) larger than shell.
10. Mantle texture (0) smooth, (1) pustulose, (2) puckered.
11. Mantle spicules (0) absent, (1) present embedded in mantle.
12. Anterior border of mantle (0) entire, (1) weakly emarginate, (2) deeply cleft.
13. Anterior mantle and oral veil (0) separated ("absent"), (1) fused ("present").
14. Oral tentacles (0) separate (*Umbraculum* only), (1) joined by oral veil.
15. Oral veil relative to body (0) very narrow, (1) narrow, (2) moderately broad, (3) very broad. [unordered]
16. Oral veil papillae (0) absent, (1) present along anterior edge.
17. Rhinophores (0) separate (*Umbraculum*), (1) together but without basal fusion (Tylodinidae), (2) together with bases fused (Pleurobranchacea). [miscoded]
18. Pedal gland (0) absent, (1) present on sole of foot of sexually mature individuals.
19. Gill location (0) in right posterior quadrant of body, (1) well back in right posterior quadrant of body, (2) extending from left antero-lateral corner of body almost to posterior midline (*Umbraculum* only), (3) in right anterior quadrant of body (*Tomthompsonia* only). [recoded states 0 and 1]
20. Gill attachment to body (0) half length, (1) less than half length, (2) almost entire length (*Umbraculum* only).

21. Gill rachis (0) smooth, (1) with row of pustules.
22. Anus location (0) middle of gill basement membrane, (1) in front of hind end of gill basement membrane, (2) above hind end, (3) well behind posterior end of gill basement membrane (*Umbraculum* only).
23. Median buccal (= dorsal accessory) gland (0) absent, (1) present.
24. Radular rachidian row (0) present, (1) absent. [coding reversed]
25. Radular lateral tooth, basal denticle (0) absent, (1) single on at least some teeth.
26. Radular lateral tooth (0) not lamellate ("absent"), (1) lamellate ("present," i.e., two or more denticles on blade).
27. Jaws (0) present, (1) absent. [characters 28 and 29 not homologous, split from a single original character]
28. Labial cuticle (0) absent or not thickened, (1) thickened ring.
29. Jaw elements (0) oval or polygonal, (1) elongate with paired lateral projections (i.e., cruciform). [coding reversed]
30. Jaw element blades (0) denticulate, (1) smooth. [coding reversed]
31. External ciliated autospermal groove on penis (0) present, (1) absent. [coding reversed]
32. Penis location (0) at base of right oral tentacle (1) on right side in front of anterior end of gill, (2) in vertical cleft in anterior midline, immediately below rhinophores and above mouth (*Umbraculum* only). [coding reversed]
33. Penis (0) not retractile, (1) retractile.
34. Penial outer surface (0) smooth, (1) with papillae.
35. Allosperm receptacles (0) two (bursa copulatrix and receptaculum seminis), (1) one (bursa copulatrix only). [coding reversed]
36. Receptaculum seminis location (0) arising low down off the vagina near female genital aperture, (1) arising high up off the vagina near base of bursa copulatrix. [coding reversed]
37. Prostate gland (0) surrounding or ensheathing autosperm duct, (1) absent, (2) distinct organ. [recoded]
38. Penial gland (0) absent, (1) present.
39. Penial sac (0) absent, (1) present and muscular.
40. Vas deferens (0) not coiling within penial sack, (1) coiling extensively within penial sack.

Fifteen of Willan's characters were omitted from the analysis as automorphies of single taxa: posterior mantle border entire or permanently cleft; mantle margin entire, slightly crenulate or serrate; mantle autotomization incapable or capable; posterior mantle and foot separated or fused; oral tentacles one or two pairs; rhythmic rhinophoral activity absent or present; foot upper surface smooth or pustulate; pedal gland small or large; caudal spur absent or present; anterior pedal cleft absent or present; anal opening flush with body or at end of anal tube; mouth location outside or within pedal cleft; buccal mass non-protrusible or protrusible during feeding (when coding reversed); accessory denticle on lateral tooth absent or single; flaps surrounding genital apertures absent or enlarged. Configuration of the reproductive system (monaulic, diaulic, or triaulic) was omitted for reasons discussed in the text.

Data matrix:

	00000000001	1111111112	2222222223	3333333334
	01234567890	1234567890	1234567890	1234567890
Outgroup	00000000000	0000000000	0000000000	0000000000
Tylodina	00001000000	0001001010	02001011--	00001-0000
Anidolyta	00010000000	0001001010	02010111--	00001-0000
Umbraculum	00012010100	0000--0022	03010011--	02-0--2000
Berthella	01000120??0	1101102101	0?11?0001?	1110?1?100
Pleurobranchus	01?0012???2	1201102101	1211?00010	1110?1?000
Berthellina	010001?1210	1?01102001	021101001?	1110000100
Pleurehdera	01000121110	-001102101	0211110010	1110010100
Bathyberthella	01100120000	0001102101	0211000000	1110010?00
Pleurobranchella	1---------1	0-1121?001	?11?000000	111?1-2011
Pleurobranchaea	1---------1	0-1121?101	0110000000	11101-2011
Euselenops	1---------0	0-1131?101	1111000001	1111101000
Tomthompsonia	0100?000000	1001302130	0101000010	111001210?

Analysis: Using 12 genus-level taxa, a hypothetical all-zero outgroup, 41 unweighted characters (28 binary, 13 multistate unordered), and the Nona/Winclada combination as in Appendix I (heuristic search, 3000 replicates); resulting in a single most-parsimonious tree of 75 steps (CI 0.74, RI 0.76).

APPENDIX III

Combined analysis

Dataset consisting of 26 genus-level taxa, including representatives from each analysis, 2 from each major clade:

Architectibranchia (Mikkelsen, 1996): *Hydatina, Ringicula.*

Cephalaspidea (Mikkelsen, 1996; supplemented with data from Vayssière, 1880 in 1879–1880; Bergh, 1901; Eales, 1938; Lemche, 1956): *Bulla, Haminoea, Cylichna, Acteocina.*

Anaspidea (Mikkelsen, 1996; supplemented with data from Eales, 1960): *Akera, Aplysia.*

Sacoglossa (Jensen, 1996; Mikkelsen, 1996, 1998a; supplemented with data from Eliot, 1906; Gascoigne and Sordi, 1980; Boss, 1982; Ortea, 1984; Jensen, 1991, 1992): *Cylindrobulla, Lobiger, Plakobranchus, Elysia, Cyerce, Placida.*

Notaspidea (Willan, 1987, reanalyzed herein; supplemented with data from Vayssière, 1885; Thompson and Slinn, 1959; MacFarland, 1966; Willan, 1977, 1983; Marcus and Gosliner, 1984; Marcus, 1985; Wägele and Willan, 1994; Rudman, 1998): *Tylodina, Umbraculum, Pleurobranchella, Euselenops, Bathyberthella, Pleurobranchus.*

Lower heterobranchs (Mikkelsen, 1996; supplemented with data from Haszprunar, 1985c): *Gegania, Acteon.*

Nudibranchia (Wägele and Willan, 2000; supplemented with data from Gohar and Soliman, 1963; Young, 1969; Thompson, 1972; Wägele, 1989, 1995; Wells and Bryce, 1993; Kolb, 1998): *Bathydoris, Hexabranchus, Armina, Tritonia.*

Outgroup is a hypothetical all-zero "prosobranch".

Characters (with source of data):

0. Shell (0) present, (1) absent (from Willan, 1987; Jensen, 1996; Mikkelsen, 1998a; Wägele and Willan, 2000).
1. Shell (0) external, (1) internalized or internal beneath mantle (from Willan, 1987; Mikkelsen, 1996).
2. Shell shape (0) coiled, (1) circular (limpet-like), (2) rectangular (from Willan, 1987).
3. Shell size relative to body (0) large, (1) small (from Willan, 1987; Jensen, 1996; Mikkelsen, 1998a).
4. Rhinophores (0) absent, (1) present, enrolled, (2) present, solid column (from Willan, 1987; Jensen, 1996; Mikkelsen, 1998a; Wägele and Willan, 2000).
5. Radular rachidian row (0) present, (1) absent (from Willan, 1987; Wägele and Willan, 2000).
6. Radular lateral teeth (0) present, (1) absent (from Wägele and Willan, 2000).
7. Jaws (0) present, (1) absent (from Willan, 1987; Mikkelsen, 1996; Wägele and Willan, 2000).
8. Labial cuticle (0) not thickened or absent, (1) thickened ring (from Willan, 1987; Mikkelsen, 1996).
9. Penis (0) not retractile, (1) retractile (from Willan, 1987; Mikkelsen, 1996).
10. Operculum (in adult) (0) present, (1) absent (from Mikkelsen, 1996).
11. Parapodia (0) absent, (1) present (from Mikkelsen, 1996).
12. Shell adductor muscle (0) absent, (1) present (from Jensen, 1996; Mikkelsen, 1996, 1998a).

13. Cerata (0) absent, (1) present (from Jensen, 1996; Mikkelsen, 1998a).
14. Anus (0) anterior, (1) posterior (from Willan, 1987; Jensen, 1996; Mikkelsen, 1996, 1998a).
15. Pharyngeal pouches (0) absent, (1) present (from Jensen, 1996; Mikkelsen, 1996, 1998a).
16. Ascus (0) absent, (1) present (from Jensen, 1996; Mikkelsen, 1996, 1998a).
17. Oral tube during feeding (0) non-protrusible, (1) protrusible (Willan, 1987; Jensen, 1996; Mikkelsen, 1998a).
18. Shape of rachidian teeth (0) plate with large central cusp, (1) dagger (= blade, chisel, sabot), (2) rounded/indented, (3) transversely elongated (from Jensen, 1996; Mikkelsen, 1996, 1998a).
19. Visceral nerve loop (0) long, (1) short (from Jensen, 1996; Mikkelsen, 1998a).
20. Penial stylet (0) absent, (1) present (from Willan, 1987; Jensen, 1996; Mikkelsen, 1998a).
21. Female genital papilla (0) absent, (1) present (from Willan, 1987; Jensen, 1996; Mikkelsen, 1998a).
22. Cerebropleural ganglia (0) separate, (1) fused (from Mikkelsen, 1996, 1998a).
23. Esophageal gizzard with gizzard plates (0) absent, (1) present (from Mikkelsen, 1996, 1998a).
24. Posterior foot (0) absent, (1) present (from Mikkelsen, 1996).
25. Enlarged sickle-shaped lateral teeth (0) not present (but other laterals/marginals present), (1) present (from Mikkelsen, 1996).
26. Gizzard plate number (0) >3, (1) 3 (from Mikkelsen, 1996).
27. Gizzard spines (0) present, (1) absent (from Mikkelsen, 1996).
28. Filter chamber (0) absent, (1) present (from Mikkelsen, 1996).
29. Stomach (0) with pouch-like chamber, (1) flow-through (from Mikkelsen, 1996).
30. Oral veil (0) absent, (1) present (from Willan, 1987; Wägele and Willan, 2000).
31. Nerve-ring location (0) pre-pharyngeal, (1) post-pharyngeal (from Mikkelsen, 1996; Wägele and Willan, 2000).
32. Internal sperm duct (“vas deferens”) (0) present, (1) absent (from Willan, 1987; Jensen, 1996; Mikkelsen, 1996, 1998a).
33. Lateral external cilated groove (0) absent, (1) present (from Mikkelsen, 1996).
34. Oral tentacles (0) absent, (1) present (from Jensen, 1996; Mikkelsen, 1998a; Wägele and Willan, 2000).
35. Propodial tentacles (0) absent, (1) present (from Wägele and Willan, 2000).

36. Typhlosole (0) present; (1) absent (Mikkelsen, 1996; Wägele and Willan, 2000).
37. Radular nerve (0) separate, one emanating from each buccal ganglion, (1) fused, a single nerve arising from the buccal commissure (Wägele and Willan, 2000).
38. Gonadal acini (0) hermaphroditic, i.e., spermatogonia and oogonia located in same acini, (1) spermatogonia and oogonia located in separate acini (Mikkelsen, 1996; Wägele and Willan, 2000).
39. Blood gland (0) absent, (1) present (Wägele and Willan, 2000).
40. Ciliated strips (0) blunt at mantle edge, (1) flexed at mantle edge, (2) exogyrous (Mikkelsen, 1996).

Data matrix:

	00000000001 01234567890	1111111112 1234567890	2222222223 1234567890	3333333334 1234567890
Outgroup	00000000000	0000000000	0000000000	0000000000
Hydatina	00000?00001	1001000?00	01000--010	0001110000
Ringicula	00000101011	1000000-00	01001--010	100010?100
Bulla	00000000011	1001000300	0011010010	0110011002
Haminoea	00000000011	1001000000	0011010010	1110001002
Cylichna	00000000011	1001000200	0011111010	0110001101
Acteocina	00000000011	1001000200	0?11111010	01100?1001
Akera	00000000011	1101000000	0011000110	1110011000
Aplysia	0111100001-	1001000000	0010000110	111101100-
Cylindrobulla	00000011111	010111?000	1100---010	1010000100
Lobiger	00001011?11	1001111110	1100---0?0	10110?0?0-
Plakobranchus	1---1011?1-	1-0?11?011	0100---0?0	10?01?000-
Elysia	1---1011?1	-1-00010110	0100---010	10?00?0?0-
Cyerce	1---1011?1	-0-1011?011	0100---0?0	10?1??0?0-
Placida	1---1011?1	-0-1?010111	0100---0?0	10000?0?0-
Tylodina	0010100111	10001001010	00000--011	11110?1??-
Umbraculum	0011110110	10001000-10	?0000--011	11010??0?-
Pleurobranchella	1---1?0001	-0-01001??0	0?000--0?1	10010???1-
Euselenops	1---110001	-0-01001-?0	0?000--0?1	10?1????1-
Bathyberthella	0120110001	-0001001-10	01000--011	1001010?1-
Pleurobranchus	0121110001	-0001001-?0	0?000--011	10010?1?1-
Gegania	000000010-	0000000000-	00000--000	0000010100
Acteon	0000010100	00000000-00	01000--000	000010?000
Armina	1---200001	-0-0100?010	11000--001	10001?100-
Bathydoris	1---200001	-0-0100?010	10000--001	100101001-
Hexabranchus	1---210001	-0-01000-?0	??000--001	100?0???1-
Tritonia	1---200011	-0-10000010	10000--001	100100110-

Analysis: Using 26 genus-level taxa, a hypothetical all-zero outgroup, 41 unweighted characters (38 binary, 3 multistate unordered), and the Nona/Winclada combination as in Appendix I (heuristic search, 3000 replicates); resulting in 45 most-parsimonious trees of 110 steps (CI 0.41, RI 0.69). The strict consensus tree (length 140 steps, CI 0.32, RI 0.55) of this result is presented in Figure 11.

The Scaphopoda

Patrick D. Reynolds

Biology Department, Hamilton College, Clinton, NY 13323, USA
e-mail: preynold@hamilton.edu

ADVANCES IN MARINE BIOLOGY VOL. 42
ISBN 0-12-026142-1

The Scaphopoda are marine infaunal carnivores that feed on foraminiferans and other microorganisms selected and manipulated by their unique feeding tentacles or captacula. Their tusk-like shell is open at both ends; the burrowing foot and captacula protrude anteriorly, while respiratory currents pass through the posterior opening. Although the scaphopods comprise one of the smallest molluskan classes in terms of species diversity, they have a worldwide distribution ranging from intertidal to depths in excess of 6000 m.

Despite detailed monographic work from the late nineteenth and early twentieth centuries, however, the biology of scaphopods is today among the least understood of mollusks. To some extent, this is related to a significant departure from more typical molluscan organization inferred from studies of the larger classes. For example, the mantle cavity is elongate and restricted in space, and several associated organs are lost, including the osphradium and ctenidia; the latter is associated with the loss of auricles from the scaphopod heart. There is also little record of ontogeny in the adult shell, as the older portion is periodically decollated to maintain the posterior aperture for passage of inhalant and exhalant currents. This and other constraints on scaphopod diversification are reflected in the range limits of shell shape within the class.

In contrast, the "Dentalium" egg has been a model system for experimental embryology, particularly in studies of cell lineage and early morphogenesis. Later larval development and most organogenesis is nonetheless poorly known, although recent studies have considerably enhanced our knowledge in this area.

Global biogeographic patterns of scaphopod diversity have been studied only preliminarily. A general decrease in diversity with depth is broken by small diversity peaks at bathyal depths. There is also evidence for a marked latitudinal diversity gradient in the world's oceans, near equatorial in the Pacific but at ~20°N in the Atlantic.

Scaphopods have a wide diversity of ecto- and endo-symbiotic associations with other organisms, including commensal bacteria and ciliates, mutualistic anemones and corals, and parasitic algae and platyhelminths. Other documented associations include predation by naticid gastropods and ratfish, and the inhabitation of empty scaphopod shells by a variety of sipunculans and hermit crabs.

Phylogeny within the class is still unresolved, although significant progress has been made recently in documenting morphological variation among families, genera, and species for application in a cladistic context. The evolutionary relationships of Scaphopoda within Mollusca are farther from resolution, with every conchiferan class a proposed sister group among modern analyses. Molecular avenues should provide significant progress in scaphopod phylogenetics, as indicated by recent comparative sequence analysis and gene expression patterns. Scaphopods comprise the most recent class of mollusks to appear in the fossil record, dating at least from the Mississippian Carboniferous (362.5 My). They appear from first-order occurrence data to have a substantially documented fossil record, with several sharp extinction and radiation events evident during their comparatively short geologic history.

The unresolved position of the Scaphopoda within molluscan phylogeny, and their relatively limited range of morphological disparity but substantial derivation from other conchiferans, renders the scaphopods an enigmatic group. They represent a significant evolutionary avenue within the Mollusca, but one that is poorly examined. Nonetheless, comprehensive phylogenetic analysis and documentation of diversity within this small, monophyletic, class-level taxon, well-preserved in the fossil record, is imminently tractable, and presents a singular opportunity in the study of molluscan evolution.

1. INTRODUCTION

Scaphopods are mollusks that are commonly known for the characteristic tusk-like shape of their shell (Plates 1 and 2, Figure 1). They have a worldwide distribution; all members of the class are marine infaunal carnivores or deposit feeders, ranging in size from a few millimeters to several centimeters. They burrow into soft sediments from intertidal to >6000 m depths using a muscular foot (Plate 1, Figure 1). The Scaphopoda is one of the smallest molluscan classes, with approximately 1000 described species (Pojeta, 1987), of which about 520 are extant (G. Steiner, pers. comm.; see also Scarabino, 1994). There are currently 61 genera (46 Recent) (Steiner and Kabat, 2001) in 14 families (Scarabino, 1995) within two orders, Dentaliida (Palmer, 1974) and Gadilida (Starobogatov, 1974) (Figure 3);

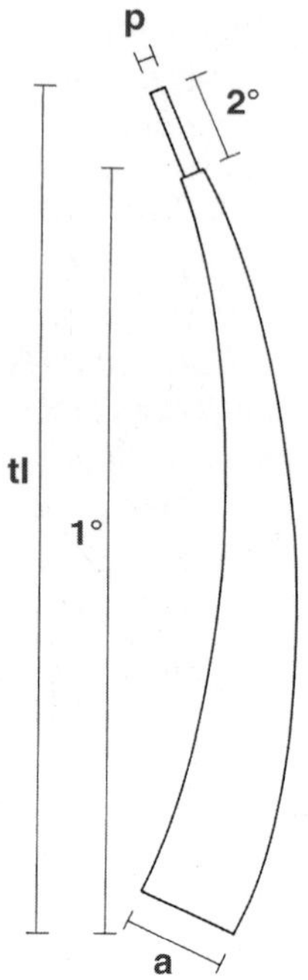

Figure 1 Schematic diagram of scaphopod shell, with terms of reference indicated: tl, total shell; 1°, primary shell; 2°, secondary shell; a, anterior aperture; p, posterior aperture.

the latter order is divided into suborders Entalimorpha and Gadilimorpha (Table 1).

Despite detailed early anatomical, developmental and behavioral investigations dating from the 1800s (Deshayes, 1825; Clark, 1849; Lacaze-Duthiers, 1856–1857, 1885; Plate, 1888, 1891a, b, c, 1892; Fol, 1885, 1889; Kowalevsky, 1883, 1889; Léon, 1894; Wilson, 1904a, b), the Scaphopoda is today one of the least studied classes of the Mollusca. One reason for this is the restriction of early investigations to one of two northeast Atlantic species (*Antalis entalis*, *A. vulgaris*), which have subsequently served as the model for the class as a whole; an explanation offered for the paucity of comparative scaphopod studies is the restriction of many species to off-shore and deeper waters (e.g., Morton, 1959; Shimek, 1988; Scarabino, 1995), and the consequent difficulty in obtaining specimens for live observation and soft-part examination. More recently, several examples of diverse, if not abundant, near-shore scaphopod populations have come to light; one example in the northeast Pacific has

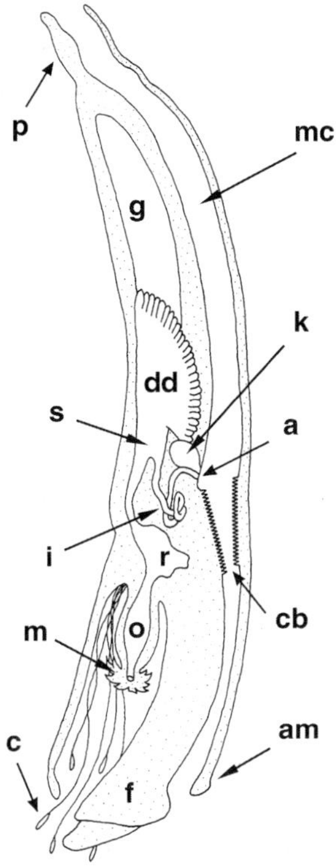

Figure 2 Diagrammatic representation of the scaphopod body plan, shell removed. Dorsal is to the left, posterior to the top. a, opening of rectum; am, anterior mantle margin; c, captacula; cb, ciliated ridges of the mantle cavity; dd, digestive diverticula; f, foot; g, gonad; i, intestine; k, kidneys; m, mouth and labial palps; mc, mantle cavity; o, esophagus; p, posterior mantle or pavilion; r, radula; s, stomach. (Redrawn with permission from Pelseneer (1906b, figure 182) © A & C Black.)

provided an opportunity to learn much about scaphopod feeding behavior and ecology (Poon, 1987; Shimek, 1988, 1990).

1.1. Phylogeny

Understanding basic scaphopod structure and function and documenting comparative anatomy is a prerequisite to significant understanding of evolution in the class. Morphology-based analyses of scaphopod phylogeny (Steiner, 1992b, 1996, 1997, 1998, 1999; Reynolds, 1997, 1998; Reynolds and Peters, 1997; Reynolds and Okusu, 1999) have been limited in their

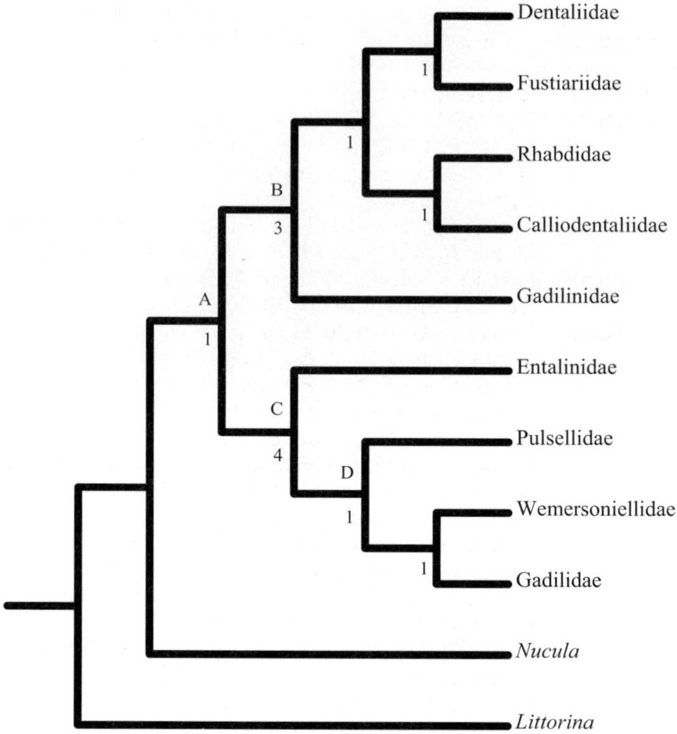

Figure 3 Single most parsimonious tree (length = 61; CI = 0.83, RC = 0.72) of nine scaphopod families. Numbers below horizontal bars represent Bremer support indices; letters above horizontal bars identify nodes/clades, which correspond to the following taxa: A = Scaphopoda, B = Dentaliida, C = Gadilida, D = Gadilimorpha; Entalinidae comprises the Entalimorpha (see also Table 1). (Reproduced with permission from Reynolds and Okusu (1999, figure 4) ©Academic Press.)

ability to resolve relationships; in part this is due to the limited availability of wet-preserved specimens for most species, and the narrow range of at least external morphological disparity among scaphopods compared to more diverse molluskan classes. Early evolutionary analyses of relationships amongst scaphopods relied almost exclusively upon shell characteristics (Emerson, 1962) or hard parts including the radula (Chistikov, 1975, 1978, 1979, 1984). Incorporation of soft-part anatomy and a cladistic approach to phylogenetic analysis was first brought to the Scaphopoda, at a family-level analysis, by Steiner (1992b, 1996), and later by Reynolds and Okusu (1999); Reynolds (1997) and Steiner (1997, 1998) have discussed methodological approaches to phylogenetic analysis and

Table 1 Classification of Scaphopoda (based on Steiner and Kabat (2001): Recent taxa from table 1, fossil taxa [F] compiled from text).

Class **SCAPHOPODA** Bronn, 1862
  Order **DENTALIIDA** da Costa, 1776
    Family ANULIDENTALIIDAE Chistikov, 1975
      Genus *Anulidentalium* Chistikov, 1975
      Genus *Epirhabdoides* Steiner, 1999
    Family BALTODENTALIIDAE Engeser & Riedel, 1992 [F]
      Genus *Baltodentalium* Engeser & Riedel, 1992
    Family CALLIODENTALIIDAE Chistikov, 1975
      Genus *Calliodentalium* Habe, 1964
    Family DENTALIIDAE [Children, 1834]
      Genus *Antalis* H. & A. Adams, 1854
      Genus *Coccodentalium* Sacco, 1896
      Genus *Compressidentalium* Habe, 1963
      Genus *Dentalium* Linnaeus, 1758
      Genus *Eodentalium* Medina & del Valle, 1985 [F]
      Genus *Eudentalium* Cotton & Godfrey, 1933
      Genus *Fissidentalium* Fischer, 1885
      Genus *Graptacme* Pilsbry & Sharp, 1897
      Genus *Paleodentalium* Gentile, 1974 [F]
      Genus *Paradentalium* Cotton & Godfrey, 1933
      Genus *Pictodentalium* Habe, 1963
      Genus *Plagioglypta* Pilsbry in Pilsbry & Sharp, 1898
      Genus *Prodentalium* Young, 1942 [F]
      Genus *Schizodentalium* Sowerby, 1894
      Genus *Striodentalium* Habe, 1964
      Genus *Tesseracme* Pilsbry & Sharp, 1898
    Family FUSTIARIIDAE Steiner, 1991
      Genus *Fustiaria* Stoliczka, 1868
    Family GADILINIDAE Chistikov, 1975
      Subfamily EPISIPHONINAE Chistikov, 1975
        Genus *Episiphon* Pilsbry & Sharp, 1897
      Subfamily GADILININAE Chistikov, 1975
        Genus *Gadilina* Foresti, 1895
      Subfamily LOBANTALINAE Chistikov, 1975 [F]
        Genus *Lobantale* Cossmann, 1888
    Family LAEVIDENTALIIDAE Palmer, 1974
      Genus *Laevidentalium* Cossmann, 1888
      Genus *Pipadentalium* Yoo, 1988 [F]
      Genus *Rhytiodentalium* Pojeta & Runnegar, 1979 [F]
      Genus *Scissuradentalium* Yoo, 1988 [F]
    Family OMNIGLYPTIDAE Chistikov, 1975
      Genus *Omniglypta* Kuroda & Habe in Habe, 1953
    Family PRODENTALIIDAE Starobogatov, 1974 [F]
      Genus *Prodentalium* Young, 1942
    Family RHABDIDAE Chistikov, 1975
      Genus *Rhabdus* Pilbry & Sharp, 1897

(contd)

Table 1 (contd)

- Dentaliida, *incertae sedis*
  - Genus *Progadilina* Palmer, 1974 [F]
  - Genus *Suevidontus* Engeser, Riedel & Bandel: 1993 [F]
- Order **GADILIDA** Starobogatov, 1974
  - Suborder **ENTALIMORPHA** Steiner, 1992
    - Family ENTALINIDAE Chistikov, 1979
      - Subfamily BATHOXIPHINAE Chistikov, 1983
        - Genus *Bathoxiphus* Pilsbry & Sharp, 1897
        - Genus *Rhomboxiphus* Chistikov, 1983
        - Genus *Solenoxiphus* Chistikov, 1983
      - Subfamily ENTALININAE Chistikov, 1979
        - Genus *Entalina* Monterosato, 1872
      - Subfamily HETEROSCHISMOIDINAE Chistikov, 1982
        - Genus *Costentalina* Chistikov, 1982
        - Genus *Entalinopsis* Habe, 1957
        - Genus *Heteroschismoides* Ludbrook, 1960
        - Genus *Pertusiconcha* Chistikov, 1982
        - Genus *Spadentalina* Habe, 1963
  - Suborder **GADILIMORPHA** Steiner, 1992
    - Family GADILIDAE Stoliczka, 1868
      - Subfamily GADILINAE Stoliczka, 1868
        - Genus *Bathycadulus* Scarabino, 1995
        - Genus *Cadulus* Philippi, 1844
        - Genus *Gadila* Gray, 1847
        - Genus *Sulcogadila* Moroni & Ruggieri, 1981 [F]
      - Subfamily SIPHONODENTALIINAE Tryon, 1884
        - Genus *Dischides* Jeffreys, 1867
        - Genus *Polyschides* Pilsbry & Sharp, 1898
        - Genus *Sagamicadulus* Sakurai & Shimazu, 1963
        - Genus *Siphonodentalium* M. Sars, 1859
        - Genus *Striocadulus* Emerson, 1962
      - Gadilidae, *incertae sedis*
        - Genus *Calstevenus* Yancey, 1973 [F]
        - Genus *Gadilopsis* Woodring, 1925 [F]
    - Family PULSELLIDAE Scarabino in Boss, 1982
      - Genus *Annulipulsellum* Scarabino, 1986
      - Genus *Pulsellum* Stoliczka, 1868
      - Genus *Striopulsellum* Scarabino, 1995
    - Family WEMERSONIELLIDAE Scarabino, 1986
      - Genus *Chistikovia* Scarabino, 1995
      - Genus *Wemersoniella* Scarabino, 1986
  - Gadilida, *incertae sedis*
    - Genus *Compressidens* Pilsbry & Sharp, 1897
    - Genus *Megaentalina* Habe, 1963
- Scaphopoda, *incertae sedis*
  - Genus *Cyrtoconella* Patrulius, 1996 [F]

consequent influence on hypotheses of scaphopod relationships. With cumulative character development and terminal taxon addition through all of these studies, resolution of phylogeny at the family level is emerging (e.g., Figure 3).

The addition of new characters and terminal taxa at the genus and species level, however, indicates that a phylogenetic tree for the Scaphopoda is far from stable below suborder level, and monophyly of family-level taxa cannot be assumed (Reynolds, 1997, 1998; Steiner, 1997, 1999; Reynolds and Okusu, 1999). While preliminary reports of molecular avenues of phylogenetic analysis in the class are promising (Reynolds and Peters, 1997; Reynolds, 1998; Steiner and Dreyer, 2001), much has yet to be learned about scaphopod comparative biology before hypotheses of relationships can be applied in a meaningful way to character evolution in the class.

A review of the hypotheses on the phylogenetic placement of Scaphopoda within Mollusca reveals a remarkably wide diversity of opinion. For example, recent analyses have placed the scaphopods as sister to the Cephalopoda (Waller, 1998), sister to a Gastropoda–Cephalopoda clade (Haszprunar, 2000), sister to the Bivalvia (among Recent taxa, Salvini-Plawen and Steiner, 1996), and as a subclade of the Rostroconchia, sister to the Conocardioida (Engeser and Riedel, 1996) or Ribeirioida (Runnegar, 1996). Wanninger and Haszprunar (in press) argue against the Diasoma (Bivalvia + Scaphopoda + Rostroconchia) concept, proposed by Runnegar and Pojeta (1974) (see Section 9.4).

This lack of resolution may be a consequence of the significant departure of several scaphopod organ systems from structural commonalities among the major molluskan classes. The absence or reduction of ctenidia, osphradium, auricles and ventricle are striking characteristics of the class, and apparently reflect the severe restriction in volume of the mantle cavity along the elongated dorso–ventral axis (functionally the posterior–anterior axis; see Edlinger, 1991; Steiner, 1992b; Shimek and Steiner, 1997; and Waller, 1998 for discussion of scaphopod body axes). In addition, while early development is well studied, there is relatively little information about scaphopod organogenesis and larval biology; further study in these areas should be informative for scaphopod relationships within Mollusca.

1.2. Scope of review

This review aims at a broad and comprehensive analysis of current understanding and historical development of scaphopod biology, largely from the perspective of functional life systems but with additional treatment of biogeographic patterns and interspecific interactions. Several reviews of the class with widely different emphases have appeared in recent years, including microscopic anatomy (Shimek and Steiner, 1997), the Australian fauna (Palmer and Steiner, 1998), functional morphology of the invertebrate skeleton (Yochelson, 1999), and larval development (Buckland-Nicks *et al.*, 2002). This review complements these; for some topics a summary of current knowledge and newly published literature is provided here, and the reader is directed to a recent source for further review.

Results from theses and dissertations (Kollman, 1908; Sahlman, 1973; Taib, 1976; Scarabino, 1979; Hebert, 1986; Reynolds, 1990a; Steiner, 1990), published abstracts, and several personal communications of work in progress have been incorporated here, where these contributions significantly enhance our understanding of a topic and peer-reviewed published sources for the data do not yet exist. The literature on broad paleontological and biogeographic patterns of scaphopod diversity is limited, but many species-specific data points on shell form, fossil occurrence, and geographic distribution are recorded throughout the literature. These primary data have been compiled for a large proportion of scaphopod species diversity, and resultant patterns are presented in a preliminary form here for the first time.

2. SHELL

2.1. General form

The tusk-shaped shell of scaphopods is a hollow, curved, conical tube that is open at both ends (Plates 1 and 2, Figure 1). Shell secretion to accommodate soft-tissue growth occurs at the anterior aperture, and the general surface of the mantle progressively thickens the shell. In most scaphopods, the maximum diameter of the shell is at the anterior, larger aperture, as in all members of the order Dentaliida and most of the order Gadilida (Plate 1). For other gadilids (subfamily Gadilinae, family Gadilidae), the maximum diameter is at a point posterior to the anterior aperture (Plate 2). This results in a much more determinate growth pattern than in other scaphopods; narrowing of the aperture marks the effective end of increase in length (Shimek, 1989). Shell sculpture is absent in many scaphopods, although longitudinal sculpture of widely varying types occurs in many dentaliids and a few gadilids, and is an important diagnostic character in scaphopod taxonomy (e.g., Palmer, 1974). Members of the Omniglyptidae and some other species, such as *Episiphon subtorquatum,* are characterized by annular sculpture (Habe, 1964; Scarabino, 1995).

Waller (1998) interprets the planospiral curvature of the shell as endogastric, based upon the relatively accelerated development of the anterior end of the larval shell figured by Lacaze-Duthiers (1856–1857). The unusually slight curvature of the scaphopod shell among coiled univalve mollusks, with small spiral angle and extremely large whorl dilation, was reviewed by Thompson (1952). Yochelson (1999) also describes the coiling as a logarithmic curve, while acknowledging that the precise description of the curve is hindered by the periodic programed loss of the curve origin, or original apex of the shell (see Section 2.2.1). This latter phenomenon has also hindered efforts to analyze scaphopod shell growth from a constructional perspective as has been done by many authors for gastropods, cephalopods, bivalves and polyplacophorans (e.g., Raup, 1965; Thomas, 1988; Watters, 1991; Dommergues *et al.*, 1996). Morphometric analysis of shell parameters has been applied with success to species diagnoses (e.g., Fantinet, 1959; Shimek, 1989; Steiner, 1995; Engeser and Riedel, 1996; Steiner and Linse, 2000). In these cases, length:width ratios and measures of curvature have been useful in distinguishing among congeneric species.

Class-wide patterns in shell form or scaphopod morphospace have not been reported to date. In a sample of 234 extant scaphopod species (~45% of extant species) from the Indo-Pacific and North Atlantic (based upon illustrations in Scarabino, 1995 and Henderson, 1920, respectively) we (Reynolds, Mammano and Schwimmer, unpublished data) calculated two parameters characteristic of scaphopod shell shape, conicality and curvature. Figure 4 shows a representation of scaphopod morphospace based on these parameters. While only very weakly correlated ($r^2 = 0.25$, $p < 0.0001$), these parameters combine to produce a range of scaphopod forms from near-straight cylinders to strongly curved cones, although the latter are considerably less frequent in this sample; straight cones and curved cylinders do not exist. While not exhaustive, the distribution of shell form in the "scaphospace" represented is intriguing, and hints at trade-offs and constraints in scaphopod shell growth. Wagner (1997) comments on the highly constrained and stereotyped nature of the scaphopod shell shape, and that patterns of scaphopod morphological diversification are likely to be quite distinct from those of ribeiroid and conocardoid rostroconchs through geologic time. Such class-wide patterns of shell form evolution would be worthy of further examination in scaphopods.

The scaphopod larval shell is discussed in Section 9.4.1.

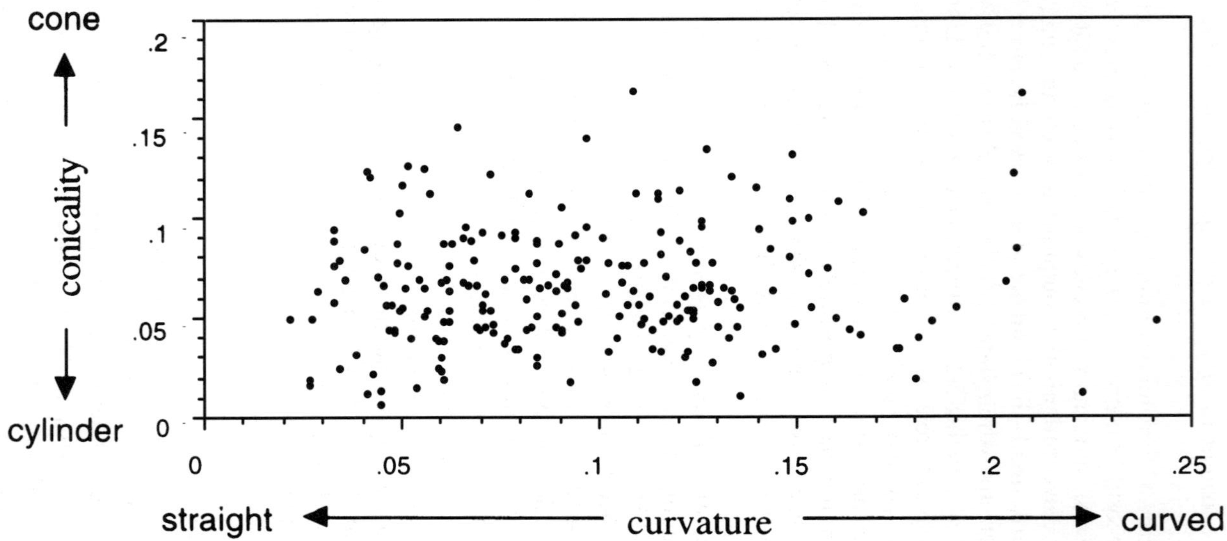

Figure 4 Scaphopod morphospace based on curvature and conicality; $n = 234$ species. Conicality, or degree of taper, was calculated as anterior-aperture/posterior aperture diameter change per unit primary-shell length, degree of curvature as ventral shell curve depth per unit primary shell length. (Reynolds, Mammano and Schwimmer, unpublished data, available for review from author.)

2.2. Shell modification

2.2.1. *Dissolution and decollation*

While shell growth occurs at the anterior end, the posterior end or apex of the shell is periodically lost. As the posterior aperture is the portal through which respiratory currents, gametes, and wastes largely pass, the functional necessity for its progressive widening throughout growth has long been noted (Lacaze-Duthiers, 1856–1857; Fischer-Piette and Franc, 1968). Mediation of posterior aperture shape by the mantle margin is also indicated by patterns of notches in, or secondary openings near, the posterior aperture that are relatively consistent within species (e.g., Scarabino, 1995; Lamprell and Healy, 1998; Palmer and Steiner, 1998; but see Shimek 1989 for discussion of notch variability in *Cadulus tolmiei*). In *Rhabdus rectius,* an increase in posterior aperture size is observed throughout the shell size-range, and the shape and location of the characteristic apical notch is also constant through all sizes (Reynolds, 1992a).

That these characteristics of scaphopod shell growth are a result of reabsorption or dissolution of shell material by the mantle was inferred by Pilsbry and Sharp (1898), Yonge (1937), Fischer-Piette and Franc (1968) and Stasek and McWilliams (1973). In *Rhabdus rectius,* aperture widening and notch maintenance is achieved through shell decollation, or removal of an apical section of the shell, as shown by direct observation (Reynolds, 1992a) (Figure 5). Mantle-mediated shell dissolution occurred at a point removed from (anterior to) the aperture, thereby truncating or decollating the shell; notch dissolution scars on the decollated portion of shell evidenced several abandoned attempts at decollation preceding the successful creation of a new apical notch (Reynolds, 1992a) (Figure 6). While known only from the family Caecidae among marine gastropods, decollation also occurs widely in terrestrial snails (Vermeij, 1993; also see Reynolds, 1992a for review).

2.2.2. *Secondary secretion*

While the major portion of the scaphopod shell is secreted by the anterior mantle margin (and referred to here as the primary shell, Figure 1), a thin secondary shell, produced by the posterior mantle margin, extends from the apex of the primary shell in several species (often referred to as a "pipe") (Plate 1, Figures 1 and 5). Such a secondary shell is found only among several dentaliid genera (Stasek and McWilliams, 1973; Palmer, 1974; Lamprell and Healy, 1998), and its form is probably related to

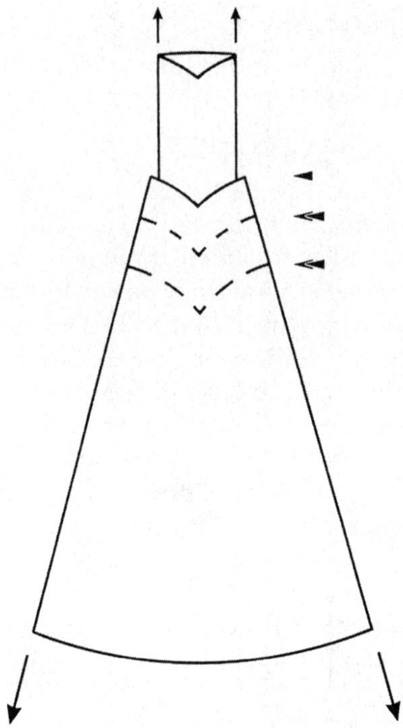

Figure 5 Schematic diagram of shell growth and decollation in *Rhabdus rectius*. The shell is viewed from the ventral side, showing notches in primary and secondary shells. Most recent (arrowhead) and possible future (double arrowheads) sites of shell decollation, and consequent posterior aperture enlargement, are indicated. Large arrows, direction of primary shell growth; small arrows, direction of secondary shell growth. (Reprinted with permission from Reynolds (1992a, figure 18) © California Malacozoological Society.)

posterior aperture increase. In *Rhabdus rectius,* this pipe does not have the conicality of the primary shell, and appears to be a separate program of shell secretion (Reynolds, 1992a).

Such continuous secretion at the posterior aperture necessitates periodic decollation in this species, and probably in other species with pipes as well; species without "pipe" growth may simply widen the posterior opening directly by gradual dissolution of the aperture edge. The possible function of the pipe is open to speculation, but may be related to affording both protection to the sensory mantle tissues at the site of inhalant respiratory currents while maintaining a widened posterior aperture (Reynolds, 1992a).

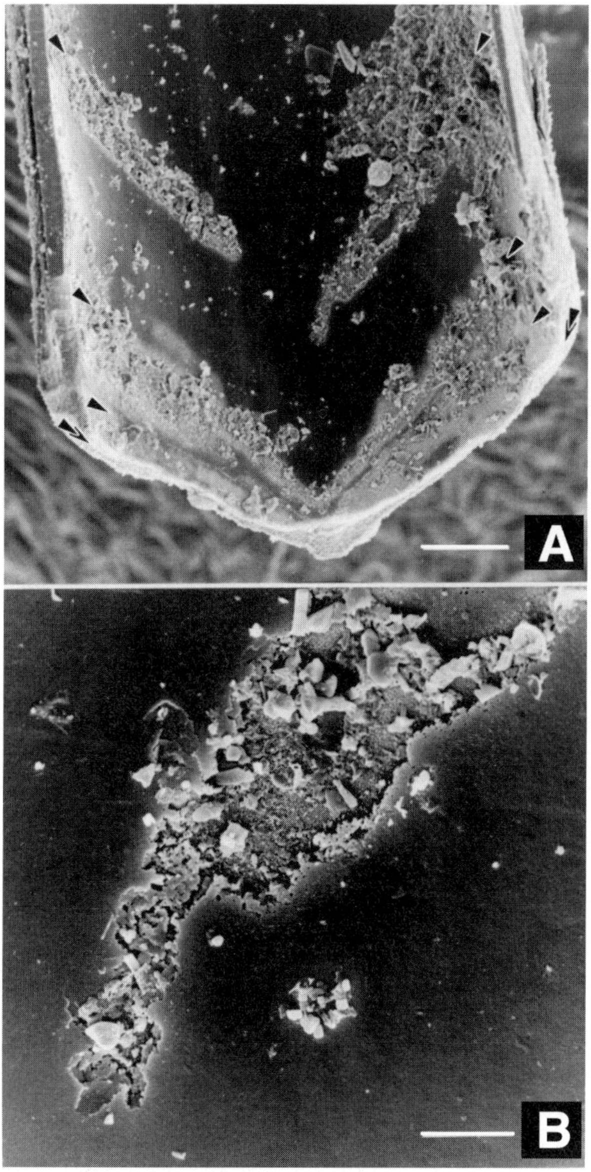

Figure 6 Shell decollation in *Rhabdus rectius.* (A) Decollated apical shell, view of internal ventral surface. Note three sets of V-shaped scars (single arrowheads) and successful decollation edge (double arrowhead), and which positionally correspond to notch left in apex of living shell. Scale bar = 0.1 mm. (B) Scar on internal ventral surface of decollated portion of shell, produced by shell dissolution. Scale bar = 20 μm. (Reprinted with permission from Reynolds (1992a, figures 14 and 15) © California Malacozoological Society.)

2.2.3. *Breakage and repair*

While repair of damage to the anterior aperture is relatively infrequent (pers. obs.), several authors have commented upon the frequency and causes of posterior shell breakage in scaphopods (e.g., Shimek, 1989; Yochelson, 1999). The disjunction between primary and secondary shell in those species with "pipes" may be confused with repairs of shell breakage, although breakage (as evidenced by repairs in primary shell) occurred in ~10% of individuals of *Rhabdus rectius* (Reynolds, 1992a; see also Shimek, 1989); Yochelson *et al.* (1983) found 20% of sub-Holocene *Dentalium laqueatum* had non-fatal shell breakage. Posterior shell breakage has been assigned to predation by fish (Shimek, 1989) or physical forces in sediments (Yochelson, 1999). Evidence of shell damage by naticid predators is usually unrepaired and presumably fatal (Yochelson *et al.,* 1983; Vermeij, 1987), although an unsuccessful naticid attack, evidenced by an incomplete bore hole, is encountered in many species (pers. obs.) and may be more common in some, such as *Antalis pretiosum* (R. Shimek, pers. comm.). (For further review of predation on scaphopods, see Section 11.2.1.)

2.3. Microstructure

The shells of scaphopods studied to date consist of aragonite (Bøggild, 1930; Haas, 1972; Alzuria, 1984; see Carter and Hall, 1990 for review), although there is one report of a calcitic fossil species (Clark and West, 1984). Scaphopod shells have been reported as consisting of two (Bøggild, 1930; Reynolds, 1992a; Steiner, 1995; Reynolds and Okusu, 1999), three (Fantinet, 1959; Haas, 1972; Alzuria, 1985a, b; Reynolds and Okusu, 1999), and four (Shimek and Steiner, 1997) calcareous layers, the latter in a generalized scheme for unpublished gadilid data. A periostracum has sometimes been reported (Bøggild, 1930; Haas, 1972; Alzuria, 1985a, b; Steiner, 1995) but appears to be lost or worn in many species; there are no reports of periostracal sculpture.

The number of shell layers observed is complicated by the appearance of an inner, third shell layer in only the posterior (older) two thirds of many, though not all, species (Reynolds and Okusu, 1999). The appearance of several sub-layers of subtle variation in crossed-lamellar structures in some species (Fantinet, 1959; Haas, 1972; Reynolds and Okusu, 1999, unpublished data) obscures layer identification and may account for some variance among shell layer reports (e.g., Haas, 1972 and Carter and Hall, 1990, for *Fissidentalium vernedei*). A comprehensive comparative survey would help clarify the degree of variability in scaphopod shell layers.

With respect to microstructure, Cayeux (1916) applied the term foliated to a dentaliid shell layer, although Carter and Hall (1990) doubted the specificity of the term. Clark and West (1984) also described a foliated structure in *Plagioglypta,* but raised questions as to the scaphopod identity of the specimens. Most reports identify prismatic and a variety of crossed-lamellar layers (Lacaze-Duthiers, 1856–1857; Simroth, 1894b; Cayeux, 1916; Bøggild, 1930; Fischer-Piette and Franc, 1968; Haas, 1972; Yochelson *et al.*, 1983; Alzuria, 1985a, b; Reynolds, 1992a; Steiner, 1995; Shimek, 1997; Shimek and Steiner, 1997; Reynolds and Okusu, 1999) (Figure 7). A homogeneous microstructure, which may be equivalent in location to the prismatic layer described by others, has been described by Bøggild (1930), Fantinet (1959), Alzuria (1985a, b), and Carter and Hall (1990).

Most shell microstructure studies with specific descriptions or illustrations report upon one (Alzuria, 1985a, b; Carter and Hall, 1990; Reynolds, 1992a) or two (Cayeux, 1916; Fantinet, 1959) species; more comparative studies include Bøggild (1930), Haas (1972), Steiner (1995), and Reynolds and Okusu (1999). Most studies have focused on dentaliid species. Haas (1972) reported no difference between dentaliid and unspecified gadilid species he examined, although the gadilid scheme of Shimek and Steiner (1997) might indicate otherwise. Neither Steiner (1995), reporting on the microstructure of larval shells from four species, three of which were gadilids, nor Reynolds and Okusu (1999), reporting on adult shells from 6 gadilids out of 17 species, found any order-level differences in microstructure.

3. MANTLE CAVITY

In mollusks, the mantle creates a sheltered enclosure adjacent to the body proper, the mantle cavity, through which pass excretory products, gametes, respiratory currents and, for many species food. The structures associated with these functions, the excretory pores, anus, ctenidia, ancillary feeding organs such as labial palps, and sensory structures to test incoming currents, are therefore all found within the cavity, as is the retracted foot in some molluskan classes. As a whole, the mantle and mantle cavity comprise an integrated system that mediates much of the animal's interaction with its environment.

Among the characters that distinguish the Scaphopoda from other molluskan classes are several unique morphological features associated

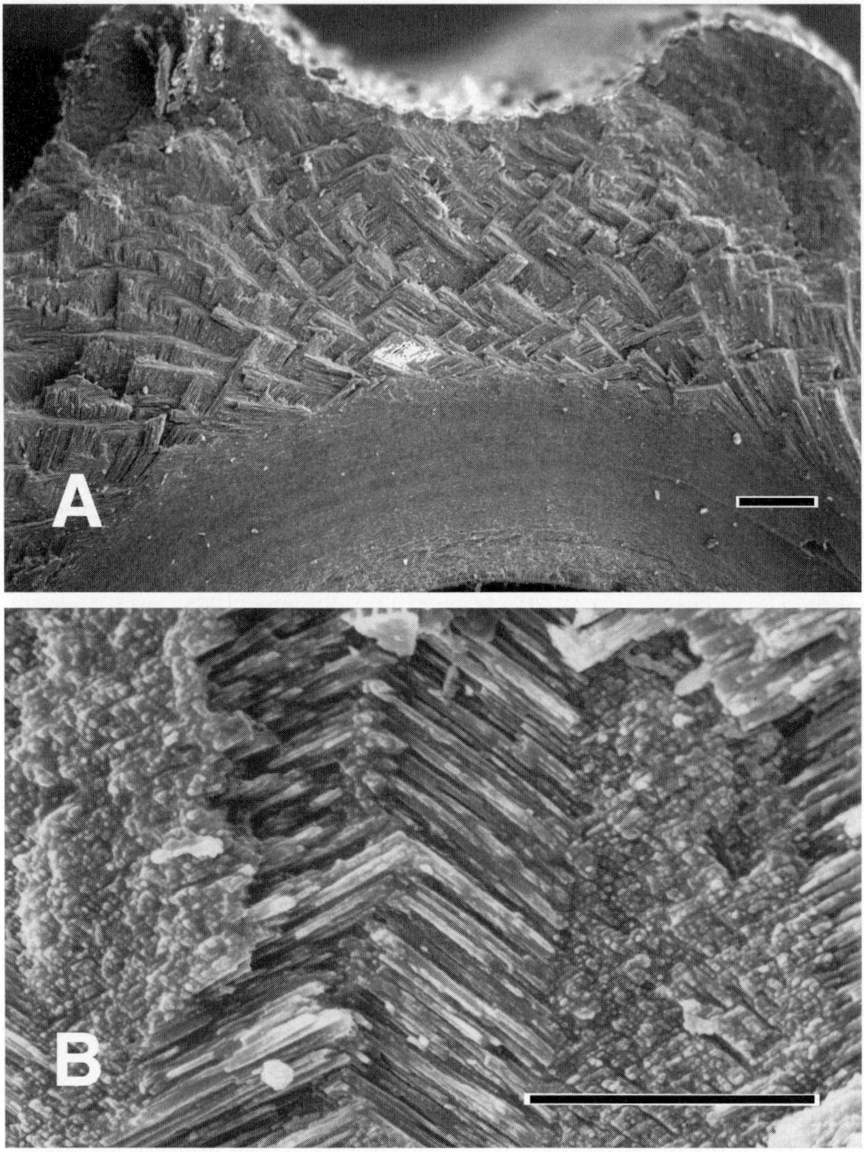

Figure 7 Shell microstructure. (A) Cross sectional view of *Dentalium laqueatum.* Note longitudinal ribs and predominant crossed-lamellar structure in two main layers, the inner one of finer, more irregular crossed-lamellar structure; a third, thin, prismatic layer is to the exterior. Scale bar = 200 μm. (B) Cross-sectional view of *Pulsellum salishorum* showing detail of crossed-lamellar layer. Scale bar = 100 μm. (Okusu and Reynolds, unpublished data.)

with the mantle cavity. Most notable is the absence of otherwise widespread features of the phylum, the sensory osphradium and the ctenidia. Associated with the loss of gills is the loss of auricles and a reduced ventricle (Reynolds, 1990b; see Section 7.1). These losses are probably related to the severe constraint in mantle cavity shape and size in the tusk-shaped shell; during development, two mantle lobes originate dorso-laterally and fuse ventrally to form a narrow space that extends the length of the animal (Lacaze-Duthiers, 1856–1857; Kowalevsky, 1883; see Section 9.4); the mantle cavity remains open at both ends (Figure 2).

The mantle also has several characteristic additions. Anteriorly, support structures at the mantle edge (Steiner, 1991) lend shape to shell secretion at the anterior aperture. In the mid-region, a series of ciliated bands and paired slit-shaped openings to the hemocoel are found. At the posterior end, the mantle cavity valve and multi-functional "pavilion" (e.g. Lacaze-Duthiers, 1856–1857) are yet other specializations found in the class.

3.1. Mantle organization

The scaphopod mantle has been studied at the gross and histological level by several workers, dating from the earliest reports on the class (including Deshayes, 1825; Lacaze-Duthiers, 1856–1857; Plate, 1892; Léon, 1894; Boissevain, 1904; Odhner, 1931; Stasek and McWilliams, 1973); a review of these early results is found in Steiner (1991), in addition to a comparative and comprehensive account at the ultrastructural level. The microscopical anatomy of the mantle is also reviewed in Shimek and Steiner (1997); with little further published work since, mantle organization is summarized here, although mantle sensory receptors are reviewed in more detail in Sections 8.1.1 and 8.1.4.

The anterior mantle margin secretes the primary shell of scaphopods, although additional shell material is added by the general exterior mantle surface in most species (see Sections 2.1 and 2.3). The supporting ring of cartilaginous tissue within the anterior mantle is associated with a central fold, which lies peripheral to an inner glandular region and serves as a site of insertion for longitudinal and radial mantle muscles (Steiner, 1991). In the order Dentaliida, an additional outer glandular region is present (Steiner, 1991, 1992b; Reynolds and Okusu, 1999). Steiner (1991) also describes a ciliary band lining the anterior mantle aperture, found in all Dentaliida studied to date except *Rhabdus rectius* (although see also Reynolds and Okusu, 1999). Sensory receptors are abundant and of several different forms (Steiner, 1991, see also Section 8.1.1).

In the mid-mantle region, there is a series of ciliated bands that runs transversely along both body and mantle walls, described by many workers (Lacaze-Duthiers, 1856–1857; Plate, 1888, 1892; Fol, 1889; Pelseneer, 1901; Boissevain, 1904; Distaso, 1905a; Odhner, 1931; Yonge, 1937; Steiner, 1991) (Figures 8, 9, 10). While such bands are present in all scaphopods, there are generally fewer in the Gadilida than in the Dentaliida. They range from a single band in *Gadila metivieri* (Steiner, 1991) to at least 30 in *Rhabdus rectius* (Reynolds, 1990a). The precise number varies among individuals within a species (Reynolds, 1990a) and is probably a function of overall body size (Steiner, 1991), both through ontogeny and among species, and related to respiratory current load (see Section 4).

The anal bulb is a large papilla that bears the anus and continually pulsates, with the anus opening and closing, drawing water into the rectum and ejecting it forcefully (Yonge, 1937; pers. obs.; see also Section 7.1). On either side and posterior to the anal bulb are the circular excretory pores (e.g., Reynolds, 1990c); anterior to these are the paired slit-shaped epithelial openings to the hemocoel (see Section 7.2) (Figure 9).

The posterior mantle opening consists of a slit- or crescent-like valve, laterally oriented in Dentaliida and vertically oriented in Gadilida (Steiner, 1991). Both valve arrangements consist of hemolymph sinuses, musculature and connective tissue that together function in regulating the size of the opening and inhalant water flow. The structure of the valves is comparatively described in Steiner (1991) and reviewed in Shimek and Steiner (1997); documentation of these structures in additional species is found in Reynolds and Okusu (1999). Extending posteriorly from the valve is the "pavilion" (dubbed "pavillon" by Deshayes, 1825, and retained by many authors since), a cylindrical tube of mantle tissue, open medially along the ventral side; sensory receptors are abundant in the pavilion (see Section 8.1.4). Steiner (1991) described the inner edges of the medial opening as ciliated in the Gadilida, which probably enhances current production. Subepithelial glands may be numerous or scarce, or variable, in the pavilion (Steiner, 1991; Reynolds and Okusu, 1999). The function of the pavilion in modifying the posterior shell aperture size and secreting secondary shell in some species (Reynolds, 1992a) is reviewed in Section 2.2.

4. GAS EXCHANGE

The ciliated bands of scaphopods have long been known to produce currents in the pallial cavity, drawing water from its posterior opening

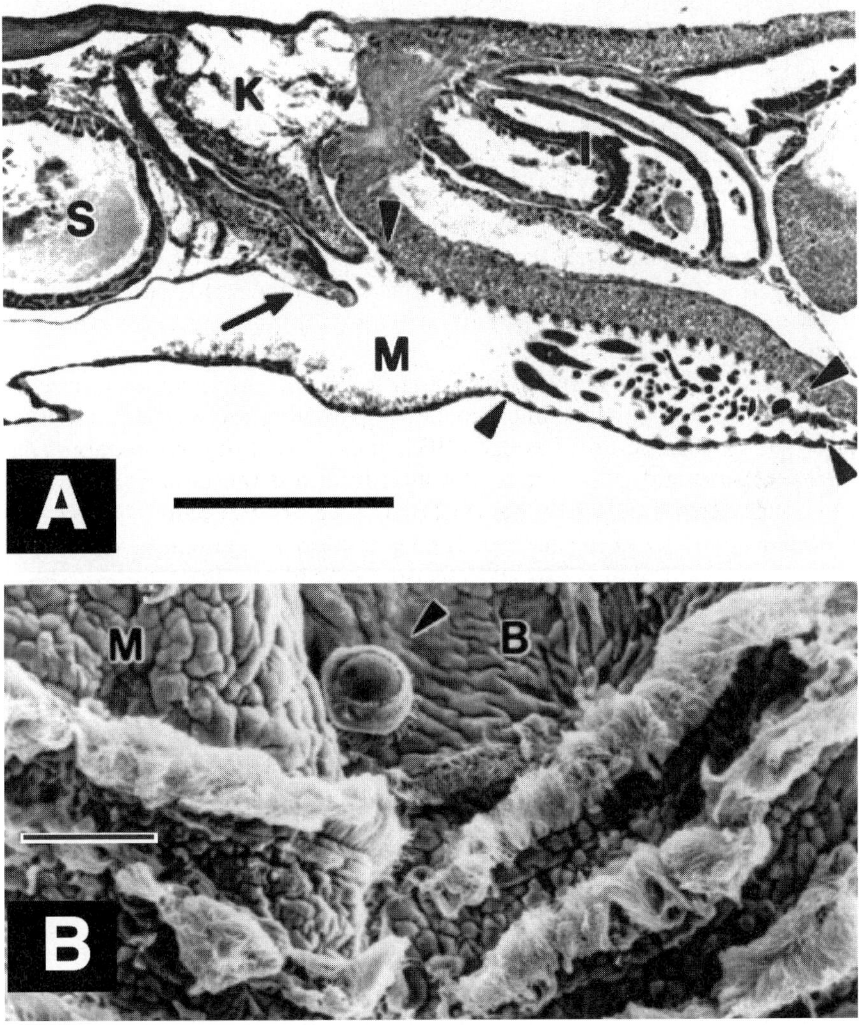

Figure 8 Mantle cavity features in *Rhabdus rectius*. (A) Longitudinal section of *Rhabdus rectius,* anterior to right. Note the ciliated ridges of the body wall (between upper arrowheads) and mantle wall (between lower arrowheads) within the mantle cavity (M). Note also the intestinal loops (I) and rectum surrounded by perianal sinus (arrow) K, kidney; S, stomach. Scale bar = 0.5 mm. (B) Ciliated bands of the body (B) and mantle (M) walls. Arrowhead indicates commensal ciliate, possibly *Trichodina* (for higher magnification, see Figure 26). Scale bar = 50 μm.

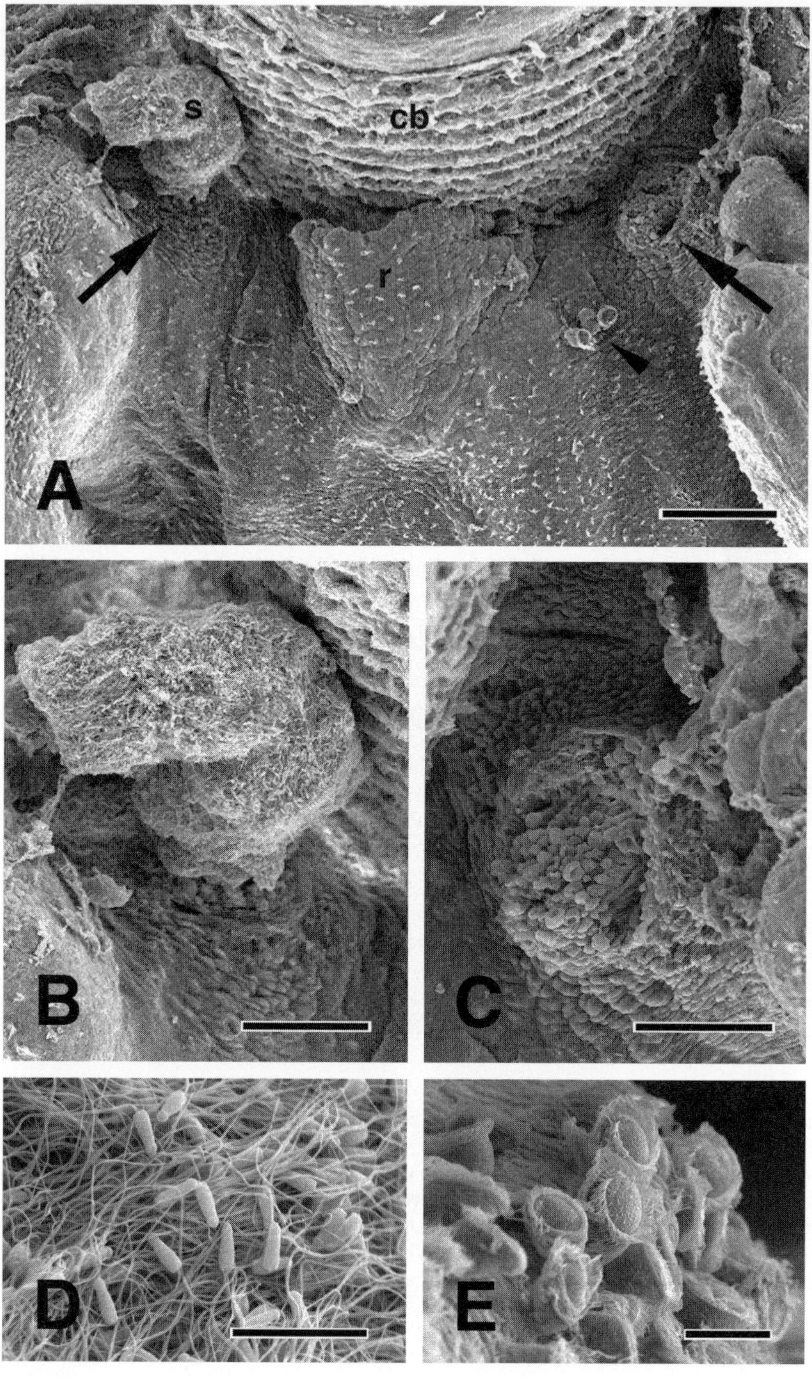
s
cb
r
A
B
C
D
E

toward the anterior (Lacaze-Duthiers, 1856–1857; Distaso, 1905a; Yonge, 1937; Taib, 1980; Steiner, 1991). Water intake may also be produced by dilation of the foot (Yonge, 1937; Gainey, 1972) and movement, perhaps periodic, of the epipodial lobes (Taib, 1980; Steiner, 1991). Water leaves the mantle cavity through the anterior mantle opening by the beating of the ciliary bands (Steiner, 1991) and through the posterior opening by periodic expulsions due to withdrawal of the foot (Clark, 1849; Yonge, 1937; Gainey, 1972; Taib, 1980; Reynolds 1992b), which also ejects feces (Yonge, 1937; pers. obs.) and typically gametes (Dinamani, 1964a; Steiner, 1993) from the mantle cavity. The ciliary bands produce a constant flow of water over the general mantle epithelium, which has been suggested as an unspecialized site of gas exchange, as has the anal bulb (Lacaze-Duthiers, 1856–1857; Plate, 1892; Simroth, 1894a; Yonge, 1937) and the ciliary bands themselves (Distaso, 1905a).

The ciliated bands are highly vascularized (Lacaze-Duthiers, 1856–1857; Distaso, 1905a; Reynolds, 1990a; Steiner, 1991). Describing the ciliated bands on the body wall in *Dentalium,* Lacaze-Duthiers (1856–1857) could not distinguish a consistent direction of blood flow, but observed that the bands received blood at various times from hemocoelic spaces anterior and posterior to the ridges, and from the anal bulb. Distaso (1905a) observed in other species of *Dentalium* that the bands received blood from the ventral vessel of the mantle which, after passing through the bands, was returned to the heart. Bidirectional blood flow has been observed through the perianal and abdominal blood sinuses of *Rhabdus rectius* (Reynolds, 1990a); contraction of the extensive musculature of the body wall and withdrawal of the foot and body within the shell would undoubtedly contribute to transport of the hemolymph supplying the ciliated bands.

Plate (1892) considered the bands to be analogous to molluscan ctenidia; while Distaso (1905a) suggested that they were homologous structures, there has been no additional support for this. In addition to their putative

Figure 9 Mantle cavity features in *Rhabdus rectius.* (A) Ventral view of the mid-mantle cavity region, anterior to top. Note ciliated bands (cb), rectum (r) with terminal anus (anterior, partially obscured), and paired excretory pores (arrows). Note slit-shaped opening to hemocoel anterior to left excretory pore, obscured on right side by sperm mass (s) exiting from right excretory pore (arrowhead, ciliates). Scale bar = 200 μm. (B) Higher magnification of sperm mass exiting from right excretory pore. Scale bar = 100 μm. (C) Higher magnification of left excretory pore and slit opening to hemocoel. Scale bar = 100 μm. (D) Spermatozoa, unmodified aquasperm type. Scale bar = 10 μm. (E) Higher magnification of peritrich ciliates, probably *Trichodina* sp. Scale bar = 50 μm.

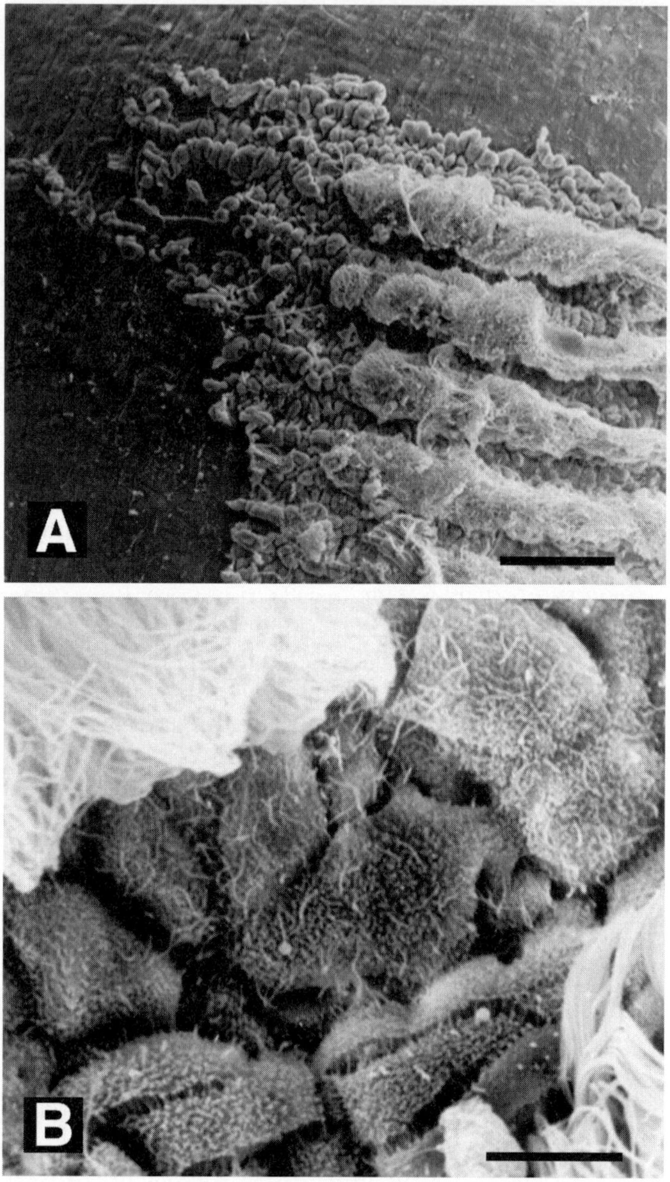

Figure 10 Mantle ciliated bands in *Rhabdus rectius.* (A) Differentiated mantle epithelium supporting the ciliated bands. Note smooth surface of indifferent mantle epithelium to the top and left. Scale bar = 0.1 mm. (B) Supporting cells of the ciliated bands. Note dense microvilli and deep, narrow crevices between cell apices; longer threads have been identified by TEM as bacteria. Scale bar = 10 μm.

role in respiration, the bands are involved in active uptake, including metals (Reynolds, 1990a), as, for example, has been shown for the ctenidia of the bivalve mollusk *Mytilus edulis* (Pentreath, 1973; George *et al.*, 1975, 1976). The bands are supplied with relatively large hemocoelic spaces (Reynolds, 1990a; Steiner, 1991) and are surrounded by microvillous supporting cells with narrow crevices between highly arched cell apices, which also distinguish the bands from the surrounding mantle epithelium (Figure 10). Bacteria are also found in association with these cells in *Rhabdus rectius* (Figure 10), confirmed with transmission electron microscopy (unpublished data).

While gas exchange may occur to some extent over the entire mantle and body wall epithelium, including the anal bulb, the ciliary bands in the scaphopod mantle cavity appear to combine current producing, gas exchange and active transport specializations into one organ system, as is found in the molluskan ctenidium. Therefore, they may be considered a functional equivalent to ctenidia, given the loss of these organs from the scaphopod mantle cavity, although further data are needed to understand fully their role in gas exchange.

The respiratory proteins in scaphopods are poorly known, as discussed further in Section 7.3.2.

5. LOCOMOTION

5.1. Foot

The anatomy of the scaphopod foot was described for two species of *Antalis* in early monographic treatments (Deshayes, 1825; Lacaze-Duthiers, 1856–1857; Fol, 1889; Plate, 1892). Steiner (1992a) extensively described pedal anatomy in a comparative approach, and scored this variability as characters in a phylogenetic analysis (Steiner, 1992b); Reynolds and Okusu (1999) similarly scored pedal variability for additional species and genera.

Pedal anatomy perhaps demonstrates the most striking differences between the two orders in the class. In the Dentaliida, the foot is a highly muscularized, cylindrical organ with a conical projection almost surrounded by epipodial lobes (Plate 1, Figure 2). Movement is effected by antagonistic contraction of longitudinal, circular, radial and oblique fibers, which surround a relatively small hemocoelic space that is characterized as a muscular hydrostat (Shimek and Steiner, 1997). The dentaliid foot in certain species extends up to the length of the shell (e.g., *Antalis vulgaris*: Steiner, 1992a). Retraction occurs fully into the anterior mantle cavity, necessitating an S-shaped folding of the cylinder in *Rhabdus rectius* (Steiner, 1992a; pers. obs.). This is in contrast to the gadilid foot, which, while cylindrical, terminates in a disk with distal and lateral papillae. In this order, discrete retractor musculature runs through an extensive hemocoelic sinus, and hydraulic pressure extends the foot, which is characterized as a hemal hydrostat (Shimek and Steiner, 1997). Foot movement in gadilids is relatively rapid (Steiner, 1992a), the foot extends up to twice the length of the shell, and retraction is through introversion. A detailed review of scaphopod pedal anatomy appears in Shimek and Steiner (1997).

5.2. Burrowing behavior

Burrowing in scaphopods occurs by a simple extension of the foot into soft sediments, anchorage through expansion of the epipodial lobes (Dentaliida) or terminal disk (Gadilida), with contraction of retractor muscles drawing the body into the sediment (Lacaze-Duthiers, 1856–1857; Morton, 1959; Dinamani, 1964a; Trueman, 1968; Steiner, 1992a). The shell is oriented with the concave (dorsal) surface uppermost during burrowing, with twisting of the shell to achieve this (Morton, 1959; Dinamani, 1964a; Gainey, 1972; Poon, 1987; pers. obs.).

Activity during burrowing (as when an individual is reintroduced to sediment in the laboratory; see Section 6.3 for separate behaviors during feeding), however, varies considerably between the orders, with dentaliids generally slower with stronger and more deliberate foot movements (Dinamani, 1964a; Steiner, 1992a). Gadilids, on the other hand, move their foot rapidly (Davis, 1968; Steiner, 1992a; Shimek and Steiner, 1997), and *Gadila aberrans* has been reported to burrow at a rate of $1\,\mathrm{cm\,s^{-1}}$ (Shimek, 1989). This higher level of activity has also been associated with the highly polished shell and narrowed anterior aperture found in some gadilids (Shimek, 1989). Upward movement has also been observed in *Graptacme eboreum, Antalis pseudohexagonum* (Gainey, 1972) and *Rhabdus rectius* (pers. obs.), and is effected by anchorage by epipodial lobes and extension of the foot through pedal circular muscle contractions (Gainey, 1972). Gainey (1972) notes that reversal is followed by downward burrowing in a new direction, thereby accessing a new feeding area.

Related also to the ordinal difference in burrowing activity may be the observations of apical, posterior shell protrusion from the sediment surface. This has been reported frequently in dentaliids (e.g., *Antalis entalis*: Yonge, 1937; Morton, 1959; *A. pseudohexagonum*: Gainey, 1972; *Rhabdus rectius*: Reynolds, 1992a; *Fissidentalium vernedei* and indirect evidence for others: Zibrowius, 1998; *Fissidentalium majorinum*: pers. obs.) and is often illustrated in textbooks, although the frequency of this behavior varies considerably among and within these species (Gainey, 1972; R. Shimek, pers. comm.), and the percentage of a population protruding from the surface at any one time may be quite small (unpublished data). Whether necessitated by efficient spawning of gametes (e.g., Steiner, 1993), or a random effect of feeding behavior, exposure to predation would appear to be an apparent consequence of apical protrusion. Gadilids, on the other hand, are reported to burrow well below the sediment surface (e.g., *Cadulus tolmiei*: Poon, 1987; *Gadila aberrans*: Shimek, 1989; *Siphonodentalium dalli antarcticus*: pers. obs.). The report of apical protrusion and burrowing convex (ventral) side uppermost in *Cadulus quadridentatus* (Davis, 1968) is very possibly a consequence of the shallow sediment provided during those observations.

Scaphopods probably burrow at least sufficiently deeply to find food, which in turn may be influenced by their feeding selectivity (see Section 6.1), and are limited by depth of aerobic sediments. The putative adaptations to rapid burrowing may render many gadilid species to be relatively wide-ranging predators; for example, *Gadila aberrans* regularly burrows to depths of 30–40 cm (Shimek, 1989, 1990).

6. FEEDING AND DIGESTION

6.1. Diet

Scaphopods are microcarnivores or microomnivores (Salvini-Plawen, 1981, 1988), feeding mainly upon foraminiferans among a suite of largely test-bearing prey (Clark, 1849; Lacaze-Duthiers, 1856–1857; Morton, 1959; Dinamani, 1964b; Fischer-Piette and Franc, 1968; Bilyard, 1974; Taib, 1980; Poon, 1987; Shimek, 1988, 1990; Steiner, 1994; Langer *et al.*, 1995; Shimek and Steiner, 1997). Species identifications of common foraminiferan prey are given in several of the above reports. Detailed analyses of gut contents and quantification of relative contribution of various prey items have been performed for *Antalis entalis stimpsoni* (Bilyard, 1974), *Cadulus tolmiei* (Poon, 1987), *Fissidentalium megathyris* (Langer *et al.*, 1995), and for *Rhabdus rectius, Gadila aberrans* and *Pulsellum salishorum* by Shimek (1988, 1990). Shimek (1990) provides the most comparative in-depth study of scaphopod diet and feeding ecology to date. While the total number of examined species is small, taxonomic and habitat breadth is wide enough for insight into the range of variability in scaphopod diet.

Various measures suggest that scaphopods are highly selective, ingesting a higher ratio of living:non-living items than is in the surrounding sediment (Bilyard, 1974; Poon, 1987; Shimek, 1990). While differences in potential prey availability among habitats and diet preferences of species preclude over-generalization, scaphopods also exhibit preferential selection among available prey species (Bilyard, 1974; Poon, 1987; Shimek, 1990). Paradoxically, while seasonality in diet was not found among the coastal species examined by Shimek (1990), an increase in dietary items during the more productive summer months, largely comprised of planktic foraminiferan species, has been found for scaphopods from 2900 m in the Rockall Trough (Davies 1987); diet seasonality is also supported in *Fissidentalium actinophorum* from 4100 m in the northeastern Pacific (R. Shimek, pers. comm.).

In all species studied to date, foraminifera were the largest single taxonomic category among food items (*Fissidentalium megathyris* ≤100%, *Antalis entalis stimpsoni* 80.9%, *Gadila aberrans* 76.19%, *Cadulus tolmiei* 65.1%, *Pulsellum salishorum* 59.20%, *Rhabdus rectius* 38.53%), and in most species dominated the diet as they do the microfauna of the sediments in which these species feed (Bilyard, 1974; Poon, 1987; Shimek, 1990; Langer *et al.*, 1995). Other small protozoans and metazoans constitute a small percentage of the diet of most species studied to date (e.g., Sahlman, 1973; Shimek, 1990). One exception to this is *Rhabdus rectius*, which takes mite eggs as a significant portion of its "live" diet (~10%); furthermore, sediment and mineral grains constitute >20% of ingested items in this species (Shimek, 1988, 1990). These observations indicate that *R. rectius* is more omnivorous than other scaphopods studied to date and, while still rejecting many or most of the potential food items encountered (Shimek, 1990), is a somewhat less selective deposit or micro-omnivorous feeder. Although detailed gut content analyses have yet to be performed, *Bathoxiphus* species also ingest significant inorganic material, possibly related to their depauperate deep-sea habitat (Steiner, 1994) (see Section 6.4.5).

6.2. Captacula

While scaphopods are radulate mollusks, they feed on foraminifera and other microorganisms selected and manipulated by the captacula, i.e., the feeding tentacles that are a characteristic feature or synapomorphy for the class. The general form and function of captacula have been reported by many early authors (e.g., Deshayes, 1825; Clark, 1849; Lacaze-Duthiers, 1856–1857; Fol, 1885, 1889; Plate, 1892). More recent histological work by Morton (1959), Gainey (1972) and Shimek (1988), and ultrastructural work by Shimek (1988) and Byrum and Ruppert (1994), currently provide the most detailed data on the functional morphology of these structures.

The captacula are thin tentacles with a bulbous ciliated tip (Figure 11) that extend to several millimeters in length and range to several hundreds in number (Morton, 1959; Shimek, 1988; Byrum and Ruppert, 1994), although only a few may be extended at one time (Davis, 1968). Captacula are frequently lost and are continually regenerated (Shimek, 1988). They arise from two dorso-lateral areas at the base of the oral proboscis. Smaller, unciliated, newly formed captacula are found along the ventral margin of this area (Figure 11A), and intermediates in size and development toward the dorso-anterior margin have led to the inference of their continual formation and gradual displacement dorsally as captacula mature and are lost (Fol, 1889; Plate, 1892). The ovoid distal bulb has a small pit, or alveolus, on the functionally ventral side. The bulb and alveolus are ciliated, the latter more heavily so (Lacaze-Duthiers, 1856–1857; Morton, 1959; Byrum and Ruppert, 1994). Stalk ciliation ranges from a continuously ciliated tract, which may or may not run to the stalk base (Fol, 1889; Dinamani, 1964b; Poon, 1987; Byrum and Ruppert, 1994; Shimek and Steiner, 1997), to equidistant tufts (Shimek, 1988), to a complete lack of ciliation (Morton, 1959) (range observed by Shimek, 1988, see Figure 11).

Apart from reports of a cuticular layer (Fol, 1889), which have not been confirmed and would be unusual in a ciliated epithelium, the captaculum generally consists of an epithelium containing nerve, muscle and glandular elements.

The alveolus is the site of sediment/food particle attachment, which is preceded by movement of the alveolar area over sediment particles by the captaculum, suggesting selectivity. A ganglion is located in the bulb, just proximal to the alveolus (Fol, 1889; Plate, 1892; Morton, 1959; Gainey, 1972; Taib, 1980; Shimek, 1988; Byrum and Ruppert, 1994). It is supplied by a nerve, and sends processes to the alveolar epithelium that probably contribute to sensory function (Byrum and Ruppert, 1994) (Figure 12). There have been no reports of specialized epithelial receptors on the bulb or within the alveolus, but their presence may be obscured by the heavy ciliation (Byrum and Ruppert, 1994).

Circular and longitudinal muscles extend throughout the stalk and bulb (Byrum and Ruppert, 1994). The number of longitudinal muscle bands varies among species studied to date, but may have different ranges within each order: 8–10 in the Dentaliida and 5–7 in the Gadilida (Shimek, 1988; Shimek and Steiner, 1997). There is no ultrastructural evidence for circular connective tissue fibers; those observed in *Antalis entalis* by Morton (1959) and Taib (1980) may in fact have been circular

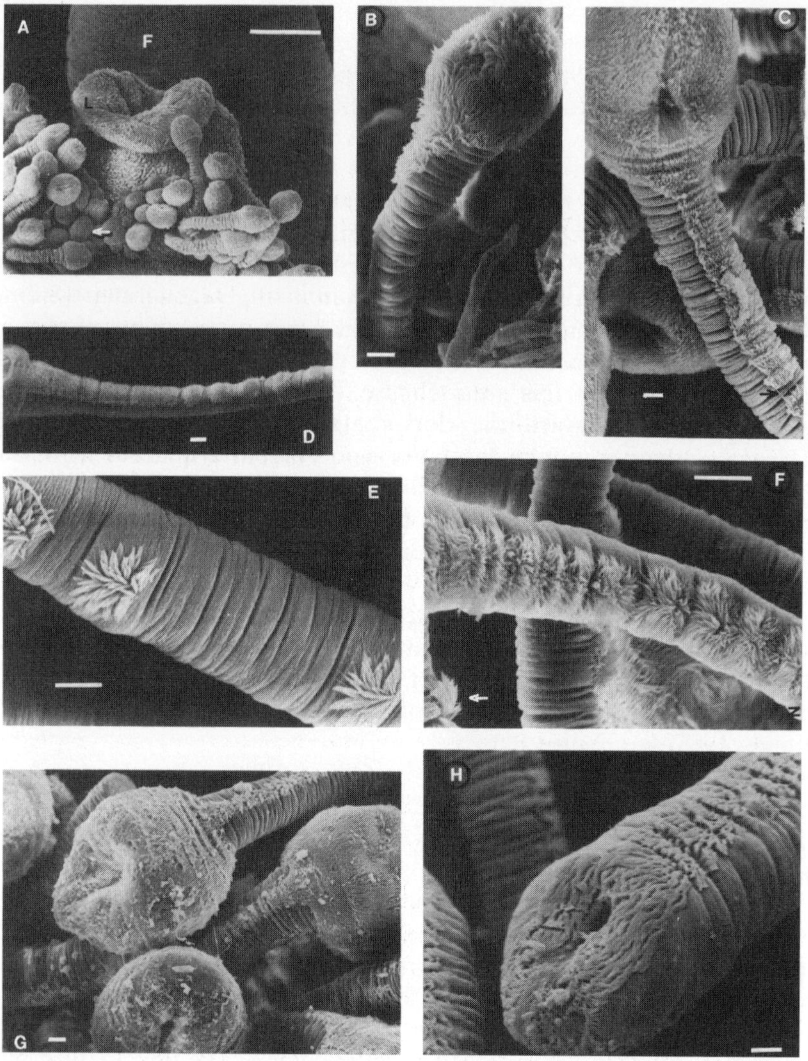

Figure 11 Captacula of several species; note bulbous tip with alveolus and stalk ciliation. Scale bar = 100 μm in A, 10 μm in B–H. (A) *Pulsellum salishorum:* dorsal view, shell and mantle removed (F, foot; L, proboscis lips; arrow, immature captaculum). (B) *Pulsellum salishorum:* note lack of stalk ciliation except for distal fringe. (C) *Cadulus tolmiei:* arrow indicates gap in stalk ciliation, note ciliary tract is complete distally (compare with F). (D) *Gadila aberrans:* note ciliary tract becomes tufted proximally. (E) *Gadila aberrans:* proximal stalk, note distinct tufts. (F) *Cadulus tolmiei:* foreground stalk is distal and contracted, where tufts form an apparently continuous band, arrow indicates an isolated ciliary tuft on an elongated stalk. (G) *Rhabdus rectius:* note complete ciliary tract and metachronal beating on bulbs. (H) *Antalis pretiosum:* note complete ciliary tract on stalk. (Reprinted with permission from Shimek (1988, figure 1) © California Malacozoological Society.)

muscle as observed by Byrum and Ruppert (1994). Extension of the captacula is therefore probably accomplished by ciliary action with a contribution from circular muscle contraction; extension by primarily hemocoelic pressure, proposed by Morton (1959), has not been subsequently supported. The relatively rapid retraction of captacula is effected by contraction of the large longitudinal muscles (Shimek, 1988; Byrum and Ruppert, 1994).

Despite the extension of muscles fibers into the bulb, musculature that could effect suction of particles to the alveolus is not present (Byrum and Ruppert, 1994). Instead, one or more of a variety of glands found in the captacular bulb are probably responsible for particle adhesion, possibly in a duo-gland system (Shimek, 1988; Byrum and Ruppert, 1994). This hypothesis is further supported by several observations: little noticeable shape change to the bulb on attachment or detachment of particles that would indicate musculature involvement (Shimek, 1988), adhesion persistence after nerve severance at captacular detachment (Shimek, 1988), and the presence of glands within the bulb (Clark, 1849; Lacaze-Duthiers, 1856–1857; Fol, 1889; Plate, 1892; Morton, 1959; Gainey, 1972; Taib, 1980; Shimek, 1988; Byrum and Ruppert, 1994).

Studies tracing the gland ducts have used different species and techniques, and results are difficult to compare directly. Most studies note a pair of prominent glands located posterior to the captacular ganglion (e.g., Fol, 1889; Shimek, 1988; Byrum and Ruppert, 1994). Byrum and Ruppert (1994) found these glands to open onto the ciliary stalk tract at the base of the bulb (Figure 12), whereas Shimek (1988) reported that they opened into the alveolus through long ducts. Byrum and Ruppert (1994) trace long ducts opening into the alveolus from a second type of gland cell, with smaller vesicles but also proximal to the ganglion (Figure 12). Additional gland bodies either originate in this area and open onto the bulb tip (Byrum and Ruppert, 1994), or are adjacent to and open onto the alveolus (Shimek, 1988).

Those glands that open onto the ciliary tract may secrete mucus for ciliary particle movement (Byrum and Ruppert, 1994); some of the variability in gland duct descriptions may be related to species differences in the presence of a stalk ciliary tract. Secretion of adhesive is a particularly likely function for glands that open into the alveolus because of the dense, truncated alveolar cilia (Lacaze-Duthiers, 1856–1857; Morton, 1959; Byrum and Ruppert, 1994) that may be specialized for spreading adhesive material and maximizing contact with the substratum (Byrum and Ruppert, 1994). Other functions proposed for the captacular glands include toxin or repugnatorial secretion (Morton, 1959), enzymatic "saliva" secretions (Clark, 1849), sensory reception and ontogenetic replacement of mature glands (Byrum and Ruppert, 1994). Little work

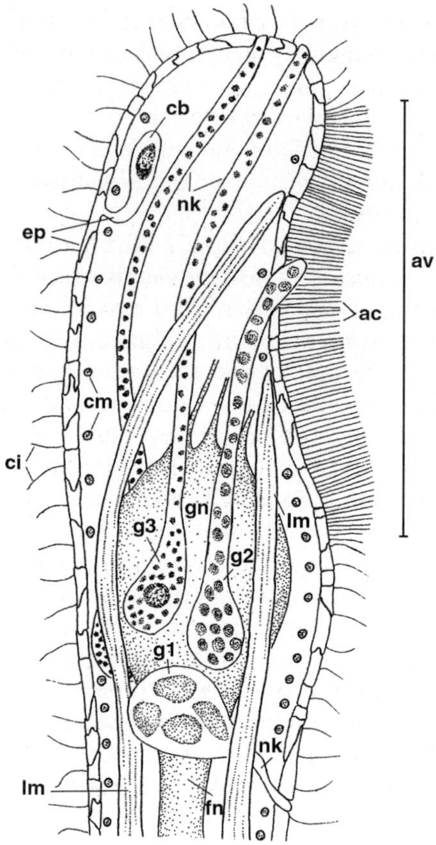

Figure 12 Internal anatomy of the captaculum in *Graptacme calamus*. ac, alveolar cilia; av, alveolus; cb, epidermal cell body; ci, cilia of multiciliated epidermal cells; cm, circular muscles; ep, epidermis; fn, filament nerve; g1–3, glands 1–3; gn, captacular ganglion; lm, longitudinal muscle; nk, gland neck. (Reproduced with permission from Byrum and Ruppert (1994, figure 22) © Blackwell Science.)

has been done on the chemical components of the glandular secretions, although histochemical reports from Sahlman (1973) and Taib (1980) are consistent with the secretion of mucosubstances.

6.3. Feeding behavior

Scaphopods have been characterized as ciliary deposit feeders (e.g., Owen, 1966) and predators (e.g., Shimek, 1990), reflecting both the action of the captacula and the specialized foraminiferan diet of most species studied to date. They are microphagous burrowers that are motile and tentaculate (*sensu* Fauchald and Jumars, 1979); their high selectivity for small heterotrophs characterizes most species as microcarnivorous (Salvini-Plawen, 1981, 1988; for diet see Section 6.1).

Feeding in scaphopods follows a several-step process that involves the foot, captacula and labial palps. Feeding behavior does vary among scaphopod species in several ways, which in some cases may reflect ordinal-level morphological synapomorphies and, in other cases, response to sediment characteristics and microfaunal communities. It is also worth noting that while many workers have observed scaphopod behavior in dishes of clean sea water or shallow sediment (e.g., Dinamani, 1964b; Davis, 1968), others have made observations in deeper sediment, sometimes using specially constructed glass aquaria and artificial transparent sediment (e.g., Dinamani, 1964b; Gainey, 1972; Poon, 1987; Shimek, 1988), which probably elicit a more accurate suite of behaviors.

The first categorical step in feeding behavior is the creation of a feeding cavity. Gainey (1972) observed individuals of two dentaliid species, *Graptacme eboreum* and *Antalis pseudohexagonum,* constructing the cavity by probing and packing the sediment surrounding the anterior aperture with the foot. The large conical proboscis of the dentaliid foot produces a fairly spacious, near-spherical cavity; a similarly spacious feeding cavity has been illustrated for *Rhabdus rectius* (Shimek and Steiner, 1997), and taphonomically preserved *in situ* in a Holocene dentaliid species by silt infilling, observed through radiographs of sediment cores (Figure 13; Shevenell *et al.,* 1996). The feeding cavity in the gadilid *Cadulus tolmiei,* on the other hand, is considerably smaller than that observed in dentaliids to date, and constructed by the foot with its terminal disk notably invaginated (Figure 14; Poon, 1987).

Once the cavity is constructed, the foot continues its feeding activity by probing the sediment, and in some dentaliids the epipodial lobes dislodge particles that are subsequently examined by the captacula (*Dentalium conspicuum*: Dinamani, 1964b). Particles are drawn directly into the mantle cavity and proximity to the mouth by repeated extension and withdrawal movements of the foot (*Graptacme eboreum* and *Antalis pseudohexagonum*: Gainey, 1972; *Cadulus tolmiei*: Poon, 1987), through a scooping action (*Rhabdus rectius*) or adherence of material to the sides of the foot (*Gadila aberrans* and *Pulsellum salishorum*) (Shimek, 1988). These feeding behaviors are by their nature non-selective and somewhat indeterminate, possibly serving other functions associated with exhalant current production through the posterior aperture. Despite these minor foot movements, there may be little apparent movement of the shell for long periods, the animal appearing somewhat sedentary while the feeding cavity remains intact.

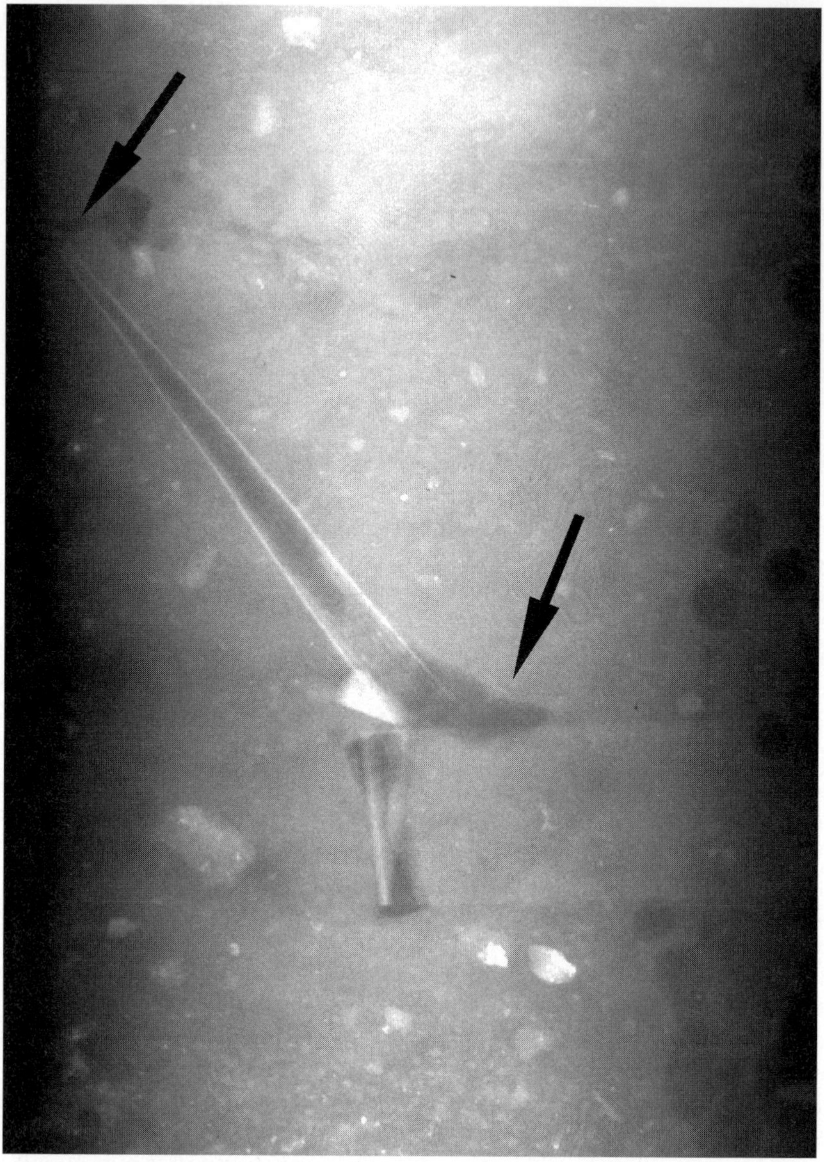

Figure 13 X-ray radiograph of portion of sediment core with fossilized dentaliid (shell length = 40 mm) preserved *in situ* from Lallemand Fjord, western Antarctic Peninsula. Note longitudinal ribbing visible in anterior half of shell, posterior shell apex (upper arrow), and darker color by silt infilling of putative feeding cavity (lower arrow). Bottom of shell is 461.25 cm below sediment surface; corrected radiocarbon date of scaphopod = 8058 years before present (Shevenell *et al.,* 1996).

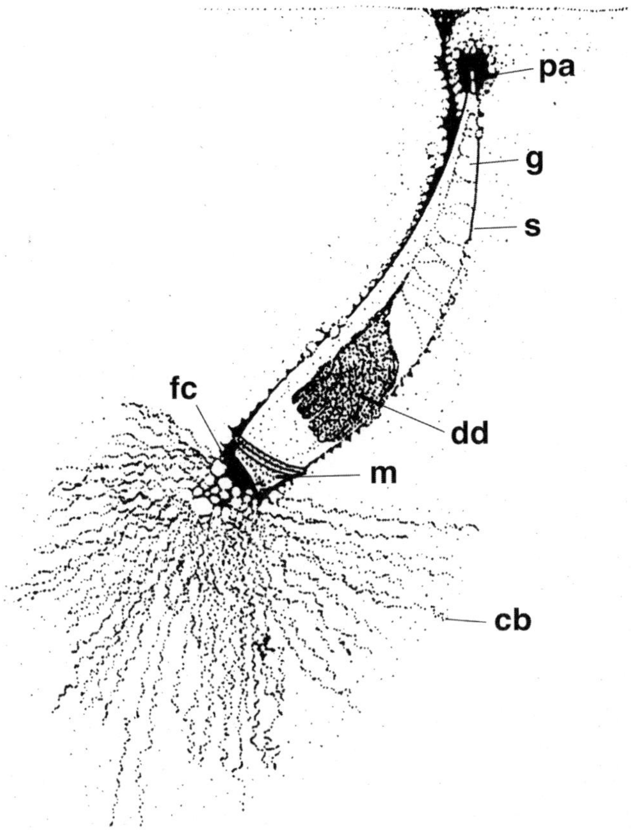

Figure 14 Feeding in the gadilid *Cadulus tolmiei.* Note small feeding cavity (fc) and radiating captacular burrows (cb). dd, digestive gland; g, gonad; m, mantle; pa, pavilion; s, shell. (Reproduced with permission from Poon (1987, figure 1).)

Directed and selective feeding in scaphopods is accomplished by the captacula. The captacula emerge from the pallial aperture and probe the cavity wall (Gainey, 1972) and extend past the wall into the sediment (Taib, 1980), or create an array of radiating captacular burrows as in *Cadulus tolmiei* (Poon, 1987). Captacula move with a rapid ciliary gliding motion (Dinamani, 1964b; Davis, 1968; Shimek, 1988; Byrum and Ruppert, 1994) ($\sim$1 mm s^{-1} in *Graptacme calamus*: Byrum and Ruppert, 1994), although Byrum and Ruppert (1994) suggest that circular musculature contraction along the length of the stalk must also play a role in captacular extension (see Section 6.2). Whereas Davis (1968) observed only 12–15 captacula extended from the anterior aperture in *Cadulus quadridentatus,* individuals of *C. tolmiei* were observed to extend many more captacula into the sediment, by moving along the taut stalks of an initial few penetrating captacula (Poon, 1987).

The captacula are involved in all stages of potential food item manipulation, both outside and within the mantle cavity. Outside the mantle cavity, captacula select and transport food items to the mantle cavity, although Gainey (1972) also observed captacula dislodging particles from the roof of the feeding cavity, which subsequently fell to the floor of the cavity and were manipulated by other captacula. Particle transport by the captacula to the mantle cavity occurs in three main ways. First, a single captaculum adheres at the site of the alveolus, and sometimes wraps around either single items such as foraminiferans (Dell, 1957; Morton, 1959; Gainey, 1972; Poon, 1987; Shimek, 1988) or several smaller particles (Poon, 1987; Shimek, 1988), and withdraws into the mantle cavity by contraction of the longitudinal muscles (Fol, 1885; Morton, 1959; Davis, 1968; Shimek, 1988; Byrum and Ruppert, 1994). Secondly, several captacula can act in concert to transport a single large foraminiferan to the mantle cavity (Morton, 1959; Dinamani, 1964b; Shimek, 1988). Third, ciliation along the captacular stalk moves small particles in a metachronal stream from bulb to the stalk base (Dinamani, 1964b; Gainey, 1972; Poon, 1987; Shimek, 1988; Byrum and Ruppert, 1994), and appears to consistently take place when a continuous ciliated path along the stalk exists (Shimek, 1988). Shimek (1988) notes the significance of this with regard to the high sediment intake of *Rhabdus rectius,* which has continuous stalk ciliation, but Poon (1987) suggests that this behavior does not contribute significantly to feeding in *Cadulus tolmiei,* in which the captacular stalks have mostly tufted ciliation (Figure 11). The degree of selectivity associated with this particular mode of captacular collection is unclear. With the hundreds of available captacula in operation, much material could potentially be collected in this manner in a short time (Dinamani, 1964b).

Within the mantle cavity, the potential food material accumulates in a longitudinal furrow along the dorsal surface of the foot, extending anteriorly from the mouth, which is surrounded by the labial palps. Upward movements of the foot direct the material toward the mouth and palps (Shimek, 1988). Captacula within the cavity may receive particles from other captacula bringing particles from the exterior (Gainey, 1972; Poon, 1987), or otherwise manipulate the particles in the foot furrow where further selection may take place (Dinamani, 1964b; Shimek, 1988). Yonge (1937) suggests that the ciliated bands of the mantle would bring particles toward the foot base for captacular manipulation, although this is probably not a major source of food. Finally, captacula deposit single particles and boluses of material on the labial palps directly

(Dinamani, 1964b; Shimek, 1988), from whence it is ingested. Ingestion accounts for only a small proportion of material in the mantle cavity, however, as most is rejected and remains in the pedal furrow (Shimek, 1988). Rejected material is expelled with fluid through contraction of the foot (Shimek, 1988), or left near the exterior of the opening by the furrow being periodically wiped along the edge of the mantle opening upon foot extension and subsequent withdrawal (Dinamani, 1964b).

Captacula have also been observed less frequently extending posteriorly along the outside of the shell (Davis, 1968) and within the mantle cavity to the extent of even emerging from the posterior aperture (Shimek, 1988). Both behaviors might reasonably be hypothesized to have a cleaning or maintenance function, removing occluding mucus, particles and fouling organisms (see also Section 11.1) from the shell, mantle ciliary bands (Shimek, 1988) and respiratory current pathways through to the posterior mantle valve.

6.4. Alimentary tract

The morphology of the digestive system in Scaphopoda is well known in terms of anatomical relationships of the associated organs (e.g., Lacaze-Duthiers, 1856–1857; Taib, 1981a; Salvini-Plawen, 1988) and in this regard is relatively consistent within the class, with the significant exception of the digestive gland, which is one-lobed in the Gadilida and two-lobed in the Dentaliida. Salvini-Plawen (1988) provides additional detail of variation in gut morphology amongst scaphopod taxa. The scaphopod gut is characterized by an essentially U-shaped arrangement, the basis for anatomical ventrality toward the larger aperture (Figure 15A). The general anatomy and histology of the gut has been comparatively examined by Salvini-Plawen (1981, 1988) and recently reviewed by Shimek and Steiner (1997). While considerable attention has been given to description at the cellular level (Lacaze-Duthiers, 1856–1857; Fol, 1885, 1889; Plate, 1888, 1892; Morton, 1959; Sahlman, 1973; Taib, 1981a, b; Shimek and Steiner, 1997), few ultrastructural data (Reynolds, 1990a) have been reported.

6.4.1. *Anterior gut and radula*

The mouth, surrounded by the labial palps, is borne on a non-retractile oral cone (Salvini-Plawen, 1988), which also bears the captacula dorso-laterally (see Section 6.2). The mouth leads to a short buccal cavity from which branch a pair of lateral, glandular, buccal pouches, which are a site

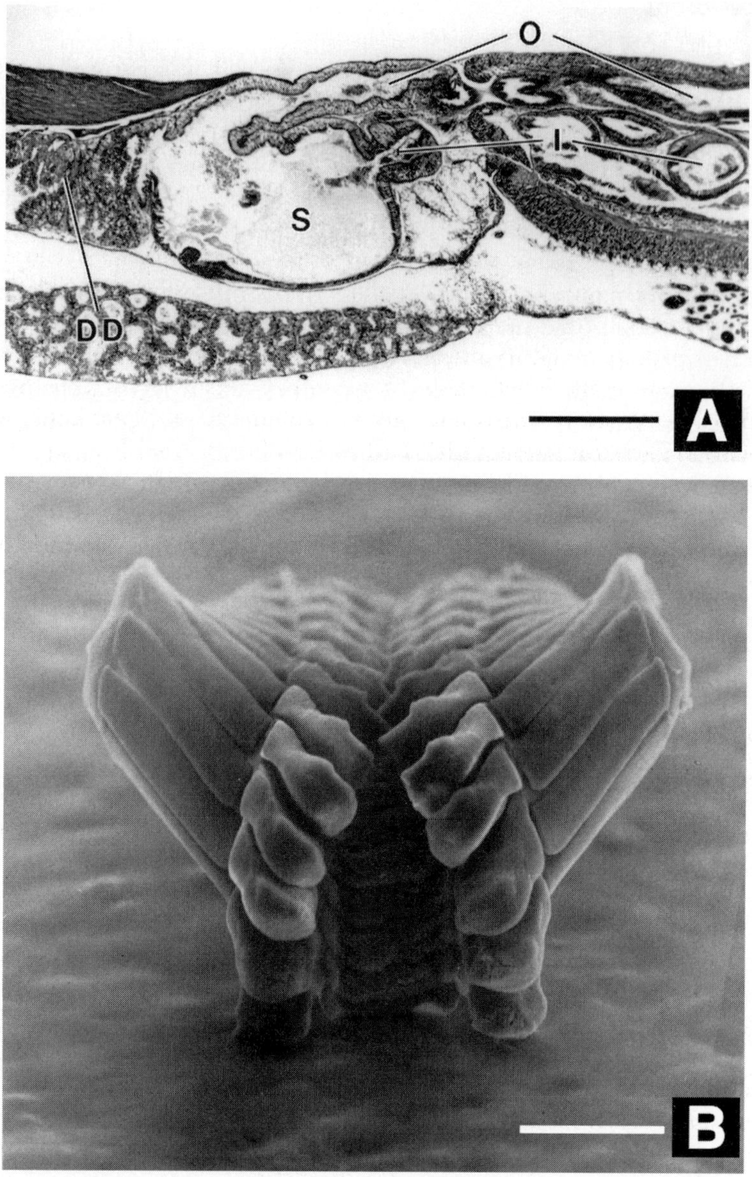

Figure 15 Gut structures in *Rhabdus rectius.* (A) Longitudinal section through the mid-region, anterior to right. Note U-shaped gut: esophagus (O) enters stomach (S) dorsally and posteriorly, and intestine (I) exits the stomach dorsally and anteriorly, ventral to the esophagus. Profiles of captacula in section can be seen in mantle cavity, adjacent to mantle ciliated bands, in lower right corner (DD, digestive diverticula in mantle lobe and main body where indicated by line). Scale bar = 0.25 mm. (B) The radula (frontal or foraminiferan view); note the paired marginal and lateral teeth, and the single, wider-than-high central tooth. Scale bar = 0.3 mm.

of food storage (e.g., Clark, 1849; Morton, 1959; Bilyard, 1974; Poon, 1987; Shimek, 1988, 1990; Langer *et al.*, 1995). The buccal cavity in turn leads to the pharynx, or foremost section of the esophagus where the radula and associated structures are located. Three components dominate the pharynx: the jaw, radular apparatus and subradular organ (Salvini-Plawen, 1988). The single chitinous jaw is found in the dorsal wall of the pharynx, which probably acts as both a substratum against which the radula can grind or move material, and protection for the cerebral ganglion from the action of the radula (Shimek and Steiner, 1997). The subradular organ is cup-shaped and found on the floor of the pharynx on a transverse ridge; it presumably functions as a gustatory sense organ (see Section 8.1.2). Adjacent to it ventro-laterally are subepithelial gland cells, which comprise the ventral foregut glands (Salvini-Plawen, 1988).

The radula has a simple arrangement common to all members of the class, but is remarkable in being among the largest in mollusks relative to body size (Morton, 1959). The radula has five teeth per row, consisting of paired marginals and laterals and a central or rachidian tooth, all supported by a radular membrane (Figure 15B). The rachidian tooth differs in shape between the orders, being wider than it is high in the Dentaliida, and vice versa (and a good deal smaller) in the Gadilida (Steiner, 1992b; Lamprell and Healy, 1998; Reynolds and Okusu, 1999). The radula has been used extensively in taxonomy of the group (e.g., Scarabino, 1979, 1995; Lamprell and Healy, 1998) and in phylogenetic analysis (Chistikov, 1984; Steiner, 1992b; Reynolds and Okusu, 1999).

Radular teeth are secreted by odontoblasts (Gabe and Prenant, 1950); in *Rhabdus rectius* the organic matrix of the teeth shows a strong histochemical reaction for protein, carbohydrates including polysaccharides but not substantial glycogen, and detectable levels of lipofuscin (Reynolds, 1990a). The teeth are highly mineralized, inorganic components including high quantities of calcium and particularly iron (Vovelle and Grasset, 1983; Reynolds, 1990a); both ferric and ferrous iron are present, and positive reactions to the von Kossa, Ebel's and ammonium molybdate histochemical tests indicate that the two iron species are complexed with phosphate (Reynolds, 1990a). Iron is a common component of molluskan radulae, and accumulated iron is utilized in the byssus of the bivalve *Mytilus edulis*, where it plays a similar hardening role (George *et al.*, 1976). In scaphopods, mineral composition of the teeth varies to some extent between orders (R. Shimek, pers. comm.).

Radular function has been observed in a few scaphopods, with representation from both orders, as reported by Shimek and Steiner (1997). Being non-eversible, the radula is used to move material from the buccal area into the esophagus in a simple cyclical motion or "rachet"-like manner (Shimek and Steiner, 1997). In dentaliids that have been observed, there is little maneuverability of the radular components; some food items are ingested to the stomach without maceration (Shimek, 1990; Shimek and Steiner, 1997). In the gadilid species reported to date, by contrast, the radula has more active movement, with lateral teeth raised and food triturated to a degree not seen in the dentaliids; *Gadila aberrans* uses the radula to break off pieces of large foraminiferans (Shimek, 1990). In *Pulsellum salishorum,* opposing halves of the radular ribbon have been observed moving with some independence (Shimek and Steiner, 1997).

6.4.2. *Esophagus*

From the pharynx, posterior to the radular apparatus, a single or paired dorsal ciliated tract marks the beginning of the esophagus (Salvini-Plawen, 1988). In Dentaliida, the continuity of the pharynx and esophagus is apparent from lateral ciliated folds that begin immediately posterior to the jaw and fuse with the dorsal ciliated tract that extends the length of the esophagus (Salvini-Plawen, 1988; Shimek and Steiner, 1997); a pair of ventral ciliated folds similarly fuses to form a ventral ciliated tract. In Gadilida, dorsal and ventral ciliated tracts are also formed, with some variability in ciliation and glandular patterns among families (Figure 16; Salvini-Plawen, 1988). This mid-region of the esophagus enlarges lateroventrally to form a crescent-shaped profile; posteriorly it separates into two lateral diverticula, the esophageal pouches, and the continuing, central, posterior esophagus with few glandular components (Shimek and Steiner, 1997). The esophagus opens through a sphincter (Salvini-Plawen, 1988) into the right-dorsal side of the stomach (Figure 15A).

Despite the absence of salivary glands, the ciliated tracts and pouches of the mid-region of the esophagus are richly endowed with glandular epithelium (Taib, 1976; Salvini-Plawen, 1988; Reynolds, 1990a; Steiner, 1990). The ciliated tracts contain mucocytes, and the esophageal pouches are lined with glandular cells. In the esophageal epithelium of *Rhabdus rectius,* numerous membrane-delimited electron opaque granules are distributed throughout the cytoplasm, structurally differentiated into two granule types based on size and consistency (Reynolds, 1990a). The esophageal cells release granules and secrete a mucus-like substance, which reacts positively to histochemical tests for carbohydrates and polysaccharides, as do cells lining other organs of the tract (Reynolds, 1990a). In addition, Taib (1976) histochemically identified a variety of enzymes from the esophageal pouches of *Antalis entalis*, thus identifying the pouches as a site of extracellular digestion.

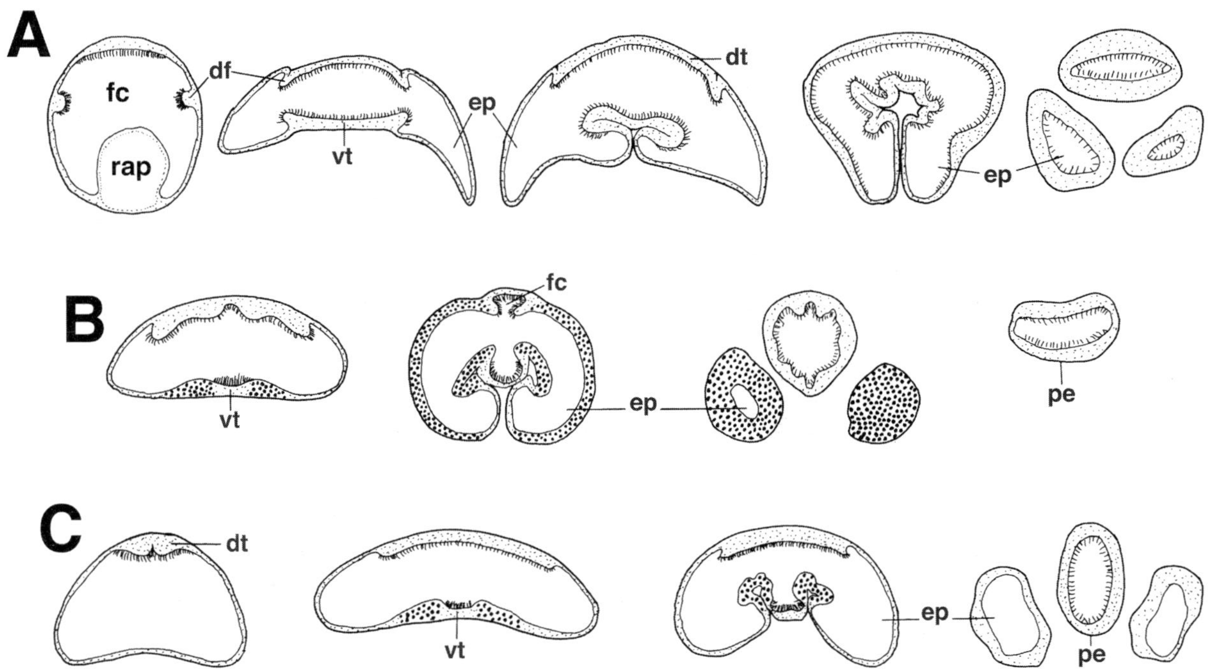

Figure 16 Anatomy of the scaphopod esophagus (left to right, subsequent cross-sections). (A) Dentaliidae species; (B) Siphonodentaliidae species; (C) *Cadulus* sp.: df, ciliated dorsal ridge/fold; dt, dorsal ciliated tract; ep, esophageal pouch; fc, food channel; pe, posterior esophagus; rap, radular apparatus; vt, ventral ciliated tract. From Salvini-Plawen, 1988; note classification follows Boss (1982) in which "Dentaliidae" includes *Fustiaria* and "Siphonodentaliidae" comprises the genera *Siphonodentalium* and *Polyschides*. (Reproduced with permission from Salvini-Plawen (1988, figure 17) © Academic Press.)

6.4.3. *Stomach*

The stomach is a relatively large, central, pyriform organ that is thin-walled and partially encircled by a muscular tunic (Salvini-Plawen, 1988). The lateral and ventral internal epithelium is lined with cuticle, forming a raised tooth or reduced gastric shield; no protostyle is present (Salvini-Plawen, 1981). Ciliated epithelium in the proximal stomach area forms several bands, which serve as a sorting area and mix ingested material (Morton, 1959; Salvini-Plawen, 1981); a rolled food/mucus strand has occasionally been observed in Dentaliida (Salvini-Plawen, 1988). In *Rhabdus rectius,* glandular cells in the wall of the stomach are dominated by small granules (~2.0 μm maximum diameter) with mixed opacity and membranous whorls (Reynolds, 1990a).

A small ciliated pit found in Dentaliida may represent the remnant of the coiled gastric caecum found in vetigastropods and bivalves (Salvini-Plawen, 1988); nearby, single (Gadilida) or paired (Dentaliida) openings lead to the digestive gland. Enzymes secreted from the digestive gland are released into the stomach (Taib, 1976) where extracellular digestion continues from the esophagus. Contractions of the muscular tunic enhance extracellular digestion, and move the chyle to the digestive gland tubules (Salvini-Plawen, 1988).

6.4.4. *Digestive gland*

All Dentaliida possess paired digestive (midgut) glands, whereas all Gadilida possess a single gland, arising from the left side of the stomach (e.g., Salvini-Plawen, 1988). The digestive gland consists of many lobes or diverticula that ramify throughout the mid-region of the body and into the mantle (e.g., Reynolds, 1990a). The digestive diverticula are lined predominantly by secretory/absorptive digestive cells, which release a variety of enzymes into the stomach (Sahlman, 1973; Taib, 1976, 1981b). Extracellular digestion continues in the diverticula (Morton, 1959), and intracellular digestion within the digestive cells (Sahlman, 1973; Taib, 1976; Shimek and Steiner, 1997). In *Rhabdus rectius,* the digestive cells are characterized by a densely microvillous apical membrane, and many intracellular lysosomes in varying stages of maturation (Reynolds, 1990a). Associated with the enzyme-secreting role of these cells is the high occurrence of Golgi and endoplasmic reticulum cytoplasmic components.

Pyramidal basophilic cells are interspersed amongst the larger digestive cells, but are much less numerous, a typical pattern of molluskan digestive glands (e.g., Salvini-Plawen, 1988). In *Rhabdus rectius,* the luminal surface of the basophilic cells possesses microvilli; the cytoplasm has an extensive rough endoplasmic reticulum and numerous vacuoles, suggesting a secretory function; granules within these vacuoles consist of alternately electron-opaque and transparent concentric bands, similar to those found in the kidney (see Section 7.5), and within amebocytes circulating in the hemocoel of the digestive diverticulum (Reynolds, 1990a).

6.4.5. *Hindgut*

Materials released from the digestive gland are rolled by the stomach ciliation and moved into the intestine by peristaltic contractions of stomach musculature. The intestine leaves the stomach anteriorly, from the dorsal-left side but ventral to the opening of the esophagus (Figure 15), and undergoes several loops and turns before opening into the mantle cavity through the anus (Salvini-Plawen, 1988; Steiner, 1994) (Figures 8A, 9A). The number of intestinal loops, typically three, has been found to vary considerably among species, ranging from two in *Cadulus aberrans* to five in *Bathoxiphus ensiculus;* corresponding intestine:body length ratios are 1.05 and 0.6, respectively (Steiner, 1994). This range may be related to diet; the bathyal *B. ensiculus* ingests considerable sediment in addition to foraminiferans, unlike the sublittoral *C. aberrans,* which is a highly specialized microcarnivore. The longer intestine of *B. ensiculus* may reflect adaptation to poorly nutritious deep-sea sediments, as has been hypothesized for the elongated hindgut of many deep-sea bivalves (Steiner, 1994).

The intestinal epithelium is simple and without mucocytes (Salvini-Plawen, 1988), although in *Rhabdus rectius,* the intestinal epithelial cells have a microvillous and ciliated apical membrane; the cells possess numerous intracellular granules and considerable mucus secretion appears to take place (Reynolds, 1990a). The mucus probably facilitates adsorption of nutrients and is incorporated into lysosomes through pinocytotic uptake, as has been described in other mollusks (Owen, 1972). The fecal material, while rolled into a rod-shaped form, is not compressed into pellets nor is there a peritrophic membrane (Salvini-Plawen, 1988; Steiner, 1994). The rectum, surrounded by the perianal sinus, projects ventrally into the mantle cavity, directed somewhat toward the anterior (Figure 8A). The rectal gland, which may secrete lipoid substances, opens into the rectum midway to the anus (Sahlman, 1973). The continuous pulsations of the perianal sinus, rectum and anus (see also Section 7.1) help in periodic voiding of the fecal threads (pers. obs.).

7. CIRCULATION AND EXCRETION

7.1. Heart and pericardium

The systems in scaphopods that have historically been the most difficult to understand are those of circulation and excretion. Loss of the auricles, associated with the loss of ctenidia from the restriction of the mantle cavity by the body orientation and form, is a considerable departure from the circulatory and excretory systems in most other mollusks. The unique body form is naturally reflected in the shape and orientation of all circulatory system components, but especially so for those remaining of the heart, the ventricle and pericardium, causing further difficulty in interpreting the system. Considerable disagreement is found in the early literature, as anatomists struggled with homology versus functional views of the scaphopod heart and to identify the ventricle from among sinuses named for position (abdominal, pedal, perianal, etc.) within a highly derived system.

Historically, investigators looked to large, spherical, central structures for a heart, Deshayes (1825) misidentifying the esophagus and Clark (1849) the stomach. But once the derived nature of the system was recognized and with careful observation, two contractile hemocoelic sinuses were separately proposed as the scaphopod ventricle. Seeking homology with the ventricle of other mollusks, Lacaze-Duthiers (1856–1857) and Fol (1889) proposed the perianal sinus, adjacent to the anterior end of the pericardial sac, based on functional, structural and positional similarities to the bivalve ventricle. Taking perhaps a more functional approach, Plate (1891b, 1892) proposed an "invagination" of the dorsal pericardial wall that enclosed a portion of the abdominal sinus ventral to the stomach, later supported by Boissevain (1904) and Distaso (1905a). While the vigorous contractions of both candidates are easily observed and have been frequently reported, most subsequent reviews of scaphopod biology have adopted Plate's more recent (1891b, 1892) interpretation of the scaphopod ventricle without discussion (Potts, 1967; Fischer-Piette and Franc, 1968; Martin, 1983; Andrews, 1988; Shimek and Steiner, 1997; Palmer and Steiner, 1998), although others have concluded that a heart is absent or reserved judgement (Purchon, 1977; Ponder and Lindberg, 1997).

In other mollusks, a single ventricle and two auricles (associated with the ctenidia) are surrounded by pericardial coelom in most species; the pericardial coelom is delimited by an endothelium referred to as the pericardium, but identified as epicardium where it overlies the ventricular or auricular musculature. The pericardium and epicardium are therefore continuous, and defined positionally, but with various functional and structural differences in adult individuals.

During molluscan development, the ventricle is formed by a variety of processes including dorsal invagination of the pericardium or fusion of paired pericardial rudiments (Raven, 1966; Moor, 1983; Salvini-Plawen, 1985). The pericardial coelom consequently surrounds the ventricle, although does not do so completely in solenogastres or some caudofoveate aplacophorans (Salvini-Plawen, 1985; Salvini-Plawen and Bartolomaeus, 1995) or some bivalves (White, 1942; Narain, 1976), where the epicardium encloses only the ventral and lateral ventricular musculature, the ventricle persisting as an incomplete dorsal invagination of the pericardium. The ventricle is usually penetrated and traversed by the rectum in bivalves (Jones, 1983; Andrews and Jennings, 1993; Morse and Zardus, 1997) and gastropod families of the Neritopsina and Vetigastropoda (Hyman, 1967; Ponder and Lindberg, 1997).

Tissue layers that comprise the molluscan ventricular wall are an epithelial layer (epicardium) and a muscular layer (myocardium) located below the basal lamina of the epicardium (Narain, 1976; Økland, 1980, 1982; Jones, 1983); an endothelium is typically found only in cephalopods (Jensen and Tjønneland, 1977). Podocytes are typical of ultrafiltration and the production of primary urine in metazoans (Kümmel, 1973), including mollusks (Andrews, 1976, 1985; Pirie and George, 1979; Økland, 1980; Morse and Reynolds, 1996). While usually interposed between the simple epithelial cells of auricular epicardia, molluscan podocytes also appear in ventricular epicardia (Andrews, 1988) and specialized elaborations of the pericardium/epicardium, such as pericardial appendages/glands of bivalves and nautiloid cephalopods (White, 1942; Meyhöfer *et al.*, 1985; Fiedler and Schipp, 1987), and the branchial heart appendages of coleoids (Fiedler and Schipp, 1987).

In scaphopods, the only ultrastructural description of the tissues comprising the perianal sinus, pericardium, stomach and the abdominal sinus to date is from *Rhabdus rectius* (Reynolds, 1990b). These ultrastructural data help clarify tissue organization of these organs and assessment of the early hypotheses on scaphopod ventricle identity.

The perianal sinus has several structural, positional and functional identities with the ventricle in other molluscan classes at anatomical, histological (Lacaze-Duthiers, 1856–1857; Fol, 1889) and ultrastructural (Reynolds, 1990b) levels:

(1) The perianal sinus musculature forms a continuous and extensive layer, enclosing hemocoel and creating a discrete chamber.
(2) Penetration of the chamber by the rectum, as in many bivalves and gastropods.

(3) Muscular trabeculae cross the lumen of the vessel between myocardium and rectal wall, as in those gastropods and bivalves with more muscular ventricles (Økland, 1982; Jones, 1983).
(4) The anterior-most end of the pericardium partially surrounds the perianal sinus musculature, the two layers separated by the basal lamina. The tissue layer organization therefore resembles that of a typical molluskan ventricle: epicardium, continuous with the pericardium, overlaying myocardial musculature (e.g., Økland, 1982; Jones, 1983).
(5) This "epicardium" contains podocytes (Reynolds, 1990b), cells characteristic of ultrafiltration and tightly associated with molluskan epicardia (e.g., Andrews, 1988; Morse and Reynolds, 1996) (for schematic orientation, see Figure 17).

In terms of its location and continuity with the large sinuses of the system, the perianal sinus is the principal hemolymph space (Shimek and Steiner, 1997); allied with its sustained strong rhythmic contractions, it may be considered the most important blood pumping site in the circulatory system.

The dorsal pericardial wall or abdominal sinus hypothesis by Plate (1891b, 1892), on the other hand, is not as well supported in either structural or positional homology criteria:

(1) A myocardium (any musculature below the basal lamina of the pericardial endothelium) is completely absent. The abdominal sinus is bounded across its width by the stomach wall dorsally and the dorsal pericardial wall ventrally (Plate, 1892; Reynolds, 1990b); there is no structural delimitation of a vessel or chamber as is sometimes represented (e.g., Potts, 1967; Shimek and Steiner, 1997).
(2) While scattered transverse muscle fibers are found in the dorsal pericardium, they are interspersed amongst the endothelial cells of the pericardial coelom, i.e., above the basal lamina, not within the hemocoel as found in molluscan myocardia. Contraction of these pericardial muscle fibers throws the dorsal pericardial wall into transitory folds, which has been frequently observed (Lacaze-Duthiers, 1856–1857; Reynolds, 1990b) and has been the basis for Plate's (1891b, 1892) hypothesis of ventricular identity.
(3) The rectum does not cross the abdominal sinus.
(4) Podocytes, which would be expected in a molluscan epicardium (Andrews, 1988) are not present (Reynolds, 1990b) (for schematic orientation, see Figure 17).

Finally, while undoubtedly pumping blood through the abdominal sinus, the dorsal pericardium plays a relatively minor role in overall circulation.

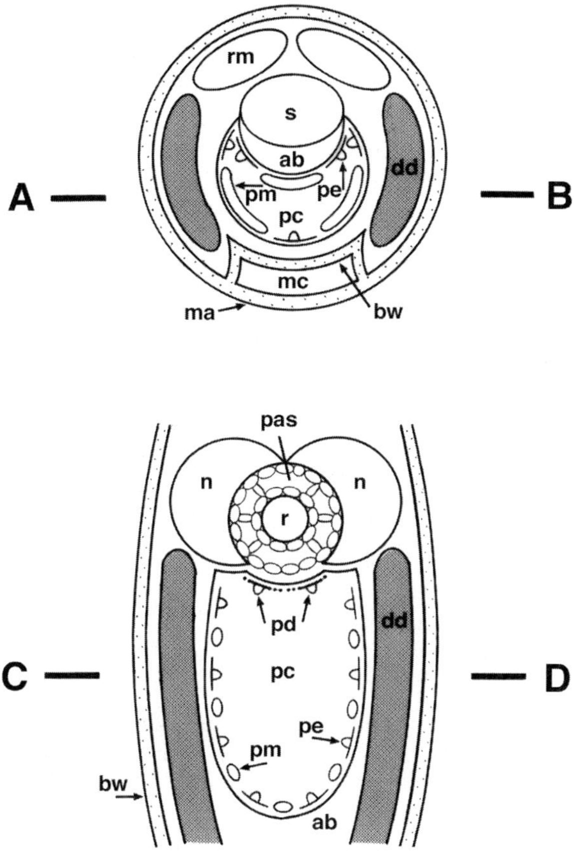

Figure 17 Diagram of scaphopod pericardial and putative ventricle structures. Top: the relative positions of the stomach (s), abdominal sinus (ab) and pericardium (pc). A–B indicates frontal section view represented below. Bottom: relative positions of the perianal sinus (pas), pericardium (pc) and rectum (r); C–D indicates cross-sectional view represented in top. ab, abdominal sinus; bw, body wall; dd, digestive diverticulum; ma, mantle; mc, mantle cavity; n, kidneys; pd, podocytes; pe, epithelial cell of the pericardium; pm, muscle cell of the pericardium; rm, retractor muscle. (Reproduced, with permission, from Reynolds, 1990b, figures 6 and 7.)

Essentially, the pericardial sac hangs within the abdominal sinus, its ventral surface closely apposed to the body wall but its dorsal surface loose in the hemocoel beneath the stomach, where contractions throw it into folds, even though myocardial musculature is absent (Reynolds, 1990b). While some early workers (Plate, 1891b, 1892) and more recent reviews (Potts, 1967; Shimek and Steiner, 1997; Palmer and Steiner, 1998) term these dorsal folds of the pericardium the "ventricle" or "heart", on the basis of current evidence this rather should be considered a "contractile pericardium". A contractile pericardium is also found in polyplacophorans, where it may aid ultrafiltration but otherwise have little impact on ventricular function (Økland, 1981), and in the more complex structure of the pericardial glands of many bivalves (White, 1942; Meyhöfer *et al.*, 1985).

The scaphopod perianal sinus, on the other hand, is better supported as the homolog of the molluscan ventricle, on the basis of identity in positional and structural detail. Given this conclusion, the partial enclosure of the perianal sinus musculature (putative ventricular myocardium) by pericardial epithelium (putative ventricular epicardium) is less than that in, for example, the bivalve *Pteria vulgaris* (White, 1942) and solenogastre *Epimenia verrucosa* (Baba, 1940), although the pericardium is not lost completely from the ventricle as in the nudibranch *Okadaia elegans* (Baba, 1937). Whether this incomplete enclosure is a reduction or a persistent invagination of the pericardium (as described for *E. verrucosa*, Salvini-Plawen, 1985) awaits developmental studies. On the basis of current evidence, the term "ventricle" would therefore be more appropriately applied to the perianal sinus, or reserved from all scaphopod structures until further resolution is brought to these long-competing hypotheses on circulatory structure and function in the Scaphopoda.

7.2. Circulatory flow

The circulatory system consists of a series of hemocoelic sinuses, lacunae and vessels first described in detail by Lacaze-Duthiers (1856–1857). Shimek and Steiner (1997) suggested that the term "vessels" be abandoned, as no endothelial linings are found in the system; this is true also of the perianal sinus, contrary to a report by Fol (1889) later refuted by Plate (1892) and Boissevain (1904). Shimek and Steiner (1997) summarized the network and communication of constituent spaces, although they follow the terminology of Plate's (1891b, 1892) currently poorly supported interpretation of ventricle and pericardial structures (see Section 7.1). Shimek and Steiner (1997) additionally provided an alternative to Plate's (1892) hypothesis on flow of blood throughout the system, based on arrangement of the sinuses and observations on the system.

The contractile centers are the pedal sinus musculature, perianal sinus musculature (=putative ventricle, see Section 7.1) and dorsal pericardium. The propulsion of blood by the pedal sinus is a consequence of burrowing and feeding behavior (see Sections 5 and 6.3); the other two centers exhibit a rhythmic, if discontinuous, contraction. Contractions of the perianal sinus, which also aid in voiding fecal strings from the rectum, which penetrates it, reach 60 contractions per minute, with periodic dilation every few minutes of up to 30 seconds (in *Rhabdus rectius,* Reynolds, 1990b). Contractions of the dorsal pericardium are independent of the perianal sinus (Plate, 1892; Reynolds, 1990b) and occur in a series of two to three anterior-to-posterior transverse constrictions; anchorage of the anterior pericardial wall (= putative ventricular epicardium, see Section 7.1) to the perianal sinus and much of the ventral pericardium to the body wall results in the wide anterior–posterior fluctuation of the posterior, freely moving end (Reynolds, 1990b). While more precise experimental approaches would contribute greatly to understanding circulation patterns in scaphopods, the reversibility of flow and disruption of contraction (Reynolds, 1990b; Shimek and Steiner, 1997) suggest a complex system of reaction and compensation among internal and external influences on local blood pressure. The paired slit-shaped openings to the hemocoel from the mantle cavity, located either side of the rectum/perianal sinus anterior to the excretory pores (Figure 9), are enigmatic; it has been surmised that they function as safety valves to release hemolymph upon rapid contraction of the foot (Lacaze-Duthiers, 1856–1857; Simroth, 1894b; Shimek and Steiner, 1997).

7.3. Blood

7.3.1. *Hemocytes*

Hemocytes were first reported by Plate (1892); Lacaze-Duthiers (1856–1857) surprisingly provided little information on scaphopod blood. Plate (1892) identified a range of cell types, grading between the two endpoints of 5-μm-diameter cells with homogeneous nuclei, and 10-μm cells with vesicular nuclei. Kollman (1908) suggested this range represented maturation of leukocytes, the smallest and youngest he termed "hyaline leukocytes", and found the largest cells to have chromatin-rich nuclei. He could not find any hemopoietic tissue.

The only other significant report on scaphopod blood cells is that of Arvy and Gabe (1951), who examined the histology and histochemistry of blood cells in a large number of scaphopods. They described three categories of cell types, leukoblasts (= proleukocytes), basophilic leukocytes and amebocytes.

Leukoblasts are spherical 12-μm-diameter cells. The cytoplasm is strongly basophilic, relatively clear with few inclusions but with a prominent chondriome, and histochemically rich in "ribonucleines" and alkaline phosphatase activity, but with little glycogen. Usually there are two nucleoli. Overall, these cells appear to be little differentiated or immature (Arvy and Gabe, 1951).

Arvy and Gabe (1951) also described basophilic leukocytes, only 8 μm in diameter and spherical to oval in shape, with regions of clear cytoplasm but with many lipid vacuoles, no birefringent granules, no alkaline phosphatase activity, weakly basophilic and less "ribonucleine". They are without nucleoli. Arvy and Gabe (1951) suggested that this cell type is the "hyaline leukocyte" of Kollman (1908). Many intermediates between the leukoblasts and this type are found, and Arvy and Gabe (1951) infer that these are probably derived from the leukoblasts.

Amebocytes are 10–15 μm in diameter and indeterminate in shape due to slow ameboid movements. Their cytoplasm is weakly basophilic, with vacuolated cytoplasm containing very small granules and small lipid droplets. They also lack nucleoli. Arvy and Gabe (1951) were uncertain as to the origin of these cells, stating that they are equally likely to derive from the basophilic leukocyte as to derive directly from the leukoblast type.

The only ultrastructural data on scaphopod blood cells are for *Rhabdus rectius* (Reynolds, 1990a). The variety in shape and size described in earlier reports is apparent from scanning electron microscopy of hemocoelic sinuses (Figure 18A). Ultrastructurally, amebocytes have a central nucleus and several pseudopodia (Figure 18B). The cells possess mitochondria and vesicles that are largely electron-transparent; glycogen and granules that histochemically test positive for iron are also characteristic cytoplasmic inclusions (Reynolds, 1990a).

In terms of relative abundance, leukoblasts make up about 5% of the blood cells, compared to equal numbers of the other two types. Arvy and Gabe (1951) also found sexual dimorphism in cell abundance, females having fewer (1400–2100 mm^{-3}) than males (2240–6500 mm^{-3}), but no significant differences in abundance throughout the size range of the animals sampled. Arvy and Gabe (1951) were unable to identify an organ with a leukopoietic function.

Unfortunately, there is no information on the functional roles of these cells in scaphopods.

7.3.2. *Hemolymph*

Little is known of scaphopod hemolymph constituents, respiratory proteins, or even whether respiratory proteins are intra- or extracellular. Two reports of hemoglobin in the pharyngeal region and radular apparatus (Manwell, 1963; Smith, 1967) are thought to represent myoglobin in the

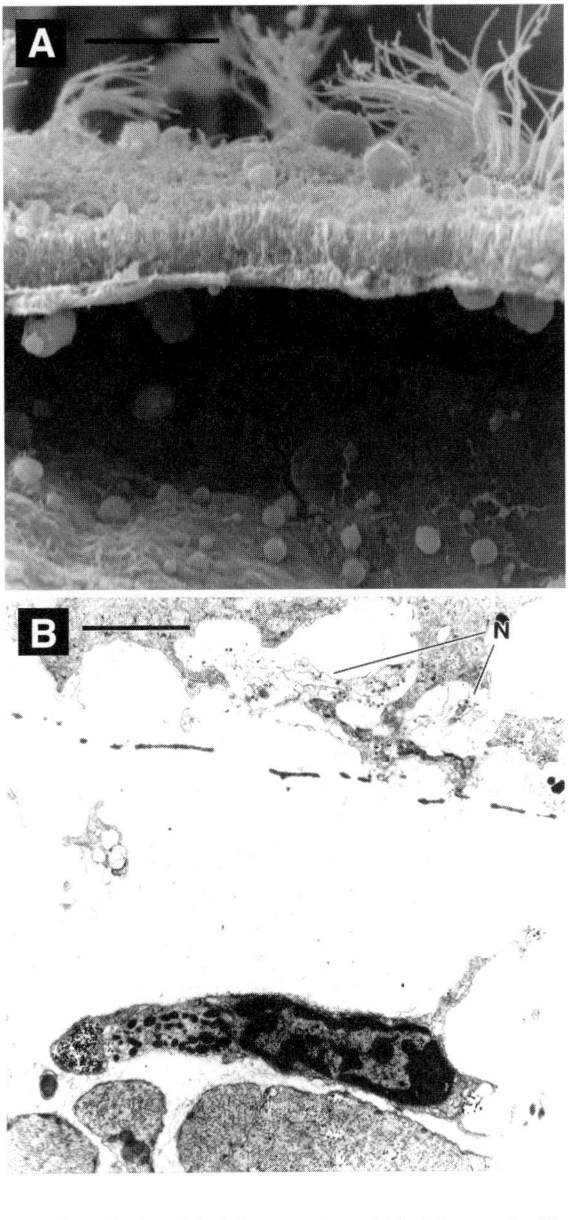

Figure 18 Blood cells in *Rhabdus rectius.* (A) View of ciliated body-wall epithelium lining the mantle cavity and underlying hemocoel in *Rhabdus rectius.* Note the numerous hemocytes within the hemocoel. Scale bar = 10 μm. (B) Amebocyte within hemocoel beneath the ciliated bands of *R. rectius.* Note granule-containing cytoplasm within the amebocyte (N, nerve profiles). Scale bar = 3 μm.

radular musculature (Bonaventura and Bonaventura, 1983). Considering the distribution of copper- and iron-based proteins among Mollusca, and especially within bivalves (Bonaventura and Bonaventura, 1983; Morse *et al.*, 1986), identification of the scaphopod oxygen carrier could be of considerable evolutionary interest.

7.4. Ultrafiltration

In mollusks, the excretory system comprises a complex interaction between the heart, pericardial coelom and kidney (Martin, 1983; Andrews, 1988; Andrews and Jennings, 1993; Morse and Reynolds, 1996; Morse and Zardus, 1997). The synergistic combination of physiological and morphological data from the better studied classes (Harrison, 1962; Little, 1965; Harrison and Martin, 1965; Martin and Aldrich, 1970; Witmer and Martin, 1973; Pirie and George, 1979; Økland, 1980; Andrews, 1981; Schipp and Hevert, 1981; Jones and Peggs, 1983; Hevert, 1984; Meyhöfer *et al.*, 1985) has defined for the Mollusca the ultrafiltration, secretion and reabsorption model that is common to all coelomic taxa (e.g., Ruppert and Smith, 1988). The first step in excretion, the production of primary urine through ultrafiltration, occurs through podocytes associated with the heart wall, while variation between ventricular and auricular sites, systemic and branchial hearts, and auricular epicardia and pericardial glands can be found amongst molluskan taxa.

In Scaphopoda, very few data on ultrafiltration are available. The early 1900s technique of ammoniacal carmine injection, which was used successfully in many molluscan taxa to localize the ultrafiltration site, produced equivocal results in scaphopods (Kowalevsky, 1889; Cuénot, 1899; Strohl, 1924), and more modern techniques have not been applied. The only ultrastructural study of heart and pericardial tissues, on *Rhabdus rectius* (Reynolds, 1990b), identified podocytes in the pericardial epithelium apposing the perianal sinus musculature (putative ventricular epicardium, see Section 7.1), auricles being absent in the class. The conservation between the podocyte structure and ultrafiltration function among those molluscan classes submitted to physiological experimentation allows strong inference that this is the site of ultrafiltration in scaphopods. Slit diaphragms, variously present or absent among mollusks (Boer and Sminia, 1976; Andrews, 1979; Økland, 1980; Schipp and Hevert, 1981; Meyhöfer *et al.*, 1985), are absent in *R. rectius* (Reynolds, 1990b).

In all extant molluscan classes, morphological or physiological evidence indicates a highly conserved ultrafiltration system, in that the ultrafiltrate passes from the pericardial coelom to one, and typically both, kidney lumina. This renopericardial connection has been difficult to discover in scaphopods; Lacaze-Duthiers (1856–1857), Fol (1885, 1889), and Plate (1888, 1892) concluded that one did not exist. Two subsequent studies reported renopericardial connections: Distaso (1905a), in *Antalis entalis*, reported a connection to the "left" kidney, although correcting to conventional orientation (convex side ventral, larger aperture anterior) renders this actually a connection to the right kidney (Distaso, 1905a, describes the retractor muscle attachment to the shell as "ventral" when it is actually on the concave, dorsal side). Reynolds (1990b: 144), studying *Rhabdus rectius,* described and illustrated a connection to the right kidney "in incidental thin (1 μm) sections" from several different specimens (not "only one section" as reported in Shimek and Steiner, 1997; Palmer and Steiner, 1998). B. Ruthensteiner and G. Haszprunar (pers. comm.), however, using light-microscopy sections, preliminarily report paired renopericardioducts from juvenile and adult specimens of *Antalis entalis;* this finding confirms the typically molluscan nature of the scaphopod pericardial cavity. It is likely that paired renopericardioducts represent the typical scaphopod condition.

7.5. Secretion and reabsorption

Before release to the exterior, the primary urine is modified by reabsorption and secretion via the cells of the kidney or nephrocytes. As for ultrafiltration, morphological characteristics have strong inferential value through complementary physiological and morphological studies on representatives of the larger molluscan classes (Little, 1965, 1979; Potts, 1967; Pirie and George, 1979; George *et al.,* 1978, 1980, 1982; George and Pirie, 1980; Andrews, 1981; George, 1982; Simkiss and Mason, 1983; Mason *et al.,* 1984; Sullivan *et al.,* 1988).

The kidneys are located either side of the rectum (Figure 9), and are separate sacs with glandular walls (Lacaze-Duthiers, 1856–1857; Fol, 1885). They were recognized early on as having accumulation and elimination functions (Lacaze-Duthiers, 1856–1857; Turchini, 1923), which was demonstrated through indigo-carmine injection to the pedal sinus (Kowalevsky, 1889). Reynolds (1990c) examined the ultrastructure of the kidney, describing two nephrocyte types distinguished on the basis of secretory products, which included mineralized granules, and which correspond well with previous histological descriptions of scaphopod kidney cells (Fol, 1889; Kowalevsky, 1889; Plate, 1892). Both cell types possess ultrastructural characteristics of cells involved in merocrine secretion (Reynolds, 1990c), which differs from, but is not unique to, the more usual apocrine secretion found in molluskan kidneys (Pirie and George, 1979; George and Pirie, 1980; Andrews, 1988). In addition, large (>20 μm) extracellular granules were frequently encountered in the kidney lumen (Reynolds, 1990c).

One of the cell types possessed microvilli and putative pinocytotic vesicles, the only evidence for reabsorption identified (Reynolds, 1990c). In scaphopods, urine release is through a pair of muscular circular pores located in the mantle cavity near the anus, just posterior to the slit-shaped hemocoelic pores (Lacaze-Duthiers, 1856–1857; Odhner, 1931; Reynolds, 1990c). The right excretory pore also serves as a conduit for the gametes (Figure 9).

8. SENSATION AND INNERVATION

8.1. Sensory receptors

Organisms interact with their environment through several different points of contact, and so must possess sensory receptors specialized to monitor those contacts in order for functional feeding, locomotion, gas exchange and other life processes. Essentially, the organism must test the environment that it moves into, its orientation in that environment, and the media that are brought into the body, such as respiratory currents and food. This suite of receptors must also operate through a range of sensory modalities, such as chemoreception and mechanoreception. Sensation in scaphopods has been almost entirely inferred from early histological survey and recent ultrastructural studies (Steiner, 1991; Reynolds, 1992b) targeted for the most part at regions of the body in contact with the surrounding environment, and with comparison to sensory receptors well studied from larger molluscan classes.

In reviewing the information available on sensory receptors in scaphopods, the sections below are divided both by region and function; the anterior mantle is the region of contact for sensing the environment into which the animal burrows (although no information is available on epithelial receptors in the foot) and where food is identified and collected, and is the site of various "exteroceptors" (sensitive to external stimuli) (Barnes *et al.*, 1993); gustatory and orientation sensation are the main "interoceptors" identified to date (sensitive to internal stimuli, insofar as food is ingested at the point of the sensory subradular organ). The posterior mantle is the site of other exteroceptors directed primarily at testing, probably in part through distance chemoreception, of the respiratory currents, thereby serving several associated functions such as gas exchange, spawning, defense and food detection.

8.1.1. *Anterior mantle*

8.1.1.1. *Ciliated epithelial receptors.* Scaphopods burrow with the anterior aperture downward, and the foot protruded through the anterior mantle opening (see Section 5.2). The areas of the mantle exposed directly to the environment, the frontal epithelium and the region just within the mantle cavity, the central fold area (Steiner, 1991), are the contact points likely to monitor the environment into which the animal moves. Ciliated epithelial receptors of these two anterior mantle regions have been described by Steiner (1991) (and reviewed in Shimek and Steiner, 1997), who identified several types of receptor among three genera (*Antalis, Fustiaria, Cadulus*). The two main types differ in ciliary ultrastructural characteristics, the first having ≤25 cilia of 2–4 μm length (type-1), found commonly in both mantle regions in members of *Antalis,* and less frequently in the frontal epithelium of *Cadulus subfusiformis* (Steiner, 1991). The second is a collar receptor, a single cilium surrounded by stereomicrovilli, more commonly found in the central fold of individuals of *Fustiaria.* Representatives of *Fustiaria* also possess a third type, similar in form and distribution to the type-1 receptor, with only 4–5 cilia but a significantly greater density of mitochondria (Steiner, 1991). A significantly different situation exists in the gadilid species of the genera *Entalina, Pulsellum* and *Cadulus*, in that the frontal epithelium is thrown into numerous papillae, each of which terminates in an epithelial receptor similar to the type-1 receptor common in *Antalis* (Steiner, 1991). In *C. subfusiformis,* on the other hand, the papillae terminate in a collar receptor (Steiner, 1991).

An annular ciliary band within the anterior mantle edge in dentaliids is probably not sensory, although may play a role in enhancing water flow over adjacent receptors of the central fold, thus aiding chemoreception; this band is absent in gadilids (Steiner, 1991).

8.1.1.2. *Slit receptors.* Steiner (1991) also described an unusual putative sensory structure in *Rhabdus rectius,* consisting of a pair of dorso-lateral slits in the anterior mantle edge, the walls of which are ciliated. Similar slits have since been found in *R. perceptum, R. dalli, Calliodentalium callipeplum, Gadilina insolita* (Reynolds and Okusu, 1999) and *Epirhabdoides ivanovi* (Steiner, 1999). The absence of an annular ciliary organ in *Rhabdus* species (Steiner, 1991; Reynolds and Okusu, 1999) originally suggested a potential structural or functional correlation or even homology between the ciliary organ and slits (e.g., Steiner, 1992b); however, both structures have been found in *Calliodentalium callipeplum* and so occur independently (Reynolds and Okusu, 1999). Nevertheless, ultrastructural studies on the slit structures are necessary to determine their sensory capability and innervation (Steiner, 1991), which may inform their homologies within the anterior mantle complex.

8.1.1.3. *Captacula.* The feeding tentacles of scaphopods, or captacula, originate from the proboscis and extend throughout the mantle cavity and along the exterior of the shell in possible cleaning activities, but operate predominantly in food collection, mostly of foraminiferans, from among sediment particles (Shimek, 1988). As such, they both test the environment that surrounds them and carry out first-order evaluation of potential food. Each captaculum is innervated by a single nerve of cerebral origin that terminates in a ganglion in the captacular bulb. From this ganglion, several nerve processes extend to the bulb epithelium, especially in the alveolar region (Byrum and Ruppert, 1994; see Section 6.2) (Figures 11, 12).

8.1.2. *Gustation*

The subradular organ probably functions as a gustatory organ, as post-contact rejection of potential food items has been observed in *Pulsellum salishorum* (Shimek and Steiner, 1997). It is a cup-shaped organ, borne on a transverse ridge, and unlike most other mollusks is found on the floor of the buccal cavity. Its unique position may reflect the non-eversive nature of the scaphopod radula (Shimek and Steiner, 1997). The general form and histology of the subradular organ is reviewed, with further data from *Antalis* species, in Shimek and Steiner (1997). Ultrastructural data would contribute substantially to our understanding of the sensory function of this organ in scaphopods.

8.1.3. *Orientation*

Orientation in scaphopods is monitored through a pair of statocysts, located posteriorly adjacent to the pedal ganglia. In scaphopods, each statocyst is a spherical capsule of innervated, densely ciliated sensory cells, within which numerous secreted calcareous particles, or statocones, move freely (Lacaze-Duthiers, 1856–1857). No ultrastructural data are available.

8.1.4. *Posterior mantle*

8.1.4.1. *Ciliated epithelial receptors.* Among mollusks, ciliated epithelial receptors and the tissues that support them reach the height of their elaboration when associated with inhalant respiratory currents. They can be distributed densely over highly branched papillae surrounding the inhalant siphon of bivalves and gastropods (e.g., Crisp, 1971; Hodgson and Fielden, 1984, 1986) or other specialized organs that have increased epithelial surface area and contact with the inhalant stream. Prominent examples of the latter include the bursicles of some gastropods (Haszprunar, 1987b) and the osphradium of many mollusks (Garton *et al.*, 1984). While little experimentation to test functional modalities has been undertaken in mollusks, such receptors are thought to confer distance chemoreception to feeding, reproduction, gas exchange and defense (e.g., Altner and Prillinger, 1980; Garton *et al.*, 1984; Hodgson and Fielden, 1984, 1986; Haszprunar, 1987a).

The inhalant respiratory stream in scaphopods enters via the posterior aperture (see Section 4), within which the pavilion, an extension of the posterior mantle edge, is located. Quite different in general form from the siphon of infaunal bivalves or gastropods, the pavilion is a synapomorphy for the class, a unique modification associated with the radically elongated body and specialized shell growth program. Another consequence of the scaphopod body form is the restriction of the mantle cavity volume, attested to by the loss of ctenidia and the absence of an osphradium (Haszprunar, 1987a, refuting a report of a scaphopod osphradium by Distaso, 1905a) (see Section 3). Nevertheless, anecdotal observations of reaction to physical and chemical stimuli introduced to the pavilion have been reported (Clark, 1849; Reynolds, 1992b).

Ciliated epithelial receptors in the posterior mantle have been observed in only two species. Steiner (1991), examining the gadilid *Entalina quinquangularis,* preliminarily reported on two rows of ciliated receptors, located proximal to the valve opening, and comprised of 5–8 rooted cilia. Reynolds (1992b) described the ciliary epithelial receptors in the pavilion of *Rhabdus rectius,* among which cilia number and length identify three receptor types, which are further differentiated by internal ultrastructural characteristics. They are distributed predominantly in two areas of the pavilion: at the rim edge, nearest the aperture (and often extended by the pavilion from the aperture) and at the base of the pavilion, on the valve that leads to the mantle cavity. The first receptor type possesses 2–5 cilia, each $\sim$1.7 μm in length, and forms a ring along the pavilion rim; the second has longer ($\sim$8.2 μm) and fewer (1–2) cilia, and was more generally distributed throughout the pavilion. The third type was more unusual, composed of a rigid bundle of 16–40+, tightly wound, long (14.4 μm) cilia, concentrated at both the pavilion rim and on the valve. The presence of striated rootlets in types 2 and 3 suggests a mechanoreceptive role; its absence in type 1 may indicate a chemosensory function (for discussion, see Reynolds, 1992b).

Two annular ciliary bands are found in the pavilion of *Rhabdus rectius*. A band running along the inside of the rim is not innervated (Reynolds, 1992b). The ciliary band at the pavilion base, described as sensory by Boissevain (1904), appears to beat in metachronal waves, and is not likely to be sensory (Reynolds, 1992b). Both bands, however, may function to enhance sensation by creating local water currents over the receptors on the adjacent rim and valve ridges, as may the annular ciliary organ of the anterior mantle edge (see Section 8.1.1), and the valve ciliary organ and ciliated ledge cells in *Entalina quinquangularis* (Steiner, 1991).

8.2. Ganglia and nerves

Shimek and Steiner (1997) have summarized the system of nerves and ganglia in scaphopods, reviewing early reports by Lacaze-Duthiers (1856–1857), Plate (1892), Biossevain (1904), and Distaso (1905b), to which the reader is directed for detail. The system is derived from a basic conchiferan tetraneury, with major cerebral, pedal and pleural ganglia, intermediate-sized visceral and buccal ganglia, and minor subradular and pavilion ganglia (Figure 19); small ganglia are present in each captaculum (see Section 6.2). Major connectives are a pair of fused cerebro- and pleuropedal nerves, and the visceral nerves that connect pleural, visceral and pavilion ganglia; minor connectives include those that run between cerebral, subradular and buccal ganglia. Among major nerves, dorsal mantle and proboscis nerves extend from the cerebral ganglia, ventral mantle nerves from the cerebral ganglia, and pedal nerves from the pedal ganglion (Figure 19). The only report on scaphopod neurosecretion (Gabe, 1956) described acidophilic cells from cerebral, pleural and subradular ganglia; the cells are much more voluminous in the latter ganglion but all possess numerous neurosecretory vesicles.

9. REPRODUCTION AND DEVELOPMENT

The number of studies and data available on reproduction and development in scaphopods are uneven to a surprising extent. The basic anatomy of the reproductive system and development is based mostly upon nineteenth century monographs. The histological record is reviewed in Shimek and Steiner (1997), who also incorporate unpublished results from Hebert (1986). Oogenesis and early development in *Dentalium*, on the other hand, have been the focus of much experimental work from the turn and latter part of the twentieth century, as reviewed by Reverberi (1971),

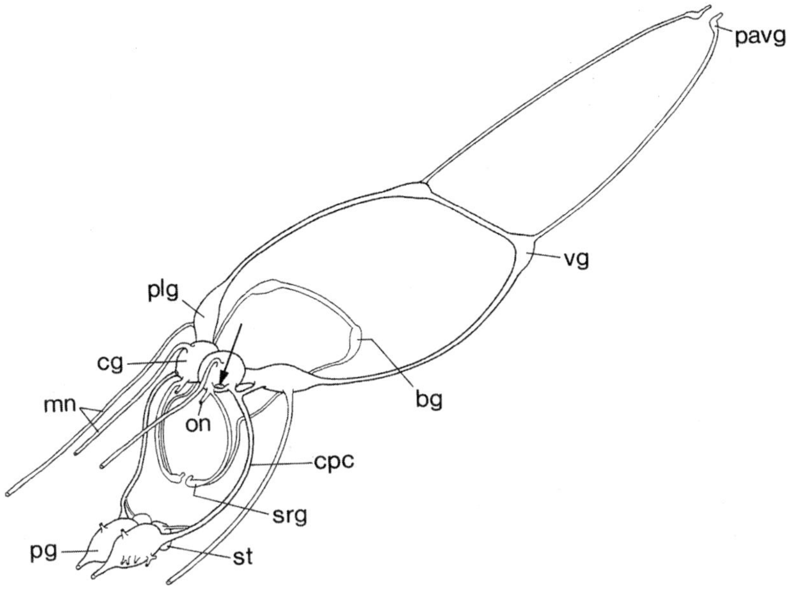

Figure 19 Diagrammatic representation of the scaphopod nervous system: bg, buccal ganglion; cg, cerebral ganglion; cpc, fused cerebro- and pleuropedal connectives; mn, mantle nerves; on, nerve of the oral cone; pavg, pavilion ganglion; pg, pedal ganglion; plg, pleural ganglion; srg, subradular ganglion; st, statocysts with nerve; vg, visceral ganglion; arrow, captacular nerve. (Reproduced with permission from Shimek and Steiner (1997, figure 57) © John Wiley and Sons.)

Van den Biggelaar and Guerrier (1983), Moor (1983), Verdonk and Van den Biggelaar (1983), and Verdonk and Cather (1983). Larval development was described by early microscopists and experimental embyologists, and recently summarized by Buckland-Nicks *et al.* (2002).

9.1. Organ systems

The anatomy of the reproductive systems is known largely from the thorough work of Lacaze-Duthiers (1856–1857), with more recent contributions from McFadien-Carter (1979) and Shimek and Steiner (1997). The scaphopod reproductive system consists of a central, variously infolded organ occupying the apex and upper half of the shell. Sexes are separate, although hermaphroditic individuals are occasionally found (Reverberi, 1971; D'Anna, 1974: *Antalis entalis*) that have been reported to be able to self-fertilize and develop normally (Reverberi, 1971).

9.2. Gametogenesis and spawning

Oogenesis, through to the ripe free egg, was described by Lacaze-Duthiers (1856–1857), Kowalevsky (1883) and Arvy (1950), and ultrastructurally in a series of detailed papers by Reverberi (1969, 1970a, b, 1972), with further contributions from Van Dongen (1977) and Dufresne-Dubé *et al.* (1983). Spermiogenesis and sperm ultrastructure have been described by Dufresne-Dubé *et al.* (1983), Hou and Maxwell (1991), and Lamprell and Healy (1998) (Figure 9). Little or no variation from the unmodified aquasperm type has been found within the class to date, although Lamprell and Healy (1998) conclude that the acrosomal pit is a unique characteristic of the class.

In all species examined to date and both sexes, a connection between the gonad and right kidney serves as a conduit for the gametes (Lacaze-Duthiers, 1856–1857; Pelseneer, 1901; Boissevain, 1904; Odhner, 1931; Steiner, 1993) (Figure 9). Sperm is subsequently spawned through the posterior aperture, as observed in *Antalis entalis* (Lacaze-Duthiers, 1856–1857) and *Pulsellum lofotensis* (Steiner, 1993). Mucus-entwined eggs are released through the posterior aperture, as in *Antalis entalis* (Lacaze-Duthiers, 1856–1857), *D. conspicuum* (Dinamani, 1964a) and *Pulsellum salishorum* (Shimek and Steiner, 1997), but released singly through the anterior aperture in *Pulsellum lofotense* and *Cadulus subfusiformis* (Steiner, 1993). Steiner (1993) describes spawning in *P. lofotense* and *C. subfusiformis* in detail, and surmises that spawning through the anterior aperture, presumably requiring withdrawal to the sediment surface, may be related to habitat and predation pressure.

Very little is known about reproductive ecology of scaphopods. Rokop (1974, 1977) found seasonal reproductive cycles in a deep-sea population of *Cadulus californicus,* based on changes in ovary length, with spawning in late summer.

9.3. Early development

Changes in gametes during fertilization at the ultrastructural level were examined by Dufresne-Dubé *et al.* (1983). Moreau *et al.* (1989) used electrophysiological techniques to demonstrate the absence of a voltage-dependent mechanism that limits sperm penetration of oocytes in *Dentalium*, as has been inferred from other approaches to be the case in mollusks generally (although such a block is found in many other metazoans). Early development is described by Délage (1899), Lacaze-Duthiers (1856–1857), Kowalevsky (1883), Wilson (1904a) and Schleip (1925) at the light microscope level (summarized in Shimek and Steiner,

1997), and detailed ultrastructural studies on the normal progress of the first cell divisions, focusing upon the formation of the polar lobes, have been carried out as part of experimental studies using "*Dentalium*" sp. (usually the species *dentalis, entalis,* or *vulgaris,* all currently referred to the genus *Antalis*) as a model organism.

The use of the *Dentalium* egg as a model system for experimental embryology dates from the studies by Délage (1899), Wilson (1904a, b) and Schleip (1925). These were the first workers to exploit the relatively large egg- and polar-lobe extrusion of scaphopods, through egg- and embryo-component isolation experiments in localization of morphogenetic factors. A sizeable literature has since developed on the experimental embryology of *Dentalium,* and includes information on cell lineage and early organogenesis (e.g., Verdonk, 1968a, b; Geilenkirchen *et al.,* 1970, 1971; Verdonk *et al.*, 1971; Van Dongen and Geilenkirchen, 1974a, b, c, 1975; Van Dongen, 1976a, b, c; Guerrier *et al.*, 1978; Cather and Verdonk, 1979; Jaffe and Guerrier, 1981). A review of these results, in the broader context of experimental embryology in mollusks generally, appears in Reverberi (1971), and there is a series of reviews of molluscan development with foci of early development (Verdonk and Van den Biggelaar, 1983), organogenesis (Moor, 1983), spatial organization (Van den Biggelaar and Guerrier, 1983) and morphogenetic processes (Verdonk and Cather, 1983); there have been relatively few contributions using the *Dentalium* model since. In one example, Render and Guerrier (1984) examined the effects of polar lobe size regulation on localization of determinative materials.

9.4. Larval development

The scaphopod larva has been referred to as a stenocalymma (Salvini-Plawen, 1980; Buckland-Nicks *et al.,* in press) and a trochophore (Wanninger and Haszprunar, in press); whereas these terms have been applied to all pre-metamorphic stages, a distinction has also been made between the early stenocalymma and later, shelled larva, sometimes referred to as a veliger (e.g., Buckland-Nicks *et al.,* in press) (Figure 20). Based on the transitory ciliary apparatus, Moor (1983) considered the scaphopod larva as intermediate between the test-cell larva and the gastropod/bivalve trochophore.

Development through to metamorphosis was described in *Dentalium* by Lacaze-Duthiers (1856–1857), Kowalevsky (1883) and Wilson (1904a), summarized by Reverberi (1971), Shimek and Steiner (1997), and Buckland-Nicks *et al.* (in press), and described in *Antalis entalis* more recently by Wanninger and Haszprunar (in press). There is general

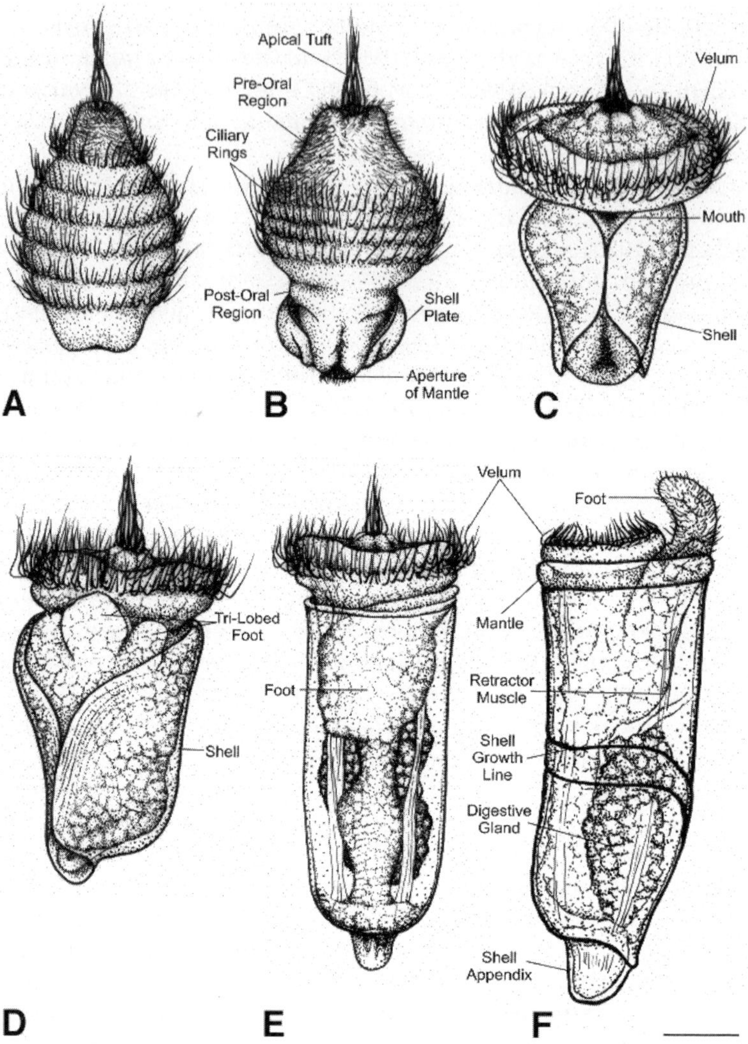

Figure 20 Sequential stages of larval development (for estimates of development time see text). (A) Early trochophore/stenocalymma larva with several ciliary bands. Scale bar = 70 μm. (B) Trochophore with three prototrochal bands. Scale bar = 70 μm. (C) Young veliger-like larva (anterior view), with ventrally growing embryonic shell and apical restriction of prototroch. Scale bar = 70 μm. (D) Late veliger-like larva. Scale bar = 105 μm. (E) Premetamorphic larva, with ventrally fused mantle and shell. Scale bar = 105 μm. (F) Metamorphically competent larva, showing tusk-like shell, prominent foot and reduced velum. Scale bar = 105 μm. (Reproduced with permission from Buckland-Nicks *et al.*, in press, figure 6; drawings from R. Koss, redrawn from Lacaze-Duthiers, 1856–1857.)

agreement in the sequence of events, although some differences in observations of structural detail and especially of timing are found (for example, see Wilson, 1904a). For comparative purposes, estimates of developmental timing given below follow Wanninger and Haszprunar (in press).

Molluskan spiralian-pattern cleavage leads to a coeloblastula; gastrulation by invagination (see also Verdonk and Van den Biggelaar, 1983) produces a blastopore that persists on the ventral side of the embryo (Wilson 1904a). Scaphopod embryos hatch by 19 h post-fertilization (hpf). Development, including fusion of several ciliated bands (Figure 20A), leads to a lecithotrophic trochophore with long apical tuft cilia, three ciliary bands forming a prototroch in the equatorial region, and a short ciliary tuft basally (Wilson, 1904a) by 32 hpf (Figure 20B). During the next 60 h, to metamorphic competency, there is a gradual reduction of the prototroch and loss of pre-trochal structures, while the post-trochal area, where the mantle fold, shell gland rudiments and definitive organogenesis are located, expands (Moor, 1983). Elongation of the post-trochal region is accompanied by secretion of the single primordium of shell by 39 hpf, with subsequent anterior and ventral growth of mantle and embryonic shell field. This post-trochal development restricts the prototroch apically into a veliger-like velum (but which lacks a muscle ring like that found in chitons, bivalves and gastropods, A. Wanninger and G. Haszprunar, pers. comm.) by about 54 h (Figure 20C, D). Ventral fusion of shell and mantle is completed by 64 hpf (Figure 20E). The foot anlage first appears by 49 hpf, as does muscle development generally (A. Wanninger and G. Haszprunar, pers. comm.); paired foot and cephalic retractors form additional fibers that function in prototroch/velum retraction (but differ structurally from larval retractors of gastropods, A. Wanninger and G. Haszprunar, pers. comm.). By metamorphic competence, the apical tuft is lost, the prototroch/velum is reduced (Figures 20F, 21C), and the foot is prominent and functional; at metamorphosis, the larva settles from a planktic to a benthic habit; the prototroch/velum is lost, its retractors are resorbed, and paired captacular anlage and anus are formed.

Ruthensteiner *et al.* (2001) describe the fine structure of the protonephridial system in the larvae and juveniles of *Antalis entalis.* The paired protonephridia consist of a terminal cell, or cyrtocyte, and a duct-releasing, glandular cell (Figure 22). The protonephridia first appear at 62 hpf (same time series as above, Wanninger and Haszprunar, in press), near the top of the slit-like mantle cavity formed by the foot and mantle fold; they are reduced and non-functional in the juvenile 13 days after metamorphosis (Ruthensteiner *et al.*, 2001). Near this stage, paired anlagen of the nephridial system are well-elaborated, and consist of a typical adult duct combined with separate protonephridial-like polyciliary cyrtocytes. The pericardium is still absent; whether this originates independently or from these terminal cyrtocytes remains uncertain (B. Ruthensteiner and G. Haszprunar, pers. comm.).

Van Loon and Van den Biggelaar (1998; see also Damen *et al.*, 1997) examined trochoblast-specific gene expression in *Dentalium vulgare*, using the *lac-Z* reporter gene under control of the *Patella* α-tubulin-4 promoter, and found conservation in expression patterns with the chiton *Acanthochiton communis* and the gastropod *Patella vulgata.* They concluded support for molluskan monophyly, and tested for this expression pattern in a nemertean and a platyhelminth for insights into spiralian relationships.

A. Wanninger and G. Haszprunar (pers. comm.) conclude that the presence of a distinct pair of cephalic retractors in the larva of *Antalis entalis*, as found in gastropods and cephalopods, is evidence for a closer phylogenetic relationship of scaphopods to those groups than to bivalves. They also consider the possibility that differences in the prototroch/velum retractor muscles and muscle ring may be secondary losses in scaphopods.

From an ecotoxicological perspective, Koster and Van den Biggelaar (1980) documented the abnormal development of eggs from *Dentalium vulgare* contaminated by hydrocarbons from the "Amoco Cadiz" oil spill in March, 1978. In addition to depressed fertilization rates, they observed several early development abnormalities including retarded and inequal cleavage, premature regression of the polar lobe constriction, and regression of cleavage furrows. Trochophores were variously malformed, and swimming was slow and uncoordinated, with elevated mortality. Those larvae that survived to metamorphosis were not affected further, and oocytes collected the following reproductive season developed normally (Koster and Van den Biggelaar, 1980).

9.4.1. *Larval shell*

During development, the anlage of the embryonic shell field appears postero-dorsally in the fully developed trochophore by about 32 hpf (Figures 20B, 21B), and the embryonic shell calcifies and proceeds to grow antero-ventrally (Wanninger and Haszprunar, in press) (Figure 20B–D). The shell is devoid of sculpture or growth lines, and forms a saddle-shaped shell with two bulbous "genae" (Engeser *et al.*, 1993); this earliest shell is termed protoconch "A" by Steiner (1995) (Figures 20C, 23A). The two lobes of the mantle and embryonic shell grow and fuse ventrally by 64 hpf (Wanninger and Haszprunar, in press), marked by faint incremental growth lines and a suture (Engeser *et al.*, 1993) (Figures 20E, 21C, 23).

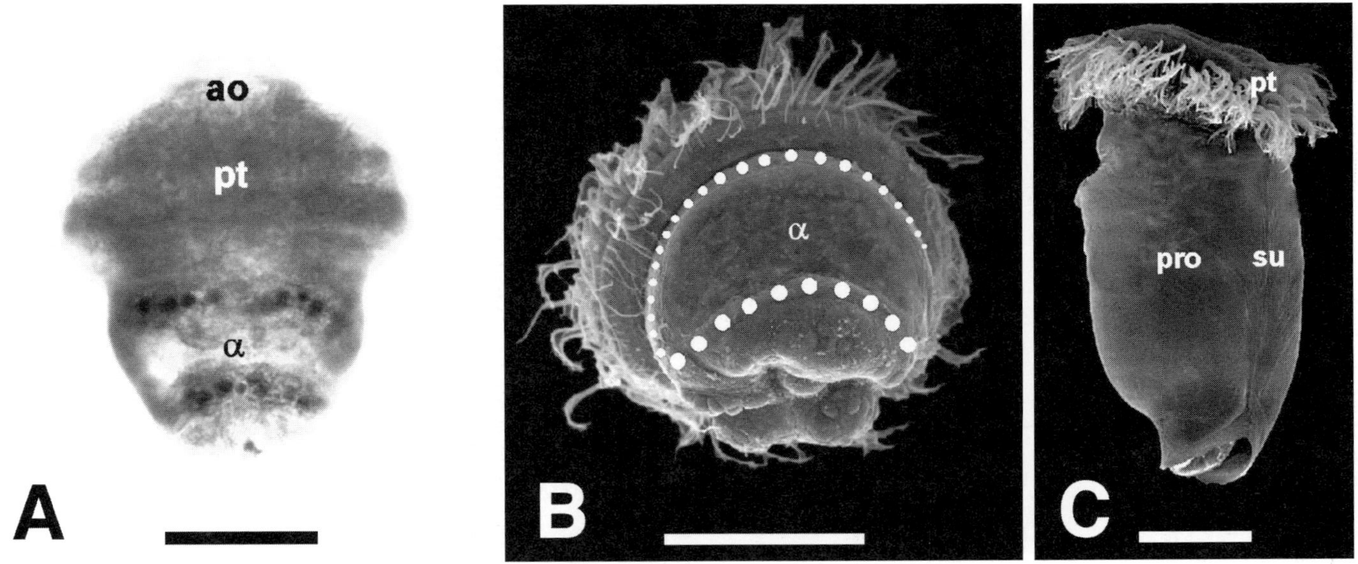

Figure 21 Shell development in the larva of *Antalis entalis.* (A) Expression pattern of 4D9-reactive cells in 39-hpf-old early trochophores of *Antalis entalis,* anterior upwards, dorsal view. 4D9-positive cells are localized at the anterior and posterior border of the embryonic shell field (α). (B) Patterning of engrailed protein expression (white dots) plotted on SEM of 48.5-hpf-old larvae with ventrally open anlage of the embryonic shell (α), postero-dorsal view. The localization of 4D9-positive cells is always found exactly at the interface of the embryonic shell field (α) and the adjacent mantle epithelium. (C) Larval development of *Antalis entalis*, SEM, anterior faces upwards. Larva at metamorphic competence, 95 hpf; ventro-lateral right view. Note the ventrally closed mantle and protoconch (pro) with distinct ventral fusion line (su, suture), as well as the completely reduced apical organ. (Reproduced with permission from Wanninger and Haszprunar, in press, figures 1i, 3C and 4C. © Blackwell Science.)

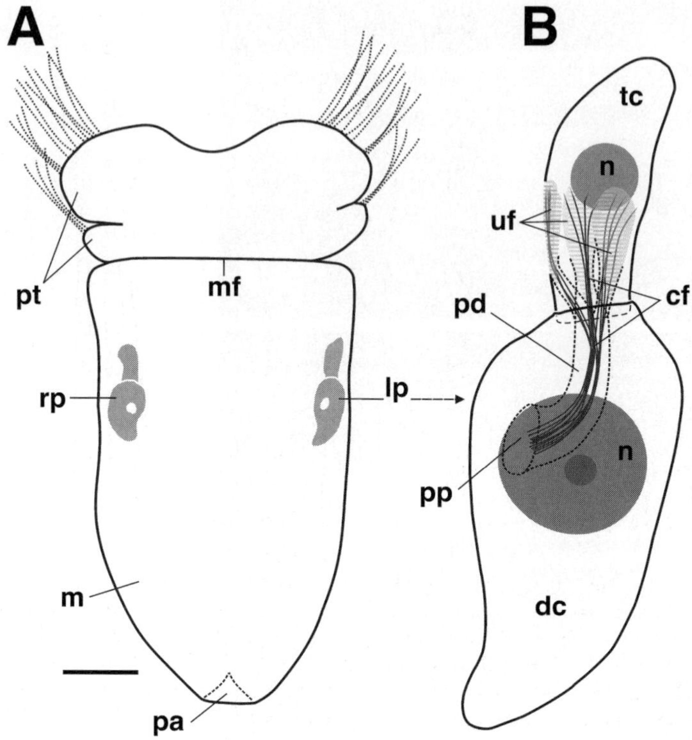

Figure 22 The larval protonephridium in *Antalis entalis.* (A) Reconstruction of a 90-hpf-old larva, dorsal view, protoconch omitted but outlined by underlying mantle (m). (B) Enlarged protonephridium, semischematic. cf, ciliary flame; dc, duct-releasing cell; lp, left protonephridium; m, mantle; mf, mantle fold; n, nucleus; pd, protonephridial duct; pa, pavilion proper; pp, protonephridial pore; pt, prototroch; rp, right protonephridium; tc, terminal cell; uf, ultrafiltration site. (Reprinted with permission from Ruthensteiner *et al.* (2001, figures 4a and 4b) © Springer-Verlag.)

After fusion, shell growth continues in a predominantly anterior direction, with a small posterior elongation termed the fumarium (Engeser *et al.*, 1993), until metamorphic competence (Wanninger and Haszprunar, in press) (Figure 21C). This stage in larval shell secretion, from fusion to metamorphosis, produces protoconch "B" which is often annulated (Steiner, 1995) (Figure 23). Metamorphosis marks the transition to the adult or teloconch shell, which in scaphopods with longitudinal sculpture can be differentiated into teloconch "A" which is smooth and precedes secretion of the ribbed teloconch "B" (Steiner, 1995); additional teloconch regions may be defined on the basis of subsequent sculptural modifications (Engeser *et al.*, 1993; Steiner, 1995).

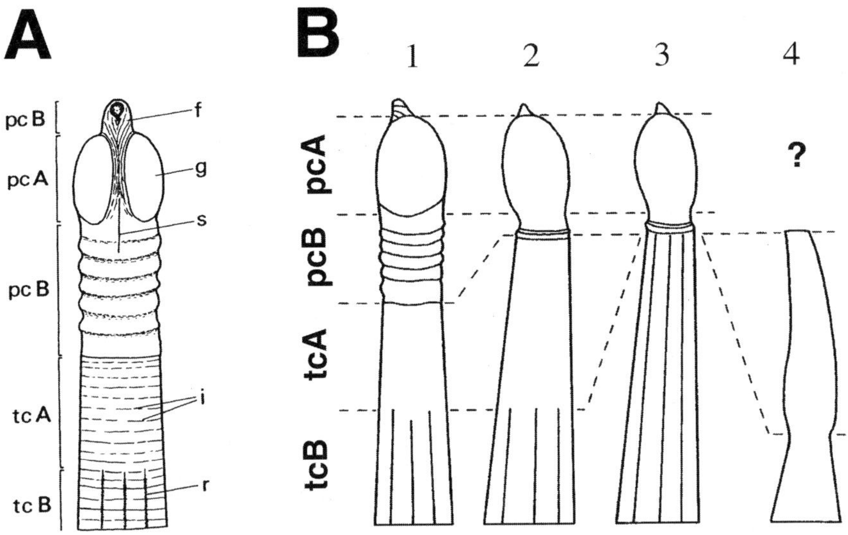

Figure 23 The scaphopod larval shell. (A) Schematic diagram of a dentalid larval shell, ventral view. pcA, protoconch A; pcB, protoconch B; tcA, teloconch A, tcB, teloconch B; f, fumarium; g, genae; i, increment lines; r, ribs; s, suture. (B) Four types of larval and juvenile shells (abbreviations as in A): 1. Dentaliida; 2. Gadilida, Entalimorpha; 3. Gadilida: Gadilimorpha; 4. *Cadulus subfusiformis.* (Reproduced with permission from Steiner, 1995, figures 1, 11.)

La Perna (1998), proposed the term "praetubulus" for the entire (pre-teloconch) scaphopod larval shell. While eventually lost through shell dissolution, and probably decollation, by the mantle (Reynolds, 1992a) (see Section 2.2.1), the larval shells or praetubuli of at least 25 species have been described or illustrated, often as incidental reports (see Steiner, 1995; Engeser and Riedel, 1996; La Perna, 1998, for species lists and sources).

Steiner (1995) compared all prior reports and material from three additional species, and identified four patterns of larval shell formation, varying largely on the basis of protoconch B and teloconch A proportions (Figure 23). In Dentaliida examined to date, roughly equal proportions are given to protoconch B and teloconch A portions (but see La Perna, 1998); within the Gadilida, two situations arise: Entalimorpha protoconch B is truncated, with teloconch A elongated, whereas in Gadilimorpha both are truncated. In *Cadulus subfusiformis,* protoconchs have not been found, but a constriction between teloconchs is evident (Steiner, 1995).

The relationship between larval shell components, their variation among scaphopods, and their development is intriguing (for discussion see

Engeser *et al.*, 1993; Steiner, 1995; Engeser and Riedel, 1996; La Perna, 1998) and comparative developmental data will be interesting in this regard. Seemingly, a comparison can be readily made between protoconch A and the embryonic shell protoconch/prodissoconch I of gastropods and bivalves, and between protoconch B and the veliger-shell protoconch/prodissoconch II of gastropods and bivalves. Steiner (1995), however, warns against inadequate evidence for homology of these structures; Wanninger and Haszprunar (in press) cite uncertainty over the existence of a "larval" (veliger) shell *per se,* applying the neutral "protoconch" to the entire praetubulus or pre-teloconch shell.

Engeser and Riedel (1996) argue the remarkable similarities between the scaphopod protoconch A and the conch of conodardiid rostroconchs, and propose heterochronic derivation from, and sister group relationship with, that clade, in contrast with the riberiid-rostroconch derivation model of Pojeta and Runnegar (1979, 1985).

Wanninger and Haszprunar (in press), however, report expression of an engrailed protein in marginal cells of the embryonic shell field, from the time of differentiation of the anlage (32 hpf), and subsequently in the mantle margin during protoconch secretion in *Antalis entalis.* Expression ceases with metamorphosis and teloconch secretion, suggesting different genetic backgrounds and evolutionary origins of protoconch and teloconch shell stages. Of greater relevance here, the *engrailed* expression pattern demonstrated a single shell-field anlage, the scaphopod larval shell being univalved throughout its morphogenesis (Figure 21). Wanninger and Haszprunar (in press) note that these data confirm earlier observations in scaphopods by Lacaze-Duthiers (1856–1857) and Kowalevsky (1883), and coincide with *engrailed* expression in the embryonic gastropod *Ilyanassa* (Moshel *et al.,* 1998). As bivalve prodissoconch I formation is initiated at two centers of calcification, each with *engrailed*-expressing cells (Jacobs *et al.,* 2000), Wanninger and Haszprunar (in press) infer a bivalved shell to be an apomorphy of the Bivalvia, and propose abandonment of the Diasoma concept (Scaphopoda + Bivalvia + Rostroconchia) (Runnegar and Pojeta, 1974; Pojeta and Runnegar, 1985).

10. BIOGEOGRAPHY

10.1. Bathymetric distribution

Scaphopods have a considerable bathymetric range in the world's oceans. Gadilida have been recorded from the low intertidal (e.g., *Cadulus vincentianus,* Australia, Lamprell and Healy, 1998) to depths of

6900–7000 m (*Siphonodentalium galatheae,* Sunda Trench, Knudsen, 1964), and a similarly wide range is found in Dentaliida (intertidal: *Laevidentalium lubricatum,* Australia, Lamprell and Healy, 1998; 4755 m: *Dentalium leptoskeles,* Watson, 1886). It has been generally considered that scaphopods are relatively diverse in the deep sea (e.g., Pilsbry and Sharp, 1898; McFadien, 1973; Scarabino, 1979, 1986a, b).

Bathymetric diversity patterns of macrofauna, including mollusks such as gastropod and bivalves, have received much attention over the past few decades (e.g., Hessler and Sanders, 1967; Rex, 1973, 1981, 1983; Grassle and Maciolek, 1992; Gage, 1996), but have been less studied in scaphopods. Henderson (1920) compiled bathymetric data for western north Atlantic scaphopods, and remarked on related provincial and habitat distribution patterns, but not bathymetric diversity patterns *per se.* McFadien (1973), studying seven Panama-Pacific species of scaphopods, found that depth ranges varied from fairly restricted (e.g., <100 m, *Cadulus austinclarki, Dentalium inversum*) to very broad (*D. splendidum, D. oerstedii*). Lamprell and Healy (1998) also comment on the wide bathymetric ranges of some scaphopod species, including the Australian scaphofauna.

Scarabino (1979), in a study of 55 western Atlantic species, defined four species groups based upon upper bathymetric limit and range; the largest group consisted of 29 bathyal species with varying range sizes (~200–3500 m); the eurybathic bathyal-abyssal group was the smallest with four species each with large ranges, the largest being *Fissidentalium meridionale* (498–4743 m). In a recent revision and catalog of Indo-Pacific scaphopods, Scarabino (1995) analyzed the bathymetric distribution of 73 species, through eight intervals: 0–100, 100–300, 300–500, 500–800, 800–1200, 1200–2000, 2000–3000 and 3000–4000 m. For both orders, diversity increased to a maximum at the 500–800 m interval, and decreased through the 800–1200 interval. Whereas Dentaliida continued to decrease in diversity through to the 3000–4000 m interval, Gadilida showed a minor diversity peak at the 1200–2000 m interval. Species of some genera, such as *Fissidentalium* and *Siphonodentalium,* were preferentially distributed at bathyal and abyssal depths. Scarabino (1995) notes that these patterns are similar to those found in the Atlantic (Scarabino, 1979).

A complementary perspective of bathymetric penetration patterns is given by a compilation of maximum depth records for scaphopods (Figure 24) (as compared to depth range, which estimates total diversity at any given depth). This sample, from 298 species, includes representation from all oceans, compiled at equal 100-m intervals (Reynolds, Gray and Hilpert, unpublished data). While there is a general trend of decreasing bathymetric penetration of scaphopod species with depth, this is marked between 0–500 m and followed by an increase between 500–900 m

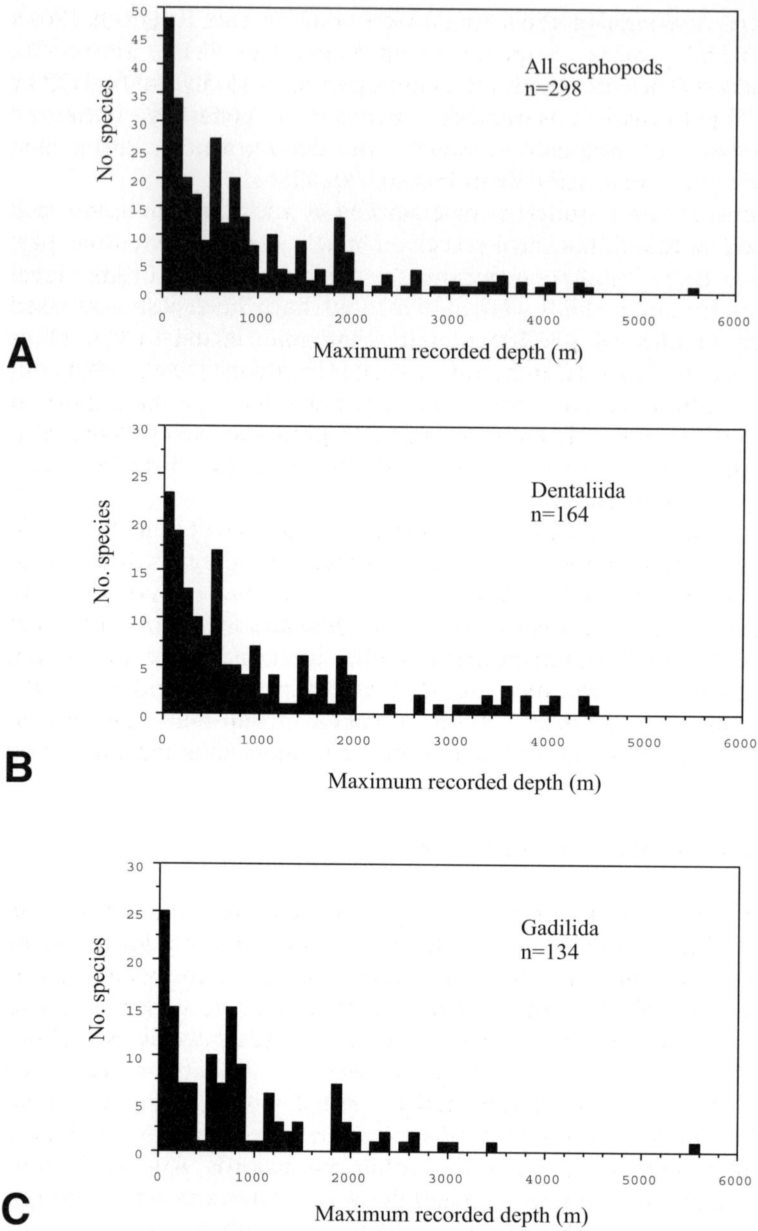

Figure 24 Distribution of maximum bathymetric records for (A) Scaphopoda, (B) Dentaliida and (C) Gadilida. (Reynolds, Gray and Hilpert, unpublished data, available for review from author.)

(Figure 24A); analysis of the orders separately (Figure 24B, C) shows that this second peak is comprised of an increase in Dentaliida species found to 500–600 m, but mostly of Gadilida species to 500–900 m. Other ordinal-level patterns in this analysis show that Gadilida have a marked penetration in the 1800–2000 m range, whereas Dentaliida are almost exclusively represented from 3000–4500 m depths.

Bernasconi (1990), studying paleocommunities from an estimated 12–500 m depth, found that scaphopod dominance varied with bathymetry, increasing to continental shelf depths (150–180 m). Several ordinal-level patterns were found: gadilids were absent at <100 m, but rapidly increased to equal dentaliid levels by 180 m, with domination again by dentaliids from that level to 500 m (Bernasconi, 1990). Bernasconi (1989, 1990) considered substratum and paleocommunity species richness to be important factors in such bathymetric patterns, and Lamprell and Healy (1998) also emphasize sediment characteristics over depth *per se* as constraining factors in scaphopod distribution.

Even given the differences in data summation and analysis in the above global and regional studies, the variance among bathymetric distribution patterns is intriguing. While a database as inclusive of scaphopod diversity as possible will provide a fuller picture, emerging patterns point to distinct radiation patterns of Dentaliida and Gadilida in the deep sea (Scarabino, 1995). Bathymetric penetration, scaphopod diversity at depth intervals, and variation in ecological role of scaphopods in neo- and paleo-communities with depth all warrant further and more detailed study.

10.2. Latitudinal diversity gradients

A latitudinal gradient in species diversity, increasing from the poles to the tropics, is well known for terrestrial fauna and flora and to a lesser extent from marine environments. Such a gradient has recently been demonstrated in marine bivalves and gastropods (Roy *et al.*, 1994, 1998). These patterns have not been extensively examined in scaphopods. Scarabino (1979) found a latitudinal gradient in western Atlantic scaphopods, which peaked at ~15–20°N, reflected in both Dentaliida and Gadilida. Among dentaliids the gradient was more marked among an infra-/circalittoral species group than in either bathyal or abyssal groups, although species numbers were generally low in the latter two groups. Among the gadilids, both infra-/circalittoral and bathyal species groups showed a marked gradient and peak at 15–20°N.

A preliminary compilation of global latitudinal species diversity for Scaphopoda is presented in Figure 25 (Reynolds, Gray and Hilpert, unpublished data). We compiled latitudinal range records for 355 species

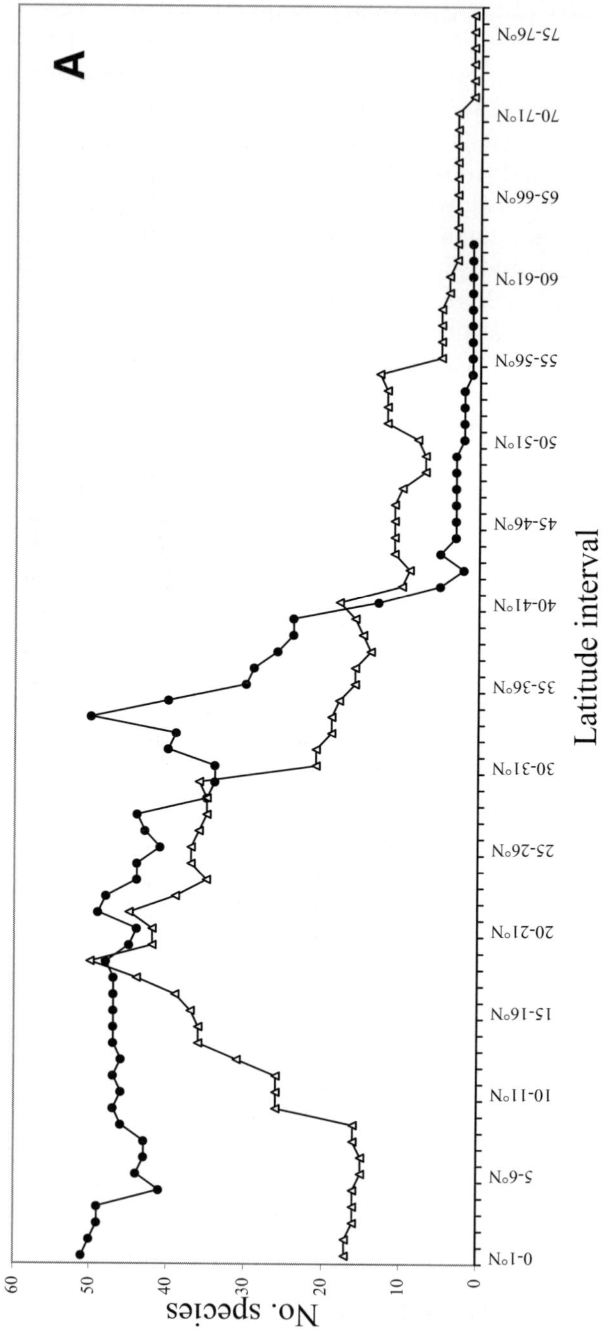
A
No. species
Latitude interval
0
10
20
30
40
50
60
0-1°N
5-6°N
10-11°N
15-16°N
20-21°N
25-26°N
30-31°N
35-36°N
40-41°N
45-46°N
50-51°N
55-56°N
60-61°N
65-66°N
70-71°N
75-76°N

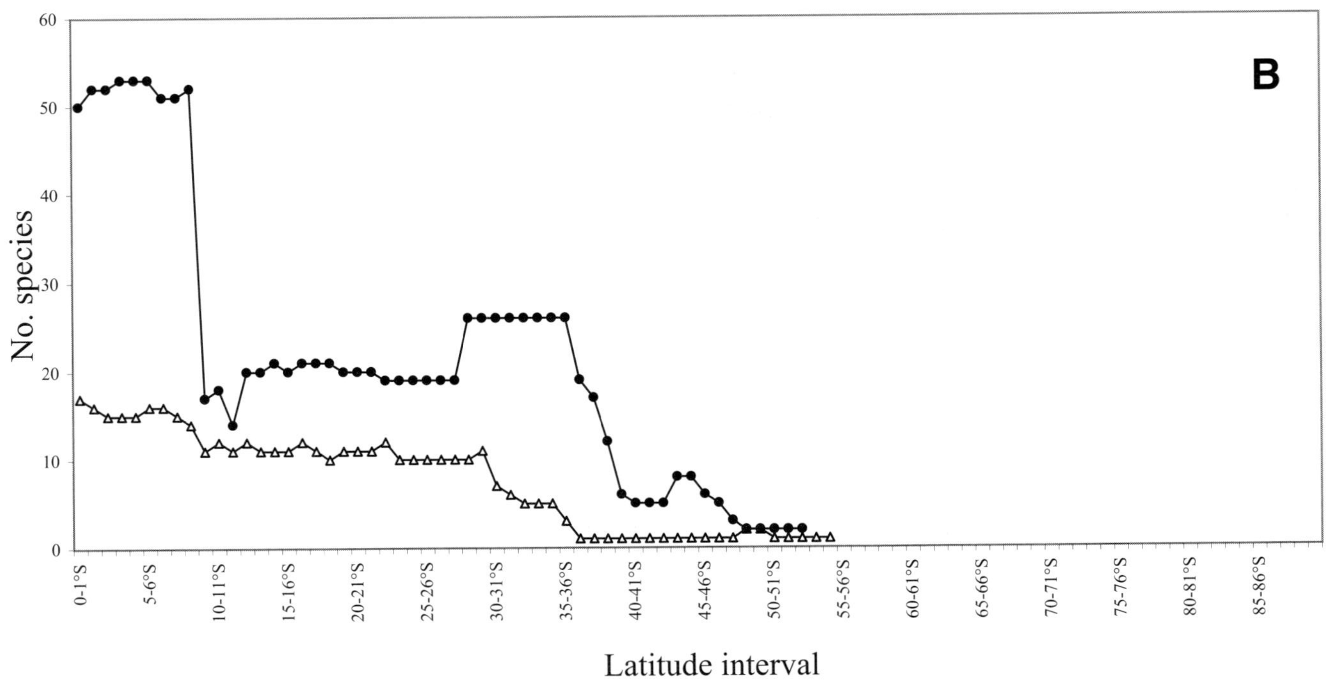

Figure 25 Latitudinal diversity of scaphopods in Atlantic (△; 145 species) and Pacific (●; 150 species) oceans. (A) Northern hemisphere. (B) Southern hemisphere. (Reynolds, Gray and Hilpert, unpublished data, available for review from author.)

(~70% of nominal extant species diversity within the class) from the primary literature, and calculated species counts for each 1° band in each hemisphere of the Atlantic and Pacific oceans. Latitudinal diversity showed marked gradients in all four major basins. Examining patterns across all latitudes in each ocean, gradient amplitude was similar in both oceans, diversity reaching a peak of ~50 species. While this occurs as expected near the equator in the Pacific, the peak is located at ~20°N in the Atlantic, confirming the pattern found in the western Atlantic by Scarabino (1979). Consequently, diversity gradients peaked near the equator in both southern hemispheres, but with a much lower maximum diversity in the southern Atlantic.

Sharp discontinuities in the otherwise gradual gradients (northern Atlantic 29–31°N, northern Pacific 34–36°N and southern Pacific 8–10°S) may reflect areas of high diversification or coastal geographic features that limit distribution, as observed in gastropods and bivalves (Roy *et al.*, 1998). For example, Scarabino (1979) also found a sharp increase in diversity at ~30°N, in the western Atlantic, which coincides with the northern limit of the Caribbean.

Possible explanations for a latitudinal diversity gradient among marine mollusks include decreasing latitudinal range size with decreasing latitude ("Rapoport's Rule": Stevens, 1989, and related literature), species-area relationships, recent geologic history of oceanic basins, and sea-surface temperatures (Roy *et al.*, 1998). For scaphopods, the latter possibility is raised by McFadien (1973), who noted the latitudinal restriction of several Panamic-Pacific species to warm waters. Preliminary analysis also suggests that Rapoport's Rule is consistent with latitudinal diversity patterns in the Atlantic, but not in the Pacific (Reynolds *et al.,* 2001). A combination of several factors is likely to influence scaphopod species distributions, and their relative influence in patterning latitudinal gradients in diversity is likely to vary among ocean basins. More detailed documentation of those patterns is required before causal hypotheses can be fully explored.

11. SPECIES-PAIR ASSOCIATIONS

In reviewing species-pair associations that include scaphopods, it is recognized that there are likely to be many anecdotal and incidental reports, which would be difficult to compile exhaustively. In many cases, the precise cost/benefit nature of the relationship can only be surmised. Several recent studies, however, give particular attention to obligate and, in some cases, species-specific relationships, and are reviewed here as a range of examples of these associations.

11.1. Symbioses

11.1.1. *Commensalism*

There are several brief reports of commensal associations with scaphopods. Ciliates, probably loricate peritrichs of the genus *Trichodina,* are frequently attached in the mantle cavity of *Rhabdus rectius,* usually near the ciliated bands (Reynolds, 1990a) (Figures 8, 9, 26). Members of *Trichodina* are common external commensals or parasites of aquatic animals (Kudo, 1954).

Although the precise nature of the symbiosis is uncertain, Timmermans *et al.* (1970) and Geilenkirchen *et al.* (1971) discovered symbiotic bacteria bound to the exterior surface of the egg at the vegetal pole of *Dentalium*

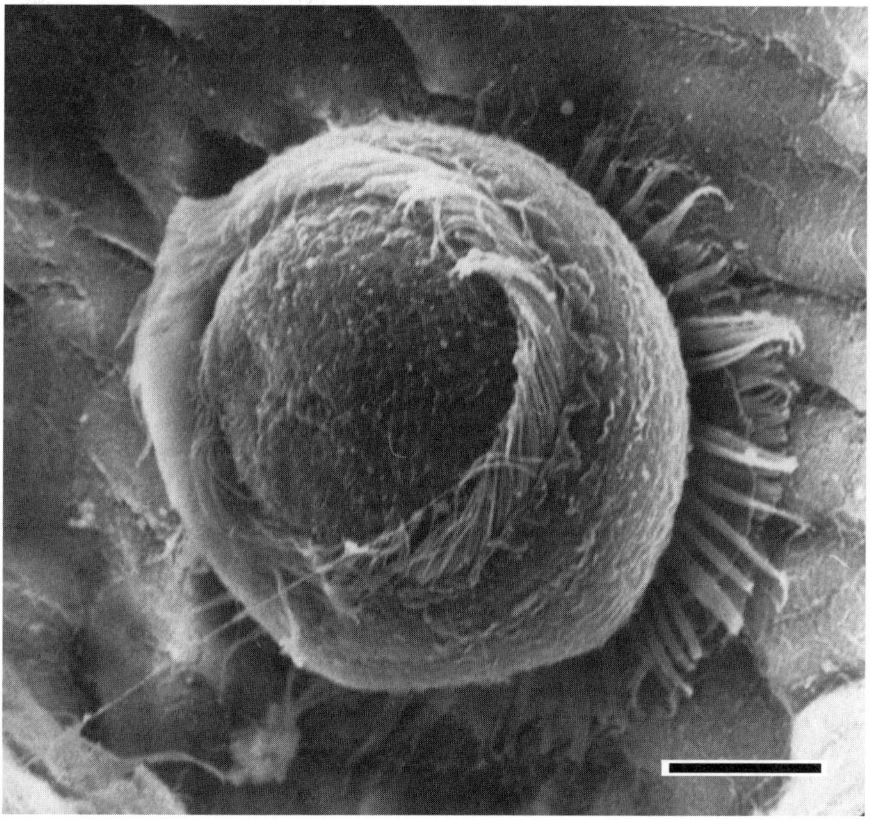

Figure 26 Commensal ciliate in the mantle cavity of *Rhabdus rectius,* probably of the genus *Trichodina.* Scale bar 10 μm.

dentalis and *D. antillarum,* respectively. The bacteria remain associated through mesoderm formation and gastrulation, successively attached to the 4d cell, the mesentoblasts, and the follicles of developing oocytes.

Bacteria have also been reported in association with the microvillous cells supporting the ciliated bands in the mid-region of the mantle cavity in *Rhabdus rectius* (Figure 10) (Reynolds, unpublished data). The presence of the bacteria, confirmed by transmission electron microscopy, is consistent throughout many individuals examined from the population and several sampling periods. The bacteria are intimately associated with the epithelial cell surfaces (Reynolds, unpublished data).

Small canals in the outer layers of the shell in living scaphopods have been the subject of somewhat enigmatic reports on *Dentalium mutabile inaequicostatum* (Alzuria, 1985b), *Fissidentalium scamnatum* (Ivanov and Memmi, 1989), *Fissidentalium floridense* and *Coccodentalium carduum* (Figure 27) (Okusu and Reynolds, unpublished data), the last with a short organic tube protruding from numerous canals opening to the shell surface. Alzuria (1985b) reviews ideas on their possible origin, and considers them to be produced by an external agent; Ivanov and Memmi (1989) suggest that a boring alga of the genus *Hyella* (Chamaesiphoneae, Pleurocapsales, Pleurosapsaceae) is responsible for the canals. As the

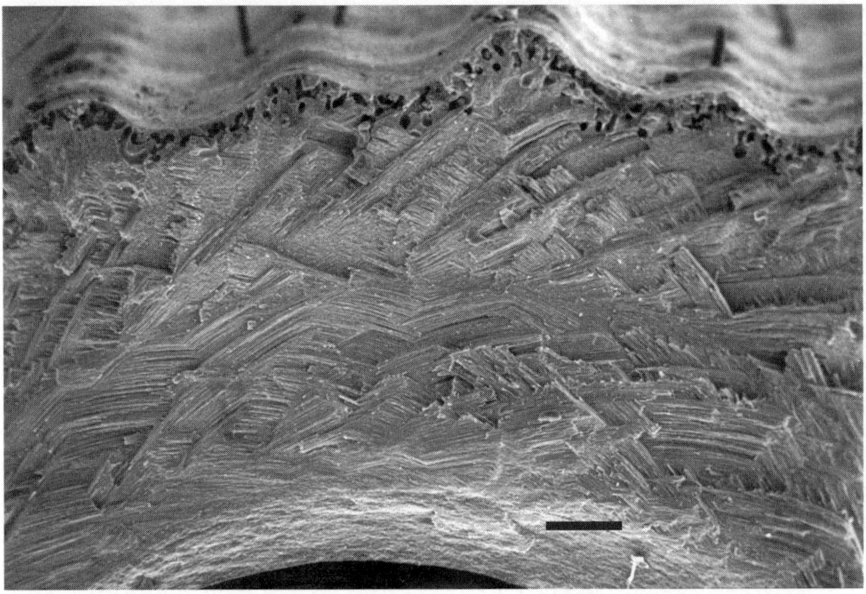

Figure 27 Canals in the outer layers of the shell in *Fissidentalium floridense,* putatively created by a boring alga. (Okusu and Reynolds, unpublished data.) Scale bar 100 μm.

degree of shell boring observed is seemingly restricted to the outer edge of the shell, detrimental effects to the scaphopod appear to be minimal, if any, and a commensal relationship is assumed.

11.1.2. *Mutualism*

Several symbiotic relationships between scaphopods and cnidarians have been described, often from deep-water habitats. The symbionts are typically found attached to the outside of the shell (Shimek, 1997; Zibrowius, 1998; White *et al.*, 1999), and consist of anemones, corals, and barnacles (Kozloff, 1990).

Anemones comprise most of the reported symbionts: *Hormathia pectinata* is occasionally found attached to a dentaliid shell (Riemann-Zürneck, 1973), and an unidentified anemone is often found in association with *Fissidentalium exuberans* (Zibrowius, 1998). The scaphopods *Fissidentalium actinophorum,* described by Shimek (1997) from depths of ~4000 m off southern California, and *Fissidentalium megathyris,* described by Shimek and Moreno (1996) from off central California, are both frequently found with the anemone *Anthosactis nomados,* a member of the Actinostolidae, which lives exclusively attached to the shell of living scaphopods (Plate 3). In most examined specimens, the anemone is attached with the oral disk roughly centered on the concave or functionally dorsal surface of the shell (White *et al.*, 1999) (Plate 3). The anemone is found much more frequently with *F. actinophora;* about 60% of living individuals of this species had an attached anemone (Shimek, 1997; White *et al.*, 1999), whereas another 30% of scaphopods showed evidence of recent anemone attachment (White *et al.*, 1999), as did most dead shells (Shimek, 1997).

The solitary coral, *Heterocyathus japonicus,* as well as attaching to gastropod and scaphopod shells inhabited by sipunculans (see Section 11.2.2), frequently attaches to the apical end of the living individual of *Fissidentalium vernedei* (Zibrowius, 1998). It is usually positioned off to one side, but may encrust the apex on all sides, changing its shape while maintaining the opening of the posterior aperture. It is typically oriented, with respect to the curvature of the shell, to maximize its height while the scaphopod burrows.

The nature of these symbioses is thought to be at least commensal, and possibly mutualistic. Shimek (1997) and White *et al.* (1999) agree that a clear benefit to the anemone *Anthosactis nomados* accrues from resuspended organic particulates generated by scaphopod feeding and burrowing behavior, and from being carried to food-rich areas by the scaphopod; the same argument can be applied to attached barnacles (illustrated as attached to the mid-region of the shell, Kozloff, 1990: figure 13.71B) and the solitary coral *Heterocyathus japonicus*. For the scaphopod, benefits may include apical protection from predation (White *et al.*, 1999) through added strength and protection through nematocyst deterrence (Zibrowius, 1998). White *et al.* (1999) further argue that this protection would thereby allow the scaphopod to avoid corrosive sub-surface sediments and feed nearer the surface, which is richer in organic nutrients and calcium carbonate. Shimek (1997) raises some potential costs to the scaphopod, noting some shell erosion at the base of the anemone and the possibility of hindrance to burrowing.

11.1.3. *Parasitism*

There are several documented instances of putative parasitism of scaphopods, although most are incidental observations. Arvy and Gabe (1951), while studying blood cells in "Dentale", found 10% (of 150 individuals) had a ciliate in the blood. Further review of protistan parasites is provided by Davies (1987).

Parasitism by platyhelminth larval stages has been reported several times; for example, Boissevain (1904) noted redia in the blood of *Antalis entalis,* and Pelseneer (1906a) found sporocysts with developing cercariae in *Antalis vulgaris.* Arvy (1949; see also Gaillard, 1956) described in detail free cercariae in the gonad of *Antalis entalis,* and the organ's associated abnormal appearance; the gonocytes were entirely absent and the sex of the scaphopod unidentifiable. Arvy (1949) considered either physical damage or toxic effects could have caused the destruction of the gonad. In a review of digenean associations, Cribb *et al.* (2001) reported that the group is primitively associated with gastropods, and that infection of scaphopods is a result of host-switching. They report only a single species occurring in scaphopods, *Ptychogonimus megastomus* (Ptchogonimidae, Hemiuroidea), and in which the cercariae are exceedingly reduced.

Adult flatworms, probably turbellarians, have also been noted replacing the gonad in *Rhabdus rectius* (R. Shimek, pers. comm.). Endoparasitic nematodes and ectoparasitic (mantle cavity) copepods have been reported by Davies (1987).

11.2. Other associations

11.2.1. *Predation*

Scaphopods are preyed upon by ratfish (Shimek, 1989), and presumably those species that burrow near the sediment surface are vulnerable to predation by bottom-feeding fish generally. It is worth noting that in response to minor sediment disturbance, some species, like *R. rectius,* will remain motionless while other species (*Antalis pretiosum, Gadilia aberrans*, and *Pulsellum salishorum*) burrow rapidly (R. Shimek, pers. comm.); severe disturbance of the animal naturally elicits withdrawal into the shell.

Naticid predation on scaphopods occurs through boring of the shell, as in other mollusks, and while often observed in shells from surface sediments, fossil populations have been the basis for extensive study of the phenomenon (e.g., Yochelson *et al.,* 1983; Vermeij, 1987; Årpád, 1993). The naticid bore-hole usually occurs in a position about halfway along the dorsal surface of the shell, due to the deep foot and mantle retraction scaphopods can attain (Yochelson *et al.*, 1983). Naticid predation rates reach over 50% in populations of *Dentalium laqueatum* from sub-Holocene coarse sediments (Yochelson *et al.,* 1983), and up to 15% in Cretaceous and Paleogene populations; a similar temporal pattern is found in gastropod families (Vermeij, 1987). The variation in predation rate with geologic time may reflect naticid diversity and abundance levels (Vermeij, 1987) or sediment type, related to naticid preference for coarser sediments (Yochelson *et al.*, 1983). Scaphopod species with ribbing are more prone to predation, but whether this is due to a vulnerability to the predator or a covarying sediment preference factor is difficult to determine (Yochelson *et al.*, 1983).

11.2.2. *Associations with empty shells*

Like gastropods, empty scaphopod shells are often used as substrata and shelters for other invertebrates. Anecdotal reports are many. Several sipunculan species, for example, are frequently found in scaphopod shells (G. Steiner, pers. comm.; pers. obs.), *Phascolion strombii* being a frequent occupant (N. Cutler, pers. comm.).

Imafuku and Ando (1999) examined the behavior and morphology of the hermit crab *Pagurus imafukui*, from Japan, which lives in empty shells of *Gadilina insolita* and *Striodentalium rhabdotum.* In this species, as in most pagurid crabs, the abdomen is asymmetrical, reflecting their normally helically spired gastropod-shell habitat; the near symmetry of the uropods in this species is an interesting reversion related to the non-helical scaphopod shell (Imafuku and Ando, 1999). Individuals of *P. imafukui* showed a significant preference for scaphopod over gastropod shells, and exhibited modified walking behavior when obliged to use gastropod shells. Courtship displays of the crab were in part mediated through the scaphopod shell. Imafuku and Ando (1999) also found *Paguristes seminudus* (infrequently) in *Fissidentalium vernedei.*

Forest (1987) reviewed the ethology and distribution of pagurid hermit crabs of the Family Pylochelidae (Decapoda, Coenobitoidea) which, rather than inhabiting gastropod shells, are found in a variety of straight tubular cavities, including decaying wood, bamboo shoots, stone fragments, sponges and scaphopod shells. Individuals of *Bathycheles incisus,* usually living in wood, often dwell within *Fissidentalium magnificum* shells; *Trizocheles sakii* is found in scaphopod shells, but also in gastropod shells and serpulid polychaete tubes (Forest, 1987).

One pylochelid species that exclusively chooses scaphopod shells is *Pomatocheles jeffreyseii,* from Japan (Forest, 1987). Its cheliped shape is modified as an operculum for the anterior shell aperture of scaphopods. Members of the Pyrochelidae are unlike other hermit crabs in having a normally developed (in most genera symmetrical) abdomen with full articulation and appendages, although they are modified somewhat in *P. jeffreyseii* to fit the scaphopod shell perfectly. This species may be further differentiated from other pyrochelids ethologically by being able to move while carrying its shelter, and having to seek a new shell as it grows; other pylochelids enlarge their spaces as needed (Forest, 1987). Another straight pugurid, *Orthopagurus minimus,* also occupies scaphopod shells almost exclusively, those of *Antalis pretiosum* in the northeast Pacific (R. Shimek, pers. comm.) (Plate 4).

Other examples of the use of empty scaphopod shells include the anemone *Paracalliactis stephensoni,* from off the Irish coast, which was found attached to a dentaliid shell inhabited by the pagurid *Parapagurus pilosimanus* (Carlgren, 1928). Sandford and Brown (1997) note that the Florida hermit-crab sponge, *Spongosorites suberitoides,* normally an encruster of gastropod shells often inhabited by hermit crabs, is occasionally found on scaphopod shells. Zibrowius (1998) reported that the solitary coral *Heterocyathus japonicus*, besides being found encrusted on living *Fissidentalium vernedei* (see Section 11.1.2), is also found encrusted on *Dentalium octangulatum* shells inhabited by sipunculans. He notes that individuals of other species in the genera *Heterocyathus* and *Heteropsammia* typically encrust gastropod shells inhabited by sipunculans, but may also attach to inhabited scaphopod shells.

12. DIVERSITY THROUGH GEOLOGIC TIME

The earliest scaphopods reported from the geologic record are frequently cited as *Plagioglypta iowensis* (Bretsky and Bermingham, 1970) and *Rhytiodentalium kentuckyensis* (Pojeta and Runnegar, 1979), both from the Ordovician, but the scaphopod and even molluscan affinities of both have been recently questioned and they may be worm tubes (for discussion see Engeser and Riedel, 1996; Lamprell and Healy, 1998; Yochelson, 1999). Several dentaliids have been reported in the Devonian (see Ludbrook, 1960; Emerson, 1962), although some of these have been subsequently reassigned to non-scaphopod taxa; Yochelson (1999) reviews pre-Carboniferous reports of scaphopods and argues that none are above question. Therefore, the Dentaliida currently dates with confidence from the Mississippian Carboniferous. The Gadilida are considerably more recent in the fossil record, with earliest undisputed reports from the Tertiary (Emerson, 1962); the Permian *Calstevenus* was assigned to the Gadilida by Yancy (1973), but Yochelson (1999) argues that its ordinal placement within the scaphopods is equivocal.

While several reviews have traced the statigraphic range of species and genera (e.g., Emerson, 1962; Skelton and Benton, 1993), there have been few attempts to represent stratigraphic or temporal species diversity. Ludbrook (1960) diagrammatically presented smoothly increasing, near logarithmic, curves to represent temporally increasing species diversity in the orders Dentaliida and Gadilida.

A compilation of stratigraphic records for 242 species of the Dentaliida, compiled from stratigraphic occurrence data from 299 primary literature papers (Reynolds, May and Maguire, unpublished data), is presented in Figure 28. Stratigraphic range-through was assumed, incorporating extant species found in the fossil record, and records of junior synonyms were collapsed to the appropriate currently valid species. Notwithstanding issues of sampling and preservational bias, and potential standardization analyses that could be applied to the dataset (e.g., Alroy *et al.*, 2001), several interesting patterns emerge. Starting with the radiation after the late Devonian extinction event (362.5 My), there is an overall increase in diversity with time, as represented diagrammatically by Ludbrook (1960). The pattern, however, shows several sharp putative extinction events, notably at the end of the Permian (247 My), during the late Liassic (187 My) with continuing decline through the Jurassic (208–145.6 My); sharp increases in diversity are seen at the beginning of the Eocene (56.5 My) and Miocene (23.3 My), with a maximum of 57 species during the Tortonian (10.4–6.7 My). The Early Cretaceous (150–100 My) appears to be a period of sustained low diversity in dentaliids (Reynolds, May and Maguire, unpublished data).

The diversity patterns of dentaliid families and geographic distribution of species through geologic time reflect these events, but provide a more complex picture than represented at the ordinal-level (Reynolds and Maguire, 1998). Sampling standardization techniques, and analysis of origination and extinction rates, is the next stage in understanding the

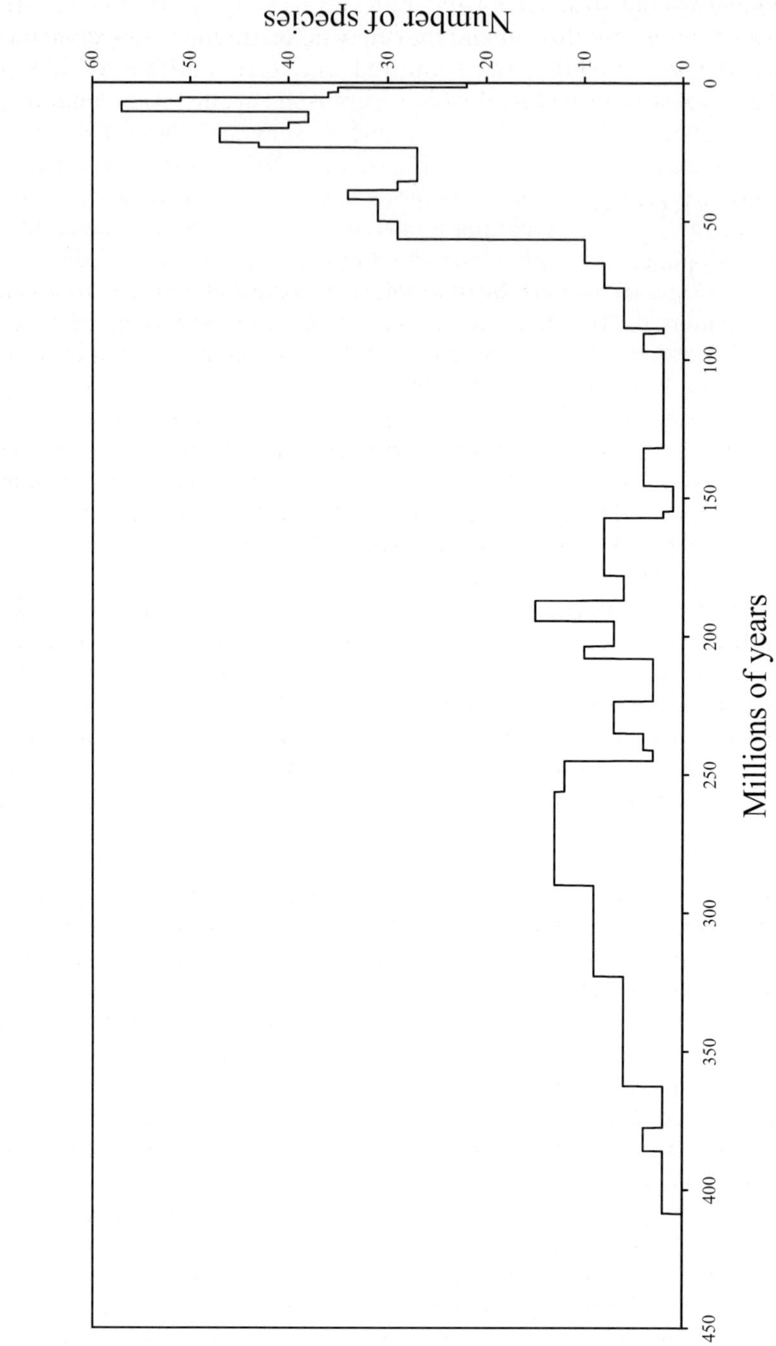

Figure 28 Diversity of Dentaliida through geologic time. (Reynolds, May and Maguire, unpublished data, available for review from author.)

scaphopod fossil record. Given the relatively recent appearance of the Scaphopoda among molluscan classes, the comparatively well-documented record and the tractable size of the group, broad patterns of change through time would provide valuable insights on scaphopod evolution.

13. CONCLUSIONS

The assumption that the class Scaphopoda is biologically homogeneous has been perpetuated by slim treatment in invertebrate biology texts, traditional reliance upon external shell characters for scaphopod taxonomy, and few comprehensive comparative studies. From a thorough review of the literature, however, it is apparent that despite constraints on morphologic disparity, scaphopod homogeneity is in many respects superficial and overestimated. There are still areas of scaphopod biology in which our knowledge is quite limited, being based either on single-species reports, anatomical work from the nineteenth century, or unpublished observations and data.

The paucity of comparative studies is due, in part, to relatively few investigators with a focus on the scaphopods. Historically, collection and preservation or experimentation on live scaphopods has been difficult due to their infaunal marine habit, dispersed diversity data and small size (in most species). As with many taxa, however, these difficulties are being alleviated indirectly through web-based museum catalogs, gradual expansion and taxonomic identification of wet-preserved museum collections, compiled distribution records in regional systematic monographs (e.g., Scarabino, 1995; Lamprell and Healy, 1998; Steiner and Linse, 2000), and molecular techniques that provide some independence from specimen size and laboratory maintenance limitations; the potential for comparative analyses on many aspects of the group's biology has been considerably enhanced in recent years.

As has been the case for other molluscan groups, it is likely that increasing the depth of detail and comparative breadth of our understanding of scaphopod biology will be motivated in large part by resolving phylogenetic relationships within the Scaphopoda and among the Mollusca. Having moved beyond dependence on shell appearance to document diversity of form and infer phylogenetic relationships, cladistic analyses are revealing a more complex picture of scaphopod evolution than previously envisioned (e.g., Steiner, 1992b; Reynolds and Okusu, 1999). While molecular approaches will probably provide significant progress in scaphopod phylogenetics, comparative non-molecular data have proven useful in resolving relationships, and are needed to adequately address questions of morphological, physiological, developmental and behavioral character evolution. An even greater challenge will be to explore mechanisms that have influenced the evolution of form, ecological role and historical biogeography in this enigmatic molluscan class.

The Scaphopoda present a singular opportunity in the study of molluscan evolution. First, they constitute a monophyletic group of mollusks at the class level, which is yet small enough for comprehensive phylogenetic analysis to be tractable and for biological diversity to be thoroughly documented. Secondly, as the most recently evolved class of mollusks, it is likely to have a more complete fossil record than other molluscan classes for examining diversification patterns since their origin. These opportunities for studying a clade evolutionarily constrained in many respects, but substantially derived from potential conchiferan sister groups, have not yet been fully exploited.

ACKNOWLEDGEMENTS

I would like to thank K. Bart, D. McHugh, D. Tewksbury, T. Gore, H. Lehman, E. Domack, D. Catlin, C. Young, M. Guyot-Bender, S. Cook and A. Raybeck for their assistance at various stages during the preparation of this review, and especial thanks to the Reference and Interlibrary Loan staff of the Burke Library, Hamilton College, for their considerable work on my behalf over several years. G. Haszprunar, B. Ruthensteiner, R. Shimek, G. Steiner and A. Wanninger graciously provided unpublished data of work in progress. I am grateful to G. Steiner, A. J. Southward, R. Shimek, D. McHugh and E. Southward for reading earlier versions of this manuscript and for making valuable suggestions for its improvement. This review, and previously unpublished data of the author, were supported by NSF grant DEB-9707400.

REFERENCES

Alroy, J., Marshall, C. R., Bambach, R. K., Bezusko, K., Foote, M., Fursich, F. T., Hansen, T. A., Holland, S. M., Ivany, L. C., Jablonski, D., Jacobs, D. K., Jones, D. C., Kosnik, M. A., Lidgard, S., Low, S., Miller, A. I., Novack-Gottshall, P. M., Olszewski, T. D., Patzkowsky, M. E., Raup, D. M., Roy, K., Sepkoski, J. J., Sommers, M. G., Wagner, P. J. and Webber, A. (2001). Effects of sampling standardization on estimates of Phanerozoic marine diversification. *Proceedings of the National Academy of Sciences of the United States of America* **98**, 6261–6266.

Altner, H. and Prillinger, L. (1980). Ultrastructure of invertebrate chemo-, thermo-, and hygroreceptors and its functional significance. *International Review of Cytology* **67**, 69–139.

Alzuria, M. (1984). Nota sobre la fraccion mineral en *Dentalium mutabile inaequicostatum* (Dautzemberg 1891) (Mollusca, Scaphopoda). *Publicaciones del Departamento de Zoología. Ediciones de la Universidad de Barcelona* **10**, 23–25.

Alzuria, M. (1985a). Ultrastructura de la concha en *Dentalium vulgare* (De Costa 1778) (Mollusca: Scaphopoda). *Iberus* **5**, 11–19.

Alzuria, M. (1985b). Ultrastructura de la concha en *Dentalium mutabile inaequicostatum* (Dautzemberg, 1891) (Mollusca; Scaphopoda). *Publicaciones del Departamento de Zoología. Ediciones de la Universidad de Barcelona* **11**, 15–22.

Andrews, E. B. (1976). The ultrastructure of the heart and kidney of the pilid gastropod mollusc *Marisa cornuarietis*, with special reference to filtration throughout the Architaenioglossa. *Journal of Zoology* **179**, 85–106.

Andrews, E. B. (1979). Fine structure in relation to function in the excretory system of two species of *Viviparus. Journal of Molluscan Studies* **45**, 186–206.

Andrews, E. B. (1981). Osmoregulation and excretion in prosobranch gastropods. Part 2: structure in relation to function. *Journal of Molluscan Studies* **47**, 248–289.

Andrews, E. B. (1985). Structure and function in the excretory system of archaeogastropods and their significance in the evolution of gastropods. *Philosophical Transactions of the Royal Society of London, Series B* **310**, 383–406.

Andrews, E. B. (1988). Excretory systems of Molluscs. *In* "The Mollusca, Volume 11, Form and Function" (E. R. Trueman and M. R. Clarke, eds), pp. 381–448. Academic Press, New York.

Andrews, E. B. and Jennings, K. H. (1993). The anatomical and ultrastructural basis of primary urine formation in bivalve molluscs. *Journal of Molluscan Studies* **59**, 223–257.

Årpád, D. (1993). Trace fossils on molluscs from the Molluscan Clay (late Oligocene, Egerian) – a comparison between two localities (Wind Brickyard, Eger, and Nyárjas Hill, Novaj, NE Hungary). *Scripta Geologica, Special Issue* **2**, 75–82.

Arvy, L. (1949). Sur la castration parasitaire chez *Dentalium entale* Deshayes. *Comptes Rendus Hebdomadaires des Séances de l'Academie des Sciences. Series D: Science Naturelles, Paris* **229**, 780–782.

Arvy, L. (1950). Donnés histologiques sur l'ovogénèse chez *Dentalium entale* Deshayes. *Archives de Biologie, Liège* **61**, 187–195.

Arvy, L. and Gabe, M. (1951). Donnés morphologiques sur le sang du Dentale. *Bulletin du Laboratoire maritime de Dinard* **35**, 15–22.

Baba, K. (1937). Contribution to the knowledge of a nudibranch, *Okadaia elegans* Baba. *Japanese Journal of Zoology* **7**, 147–190.

Baba, K. (1940). The mechanisms of absorption and excretion in a solenogastre, *Epimenia verrucosa* (Nierstrasz), studied by means of injection methods. *Journal of the Department of Agriculture, Kyushu Imperial University* **6**, 119–166.

Barnes, R. S. K., Calow, P. and Olive, P. J. W. (1993). "The Invertebrates: a New Synthesis". Blackwell Scientific, Oxford.

Bernasconi, M. P. (1989). Studi paleoecologici Pliocene ligure 5. Il Pliocene di Savona. *Bolletino della Museo Regionale di Scienze Naturali, Turin* **7**, 49–116.

Bernasconi, M. P. (1990). Some Pliocene scaphopods: controlling factors and significance within paleocommunities. *In* "Proceedings of the Fourth Symposium on Ecology and Paleoecology of Benthic Communities" (E. Robba, ed.), pp. 357–372. Museo Regionale di Scienze Naturali, Turin.

Bilyard, G. R. (1974). The feeding habits and ecology of *Dentalium entale stimpsoni* Henderson (Mollusca: Scaphopoda). *The Veliger* **17**, 126–138.

Boer, H. H. and Sminia T. (1976). Sieve structure of slit diaphragms of podocytes and pore cells of gastropod molluscs. *Cell and Tissue Research* **170**, 221–229.

Bøggild, O. B. (1930). The shell structure of the mollusks. *Det Kongelige Danske Videnskabernes Selskabs Skrifter, Naturvidenskabelig og Mathematisk Afdeling,* **9** 2, 231–326, 15 plates.

Boissevain, M. (1904). Beiträge zur anatomie und Histologie von *Dentalium. Jenaische Zeitschrift für Naturwissenschaft* **38**, 553–572.

Bonaventura, C. and Bonaventura, J. (1983). Respiratory pigments: structure and function. *In* "The Mollusca, Volume 2, Environmental Biochemistry and Physiology" (P. W. Hochachka, ed.), pp. 1–50. Academic Press, New York.

Boss, K. J. (1982). Phylum Mollusca. *In* "Synopsis and Classification of Living Organisms" (S. P. Parker, ed.), pp. 945–1166. McGraw-Hill, New York.

Bretsky, P. W. and Bermingham, J. J. (1970). Ecology of the Paleozoic scaphopod genus *Plagioglypta* with special reference to the Ordovician of eastern Iowa. *Journal of Paleontology* **44**, 908–924.

Buckland-Nicks, J. A., Gibson, G. and Koss, R. (2002). Phylum Mollusca: Polyplacophora, Aplacophora, Scaphopoda. *In* "An Atlas of Invertebrate Larvae" (C. M. Young, ed.). Academic Press, London.

Byrum, C. A. and Ruppert, E. E. (1994). The ultrastructure and functional morphology of a captaculum in *Graptacme calamus* (Mollusca, Scaphopoda). *Acta Zoologica* **75**, 37–46.

Carlgren, O. (1928). Zur Symbiose zwischen Actinien und Paguriden. *Zeitschrift für Morphologie und Ökologie der Tiere* **12**, 165–173.

Carter, J. G. and Hall, R. M. (1990). Polyplacophora, Scaphopoda, Archaeogastropoda and Paragastropoda (Mollusca). *In* "Skeletal Biomineralization: Patterns, Processes and Evolutionary Trends" (J. G. Carter, ed.), pp. 29–52, plates 122–134. Van Nostrand Reinhold, New York.

Cather, J. N. and Verdonk, N. H. (1979). Development of *Dentalium* following removal of D-quadrant blastomeres at successive cleavage stages. *Wilhelm Roux's Archives of Developmental Biology* **187**, 355–366.

Cayeux, L. (1916). "Introduction à l'étude pétrographique des roches sédimentaires: Mémoire pour Servir à l'Explication de la Carte Géologique Détaillée de la France". Ministere des Travaux Publics, Imprimerie Nationale, Paris.

Chistikov S. D. (1975). Some problems of scaphopod taxonomy. *In* "U.S.S.R. Academy of Sciences Zoological Institute. Fifth Meeting on the Investigation of Molluscs. Molluscs: Their System, Evolution and Significance in Nature. Theses of communications" (I. M. Likharev, ed.), pp. 21–23, Nauka, Leningrad [In Russian].

Chistikov, S. D. (1978). Some problems in the classification of the order Dentaliida (Mollusca; Scaphopoda). *Malacological Review* **11**, 71–73.

Chistikov S. D. (1979). Phylogenetic relations of the scaphopods. *In* "U.S.S.R. Academy of Sciences Zoological Institute. Sixth Meeting on the Investigation of Molluscs. Molluscs: Main Results of Their Study. Abstracts of communications." (I. M. Likharev, ed.), pp. 20–22. Nauka, Leningrad [in Russian].

Chistikov, S. D. (1984). Phylogenetic relationships of the Scaphopoda. *Malacological Review* **17**, 114–115.

Clark, G. R. I. and West, R. R. (1984). Microstructure and morphology of Pennsylvanian specimens of the "scaphopod" *Plagioglypta. Geological Society of America Abstracts* **16**, 472.

Clark, W. (1849). On the animal of *Dentalium tarentinum. Annals and Magazine of Natural History, 2nd Series* **4,** 321–330.

Cribb, T. H., Bray, R. A. and Littlewood, D. T. J. (2001). The nature and evolution of the association among digeneans, molluscs and fishes. *International Journal for Parasitology* **31**, 997–1011.

Crisp, M. (1971). Structure and abundance of receptors of the unspecialized external epithelium of *Nassarius reticulatus* (Gastropoda, Prosobranchia). *Journal of the Marine Biological Association of the United Kingdom* **51**, 865–890.

Cuénot, L. (1899). L'excretion chez les Mollusques. *Archives de Biologie, Liège* **16**, 49–96.

Damen, W. G. M., Klerkx, A. H. E. M. and Van Loon, A. E. (1997). Cell-specific regulation in early molluscan development. *Invertebrate Reproduction and Development* **31**, 1–3.

D'Anna, T. (1974). Ermafroditismo in *Dentalium entalis. Atti della Accademia Nazionale dei Lincei. Classe di Scienze Fisiche, Matematiche e Naturali. Rendiconti, Serie 8* **57**, 673–677.

Davies, G. J. (1987). Aspects of the Biology and Ecology of Deep-Sea Scaphopoda (Mollusca). PhD dissertation, Heriot-Watt University, Edinburgh.

Davis, J. D. (1968). A note on the behavior of the scaphopod, *Cadulus quadridentatus* (Dall) 1881. *Proceedings of the Malacological Society of London* **38**, 135–138.

Délage, Y. (1899). Études sur la merogonie. *Archives de Zoologie Expérimentale et Générale* **7**, 383–417.

Dell, R. K. (1957). A revision of the recent scaphopod Mollusca of New Zealand. *Transactions of the Royal Society of New Zealand* **84**, 561–576.

Deshayes, G. P. (1825). Anatomie et monographie du genre Dentale. *Mémoires de la Society d'histoire naturelle de Paris* **2**, 321–378.

Dinamani, P. (1964a). Burrowing behaviour of *Dentalium. Biological Bulletin* **126**, 28–32.

Dinamani, P. (1964b). Feeding in *Dentalium conspicuum. Proceedings of the Malacological Society of London* **36**, 1–5.

Distaso, A. (1905a). Sull' Anatomia degli scafopodi. *Zoologischer Anzeiger* **29**, 271–278.

Distaso, A. (1905b). Sull sistema nervoso di *Dentalium entalis* Desh. (= *vulgare* Costa). *Bollettino della Società dei Naturalisti in Napoli* **18**, 177–184.

Dommergues, J. L., Laurin, B. and Meister, C. (1996). Evolution of ammonoid morphospace during the Early Jurassic radiation. *Paleobiology* **22**, 219–240.

Dufresne-Dubé, L., Picheral, B. and Guerrier, P. (1983). An ultrastructural analysis of *Dentalium vulgare* (Mollusca, Scaphopoda) gametes with special reference to early events at fertilization. *Journal of Ultrastructural Research* **83**, 242–257.

Edlinger, K. (1991). Zur Evolution der Scaphopoden-Konstruktion. *Natur und Museum* **121**, 116–122.

Emerson, W. K. (1962). A classification of the scaphopod mollusks. *Journal of Paleontology* **36**, 461–482.

Engeser, T. S. and Riedel, F. (1996). The evolution of the Scaphopoda and its implications for the systematics of the Rostroconchia (Mollusca). *Mitteilungen aus dem Geologisch-Paläontologischen Institut der Universität Hamburg* **79**, 117–138.

Engeser, T. S., Riedel, F. and Bandel, K. (1993). Early ontogenetic shells of Recent and fossil Scaphopoda. *Scripta Geologica* Special Issue **2**, 83–100.

Fantinet, D. (1959). Contribution à l'étude des Scaphopodes fossiles de l'Afrique du Nord. *Publications du Service de la Carte Géologique de l'Algérie (Nouvelle Série), Paléontologie, Mémoire* **1**, 1–112.

Fauchald, K. and Jumars, P. A. (1979). The diet of worms: a study of polychaete feeding guilds. *Oceanography and Marine Biology: an Annual Review* **17**, 193–284.

Fiedler, A. and Schipp R. (1987). The role of the branchial heart complex in circulation of coleoid cephalopods. *Experientia* **43**, 544–553.

Fischer-Piette, E. and Franc, A. (1968). Classe des Scaphopodes. Scaphopoda (Bronn 1862). *In* "Mollusques, Gastéropodes et Scaphopodes" (P.-P. Grassé, ed.), pp. 987–1017. Masson et Cie, Paris.

Fol, H. (1885). Sur l'anatomie microscopique du Dentale. *Comptes Rendu Hebdomadaires des Seances de l'Academie des Sciences. Serie D, Sciences Naturelles* **1885**, 1352–1355.

Fol, H. (1889). Sur l'anatomie microscopique du Dentale. *Archives de Zoologie Expérimentale et Générale, 2nd Série* **7**, 91–148.

Forest, J. (1987). Ethology and distribution of Pylochelidae (Crustacea Decapoda Coenobitoidea). *Bulletin of Marine Science* **41**, 309–321.

Gabe, M. (1956). La neuro-sécrétion chez les invertébrés. *Année Biologique* **30**, 5–62.

Gabe, M. and Prenant, M. (1950). Reserches sur la gaine radulaire des Mollusques. I. – La gaine radulaire de *Dentalium entale* Deshayes. *Archives de Zoologie Expérimentale et Générale* **86**, 487–498.

Gage, J. D. (1996). Why are there so many species in deep-sea sediments? *Journal of Experimental Marine Biology and Ecology* **200**, 257–286.

Gaillard, J. M. (1956). Variations des peuplements de *Philine aperta* L. et de *Dentalium vulgare* da Costa. Relations avec les parasites de ces espèces. *Bulletin du Laboratoire maritime de Dinard* **42**, 86.

Gainey, L. F. J. (1972). The use of the foot and the captacula in the feeding of *Dentalium* (Mollusca: Scaphopoda). *Veliger* **15**, 29–34.

Garton, D. W., Roller, R. A. and Caprio, J. 1984. Fine structure and vital staining of osphradium of the southern oyster drill, *Thais haemastoma canaliculata* (Gray) (Prosobranchia: Muricidae). *Biological Bulletin, Marine Biological Laboratory, Wood's Hole* **167**, 310–321.

Geilenkirchen, W. L. M., Verdonk, N. H. and Timmermans, L. P. M. (1970). Experimental studies on morphogenetic factors localized in the first and the second polar lobe of *Dentalium* eggs. *Journal of Embryology and Experimental Morphology* **23**, 237–243.

Geilenkirchen, W. L. M., Timmermans, L. P. M., Van Dongen, C. A. M. and Arnolds, W. J. A. (1971). Symbiosis of bacteria with eggs of *Dentalium* at the vegetal pole. *Experimental Cell Research* **67**, 477–478.

George, S. G. (1982). Subcellular accumulation and detoxication of metals in aquatic animals. *In* "Physiological Mechanisms of Marine Pollutant Toxicity" (W. B. Vernberg, A. Calabrese, F. P. Thurberg and F. J. Vernberg, eds), pp. 3–52. Academic Press, New York.

George, S. G. and Pirie, B. J. S. (1980). Metabolism of zinc in the mussel, *Mytilus edulis* (L.): a combined ultrastructural and biochemical study. *Journal of the Marine Biological Association of the United Kingdom* **60**, 575–590.

George, S. G., Pirie, B. D. S. and Coombs, T. L. (1975). Transport of iron complexes in shellfish. *In* "International Conference on Heavy Metals in the Environment" (T. C. Hutchinson, ed.), pp. C37–C39. Electrical Power Research Institute, Toronto.

George, S. G., Pirie, B. J. S. and Coombs, T. L. (1976). The kinetics of accumulation and excretion of ferric hydroxide in *Mytilus edulis* (L.) and its distribution in the tissues. *Journal of Experimental Marine Biology and Ecology* **23**, 71–84.

George, S. G., Pirie, B. J. S., Cheyne, A. R., Coombs, T. L. and Grant, P. T. (1978). Detoxification of metals by marine bivalves: an ultrastructural study of the compartmentation of copper and zinc in the oyster *Ostrea edulis*. *Marine Biology* **45**, 147–156.

George, S. G., Pirie, B. J. S. and Coombs, T. L. (1980). Isolation and elemental analysis of metal rich granules from the kidney of the scallop, *Pecten maximus* (L.). *Journal of Experimental Marine Biology and Ecology* **42**, 143–156.

George, S. G., Coombs, T. L. and Pirie, B. J. S. (1982). Characterization of metal-containing granules from the kidney of the common mussel, *Mytilus edulis*. *Biochimica et Biophysica Acta* **716**, 61–71.

Grassle, J. F. and Maciolek, N. J. (1992). Deep-sea species richness: regional and local diversity estimates from quantitative bottom samples. *American Naturalist* **139**, 313–341.

Guerrier, P., Van Den Biggelaar, J. A. M., Van Dongen, C. A. M. and Verdonk, N. H. (1978). Significance of the polar lobe for the determination of dorso ventral polarity in *Dentalium vulgare*. *Developmental Biology* **63**, 233–242.

Haas, W. (1972). Micro- and ultrastructure of Recent and fossil Scaphopoda. *In* "International Geological Congress, 24th Session. Section 7 Paleontology [Proceedings]" (J. E. Gill, ed.), pp. 15–19. Harpell's Press, Québec.

Habe, T. (1964). "Fauna Japonica. Scaphopoda (Mollusca)". Biogeographical Society of Japan, Tokyo.

Harrison, F. M. (1962). Some excretory processes in the abalone, *Haliotis rufescens*. *Journal of Experimental Biology* **39**, 179–192.

Harrison, F. M. and Martin, A. W. (1965). Excretion in the cephalopod *Octopus dofleini*. *Journal of Experimental Biology* **42**, 71–98.

Haszprunar, G. (1987a). The fine morphology of the osphradial sense organs of the Mollusca. III. Placophora and Bivalvia. *Philosophical Transactions of the Royal Society of London. B* **315**, 37–61.

Haszprunar, G. (1987b). The fine structure of the ctenidial sense organs (bursicles) of Vetigastropoda (Zeugobranchia, Trochoidea) and their functional and phylogenetic significance. *Journal of Molluscan Studies* **53**, 46–51.

Haszprunar, G. (2000). Is the Aplacophora monophyletic? A cladistic point of view. *American Malacological Bulletin* **15**, 115–130.

Hebert, A. (1986). Reproductive behavior and anatomy of three central Californian scaphopods. MS thesis, California State University, Hayward, USA.

Henderson, J. B. (1920). "A monograph of the East American scaphopod mollusks". *Smithsonian Institution, U.S. National Museum Bulletin* **111**, 1–177.

Hessler, R. and Sanders, H. L. (1967). Faunal diversity in the deep-sea. *Deep-Sea Research* **14**, 65–78.

Hevert, F. (1984). Urine formation in the lamellibranchs: evidence for ultra-filtration and quantitative description. *Journal of Experimental Biology* **111**, 1–12.

Hodgson, A. N. and Fielden, L. J. (1984). The structure and distribution of peripheral ciliated receptors in the bivalve molluscs *Donax serra* and *D. sordidus*. *Journal of Molluscan Studies* **50**, 104–112.

Hodgson, A. N. and Fielden, L. J. (1986). The ultrastructure of the ciliated cells from the siphon of *Solen capensis* (Mollusca, Bivalvia). *Journal of Molluscan Studies* **52**, 161–168.

Hou, S. T. and Maxwell, W. L. (1991). Ultrastructural studies of spermatogenesis in *Antalis entalis* (Scaphopoda, Mollusca). *Philosophical Transactions of the Royal Society of London.* B **333**, 101–110.

Hyman, L. H. (1967). "The Invertebrates: Mollusca I". McGraw-Hill, New York.

Imafuku, M. and Ando, T. (1999). Behaviour and morphology of the paguird hermit crabs (Decapoda, Anomura) that live in tusk shells (Mollusca, Scaphopoda). *Crustaceana* **72**, 129–142.

Ivanov, D. L. and Memmi, M. P. (1989). Origin of canals in shell of Scaphopoda (Mollusca). *Doklady Akademii Nauk SSSR* **306**, 247–249 [in Russian].

Jacobs, D. K., Wray, C. G., Wedeen, C. J., Kostriken, R., DeSalle, R., Staton, J. L., Gates, R. D. and Lindberg, D. R. (2000). Molluscan *engrailed* expression, serial organization, and shell evolution. *Evolution and Development* **2**, 340–347.

Jaffe, L. A. and Guerrier, P. (1981). Localization of electrical excitability in the early embryo of *Dentalium. Developmental Biology* **83**, 370–373.

Jensen, H. and Tjønneland A. (1977). Ultrastructure of the heart muscle cells of the cuttlefish *Rossia macrosoma. Cell and Tissue Research* **185**, 147–158.

Jones, H. D. (1983). The circulatory system of gastropods and bivalves. *In* "The Mollusca, Volume 5, Physiology, Part 2" (A. S. M. Salueddin and K. M. Wilbur, eds), pp. 189–238. Academic Press, New York.

Jones, H. D. and Peggs, D. (1983). Hydrostatic and osmotic pressures in the heart and pericardium of *Mya arenaria* and *Anodonta cygnea. Comparative Biochemistry and Physiology* **76A**, 381–385.

Knudsen, J. (1964). Scaphopoda and Gastropoda from depths exceeding 6,000 metres. *Galathea Report* **7**, 125–136.

Kollman, M. (1908). Recherches sur les leucocytes et le tissue lymphoïde des Invertebrés. PhD dissertation, Paris.

Koster, A. S. J. and Van den Biggelaar, J. A. M. (1980). Abnormal development of *Dentalium* due to the *Amoco Cadiz* oil spill. *Marine Pollution Bulletin* **11**, 166–169.

Kowalevsky, A. (1883). Étude sur l'embryogénie du Dentale. *Annales de la Museum d'Histoire Naturelle de Marseille* **1**, 1–54.

Kowalevsky, A. (1889). Ein Beitrag zur Kenntnis der Exkretionsorgane. *Biologisches Zentralblatt (Leipzig)* **9**, 65–76.

Kozloff, E. N. (1990). "Invertebrates". Saunders, Philadelphia.

Kudo, R. R. (1954). "Protozoology", 4th edition. Charles C. Thomas, Springfield, Il.

Kümmel, G. (1973). Filtration structures in excretory systems. A comparison. *In* "Comparative Physiology: Locomotion, Respiration, Transport and Blood" (L. Bolis, K. Schmidt-Nielsen and S. H. P. Maddrell, eds), pp. 221–240. North-Holland, Amsterdam.

La Perna, R. (1998). The larval shells of *Graptacme agilis, Entalina tetragona* and *Pulsellum lofotense* (Scaphopoda) from the Mediterranean. *Bollettino Malacologico* **33**, 123–126.

Lacaze-Duthiers, H. (1856–1857). Histoire de l'organisation et du développement du Dentale. *Annales des Sciences Naturelles, Quatrième Serie, Paris* Tome **6**,

225–281, plates 8–10; 319–385, plates 11–13; Tome **7**, 5–51, plates 2–4; 171–255, plates 5–9; Tome **8**, 18–44.

Lacaze-Duthiers, H. de (1885). Note sur l'anatomie du Dentale. *Comptes Rendus Hebdomadaires des Sceances de l'Academie des Sciences, Paris* **1885**, 296–300.

Lamprell, K. L. and Healy, J. M. (1998). A revision of the Scaphopoda from Australian waters (Mollusca). *Records of the Australian Museum* Supplement **24**, 1–189.

Langer, M. R., Lipps, J. H. and Guillermo, M. (1995). Predation on foraminifera by the dentaliid deep-sea scaphopod *Fissidentalium megathyris. Deep-Sea Research, Part I. Oceanographic Research Papers* **42**, 849–857.

Léon, N. (1894). Zur Histologie des *Dentalium*-mantels. *Jenaische Zeitschrift Medizin und Naturwissenschaft* **29**, 411–416.

Little, C. (1965). The formation of urine by the prosobranch gastropod mollusc *Viviparus viviparus* Linn. *Journal of Experimental Biology* **43**, 39–54.

Little, C. (1979). Reabsorption of glucose in the renal system of *Viviparus. Journal of Molluscan Studies* **45**, 207–208.

Ludbrook, N. H. (1960). Scaphopoda. *In* "Treatise on Invertebrate Paleontology" (R. C. Moore, ed.), pp. I37–I41. Geological Society of America, Lawrence, Kansas.

Manwell, C. (1963). The chemistry and biology of hemoglobin in some marine clams, 1, Distribution of the pigment and properties of the oxygen equilibrium. *Comparative Biochemistry and Physiology* **8**, 209–218.

Martin, A. W. (1983). Excretion. *In* "The Mollusca, Volume 5, Physiology, Part 2" (A. S. M. Saleuddin and K. M. Wilbur, eds), pp. 353–405. Academic Press, New York.

Martin, A. W. and Aldrich, F. A. (1970). Comparison of hearts and branchial heart appendages in some cephalopods. *Canadian Journal of Zoology* **48**, 751–756.

Mason, A. Z., Simkiss, K. and Ryan, K. P. (1984). The ultrastructural localization of metals in specimens of *Littorina littorea* collected from clean and polluted sites. *Journal of the Marine Biological Association of the United Kingdom* **64**, 699–720.

McFadien, M. S. (1973). Zoogeography and ecology of seven species of Panamic-Pacific Scaphopoda. *The Veliger* **15**, 340–347.

McFadien-Carter, M. (1979). Scaphopoda. *In* "Reproduction of Marine Invertebrates. Volume 5. Molluscs: Pelecypods and Lesser Classes" (A. C. Giese and J. S. Pearse, eds), pp. 95–111. Academic Press, New York.

Meyhöfer, E., Morse, M. P. and Robinson, W. E. (1985). Podocytes in bivalve molluscs: morphological evidence for ultrafiltration. *Journal of Comparative Physiology* B **156**, 151–161.

Moor, B. (1983). Organogenesis. *In* "The Mollusca, Volume 3, Development." (N. H. Verdonk, J. A. M. Van den Biggelaar and A. S. Tompa, eds), pp. 123–177. Academic Press, New York.

Moreau, M., Guerrier, P. and Dufresne, L. (1989). Absence of an electrical block to polyspermy in the scaphopod mollusk *Dentalium vulgare. Journal of Experimental Zoology* **249**, 76–84.

Morse, M. P. and Reynolds, P. D. (1996). Ultrastructure of the heart–kidney complex in smaller classes supports symplesiomorphy of molluscan coelomic characters. *In* "Origin and Evolutionary Radiation of the Mollusca" (J. D. Taylor, ed.), pp. 89–97. Oxford University Press, Oxford.

Morse, M. P. and Zardus, J. D. (1997). Bivalvia. *In* "Microscopic Anatomy of Invertebrates, Volume 6A, Mollusca I" (F. W. Harrison and A. J. Kohn, eds), pp. 7–118. Wiley-Liss, New York.

Morse, M. P., Meyhöfer, E., Otto, J. J. and Kuzirian, A. M. (1986). Hemocyanin respiratory pigment in bivalve molluscs. *Science* **231**, 1302–1304.

Morton, J. E. (1959). The habits and feeding organs of *Dentalium entalis*. *Journal of the Marine Biological Association of the United Kingdom* **38**, 225–238.

Moshel, S. M., Levine, M. and Collier, J. R. (1998). Shell differentiation and *engrailed* expression in the *Ilyanassa* embryo. *Development, Genes and Evolution* **208**, 135–141.

Narain, A. S. (1976). A review of the structure of the heart of molluscs, particularly bivalves, in relation to cardiac function. *Journal of Molluscan Studies* **42**, 46–62.

Odhner, N. H. (1931). Die Scaphopoden. *Further Zoological Results of the Swedish Antarctic Expedition, 1901–1903* **2**, 1–8, 2 plates.

Økland, S. (1980). The heart ultrastructure of *Lepidopleurus asellus* (Spengler) and *Tonicella marmorea* (Fabricus) (Mollusca: Polyplacophora). *Zoomorphology* **96**, 1–19.

Økland, S. (1981). Ultrastructure of the pericardium in chitons (Mollusca: Polyplacophora), in relation to filtration and contraction mechanisms. *Zoomorphology* **97**, 193–203.

Økland, S. (1982). The ultrastructure of the heart complex in *Patella vulgata* L. (Archaeogastropoda, Prosobranchia). *Journal of Molluscan Studies* **48**, 331–341.

Owen, G. (1966). Feeding. *In* "Physiology of Mollusca, Volume II" (K. M. Wilbur and C. M. Yonge, eds), pp. 1–51. Academic Press, New York.

Owen, G. (1972). Lysosomes, peroxisomes and bivalves. *Science Progress Series, Oxford* **60**, 299–318.

Palmer, C. P. (1974). A supraspecific classification of the scaphopod Mollusca. *The Veliger* **17**, 115–123.

Palmer, C. P. and Steiner, G. (1998). Class Scaphopoda. *In* "Mollusca: The Southern Synthesis, Volume 5" (P. L. Beesley, G. J. B. Ross and A. Wells, eds), pp. 431–450. Australian Biological Resources Study, Canberra.

Pelseneer, P. (1901). Recherches morphologiques et phylogenetiques sur les mollusques archaiques. *Academie Royale des Sciences, des Lettres et des Beaux-Arts de Belgique. Memoires Couronnes et Memoires des Savants Etrangers. Brussels* **57**, 1–113, 24 plates.

Pelseneer, P. (1906a). Trématodes parasites de mollusques marins. *Bulletin Biologique de la France et de la Belgique* **40**, 161–186.

Pelseneer, P. (1906b). Mollusca. In "A Treatise on Zoology, Part V" (E. R. Lankester, ed.), pp. 1–355. Adam & Charles Black, London.

Pentreath, R. J. (1973). The accumulation from water of ^{65}Zn, ^{54}Mn, ^{58}Co and ^{59}Fe by the mussel, *Mytilus edulis*. *Journal of the Marine Biological Association of the United Kingdom* **53**, 127–143.

Pilsbry, H. A. and Sharp, B. (1898). Scaphopoda, Aplacophora. *Tryon's Manual of Conchology* **17**, 145–348.

Pirie, B. J. S. and George S. G. (1979). Ultrastructure of the heart and excretory system of *Mytilus edulis* (L.). *Journal of the Marine Biological Association of the United Kingdom* **59**, 819–829.

Plate, L. (1888). Bemerkungen zur Organisation der Dentalien. *Zoologischer Anzeiger* **11**, 509–515.

Plate, L. (1891a). Über einiger Organizations verhältnisse der Dentalien. *Sitzungsberichte der Gesellschaft zur Beförderung der gesammten Naturwissenschaften zu Marburg* **1890**, 26–29.

Plate, L. H. (1891b). Über das Herz der Dentalien. *Zoologischer Anzeiger* **14**, 78–80.

Plate, L. (1891c). Über den Bau und die systematische Stellung der Solenoconchen. *Verhandlungen Deutsche Zoologische Gesellschaft* **1**, 60–66.
Plate, L. H. (1892). Über den Bau und die Verwandtschaftsbeziehungen der Solenoconchen. *Zoologische Jahrbücher. Abteilung für Anatomie und Ontogenie der Tiere* **5**, 301–386.
Pojeta, J., Jr. (1987). Class Scaphopoda. *In* "Fossil Invertebrates" (R. S. Boardman, A. H. Cheetham and A. J. Rowell, eds), pp. 380–386. Blackwell Scientific Publications, Palo Alto.
Pojeta, J., Jr. and Runnegar, B. (1979). *Rhytiodentalium kentuckyensis*, a new genus and new species of Ordovician scaphopod, and the early history of scaphopod mollusks. *Journal of Paleontology* **53**, 530–541.
Pojeta, J., Jr. and Runnegar, B. (1985). The early evolution of diasome molluscs. *In* "The Mollusca, Volume 10, Evolution" (E. R. Trueman and M. R. Clark, eds), pp. 295–336. Academic Press, New York.
Ponder, W. F. and Lindberg, D. R. (1997). Towards a phylogeny of gastropod molluscs: an analysis using morphological characters. *Zoological Journal of the Linnean Society* **119**, 83–265.
Poon, P. A. (1987). The diet and feeding behavior of *Cadulus tolmiei* Dall, 1898 (Scaphopoda: Siphonodentalioida). *Nautilus* **101**, 88–92.
Potts, W. T. W. (1967). Excretion in the Molluscs. *Biological Reviews of the Cambridge Philosophical Society* **42**, 1–41.
Purchon, R. D. (1977). "Biology of the Mollusca", 2nd edition. Pergamon, Oxford.
Raven, C. P. (1966). "Morphogenesis: the Analysis of Molluscan Development", 2nd edition. Pergamon, Oxford.
Raup, D. (1965). Theoretical morphology of the coiled shell. *Science* **147**, 1294–1295.
Render, J. A. and Guerrier, P. (1984). Size regulation and morphogenetic localization in the *Dentalium* polar lobe. *Journal of Experimental Zoology* **232**, 79–86.
Reverberi, G. (1969). Il primo lobo polare dell'uovo di *Dentalium* al microscopio Elettronico. *Atti della Accademia Nazionale dei Lincei. Classe di Scienze Fisiche, Matematiche e Naturali. Rendiconti, Serie 8* **47**, 557–560.
Reverberi, G. (1970a). The ultrastructure of *Dentalium* egg at the trefoil stage. *Acta Embryologiae Experimentalis* **1970**, 31–43.
Reverberi, G. (1970b). The ultrastructure of the ripe oocyte of *Dentalium. Acta Embryologiae Experimentalis* **1970**, 255–279.
Reverberi, G. (1971). *Dentalium. In* "Experimental Embryology of Marine and Fresh-Water Invertebrates" (G. Reverberi, ed.), pp. 248–264. North-Holland, Amsterdam.
Reverberi, G. (1972). The fine structure of the ovaric egg of *Dentalium. Acta Embryologiae Experimentalis* **1972**, 135–166.
Rex, M. A. (1973). Deep-sea species diversity: decreased gastropod diversity at abyssal depths. *Science* **181**, 1051–1053.
Rex, M. A. (1981). Community structure in the deep-sea benthos. *Annual Review of Ecology and Systematics* **12**, 331–353.
Rex, M. A. (1983). 11. Geographic patterns of species diversity in the deep-sea benthos. *In* "Deep-Sea Biology" (G. T. Rowe, ed.), pp. 453–472. John Wiley and Sons, New York.
Reynolds, P. D. (1990a). "Functional morphology of the heart/kidney complex, digestive system and mantle of *Dentalium rectius* (Mollusca, Scaphopoda)". PhD dissertation, University of Victoria, Canada.

Reynolds, P. D. (1990b). Functional morphology of the perianal sinus and pericardium of *Dentalium rectius* (Mollusca: Scaphopoda) with a reinterpretation of the scaphopod heart. *American Malacological Bulletin* **7**, 137–149.

Reynolds, P. D. (1990c). Fine structure of the kidney and characterization of secretory products in *Dentalium rectius* (Mollusca, Scaphopoda). *Zoomorphology* **110**, 53–62.

Reynolds, P. D. (1992a). Mantle-mediated shell decollation increases posterior aperture size in *Dentalium rectius* (Scaphopoda: Dentaliida). *The Veliger* **35**, 26–35.

Reynolds, P. D. (1992b). Distribution and ultrastructure of ciliated sensory receptors in the posterior mantle epithelium of *Dentalium rectius* (Mollusca, Scaphopoda). *Acta Zoologica* **73**, 263–270.

Reynolds P. D. (1997). The phylogeny and classification of Scaphopoda (Mollusca): an assessment of current resolution and cladistic reanalysis. *Zoologica Scripta* **26**, 13–26.

Reynolds, P. D. (1998). Contribution of morphological, molecular, and fossil data toward an understanding of scaphopod phylogeny. *In* "Abstracts, World Congress of Malacology, Washington, D.C., 1998" (E. Bieler and P. M. Mikkelsen, eds), p. 274. Unitas Malacologica, Chicago [abstract].

Reynolds, P. D. and Maguire, K. (1998). Stratigraphic and biogeographic patterns of the Order Dentaliida, Class Scaphopoda (Mollusca). *American Zoologist* **38**, 139A [abstract].

Reynolds, P. D. and Okusu, A. (1999). Phylogenetic relationships among families in the Class Scaphopoda (Phylum Mollusca). *Zoological Journal of the Linnean Society* **126**, 131–154.

Reynolds, P. D. and Peters, J. (1997). Molecular and morphological phylogenetic systematics of the Scaphopoda (Mollusca). *American Zoologist* **37** (5), 54A [abstract].

Reynolds, P. D., Gray, N. and Hilpert, L. P. (2001). Latitudinal species diversity and range-size gradients in the Scaphopoda. *In* "Abstracts, World Congress of Malacology, Vienna, Austria, 2001" (L. Salvini-Plawen, J. Voltzow, H. Sattmann and G. Steiner, eds), p. 287. Unitas Malacologica, Vienna [abstract].

Riemann-Zürneck, K. (1973). Actiniaria des Südwestatlantik I. Hormathiidae. *Helgoländer Wissenschaftliche Meeresuntersuchungen* **25**, 273–325.

Rokop, F. J. (1974). Reproductive patterns in the deep-sea benthos. *Science* **186**, 743–745.

Rokop, F. J. (1977). Seasonal reproduction of the brachiopod *Frieleia halli* and the scaphopod *Cadulus californicus* at bathyal depths in the deep sea. *Marine Biology* **43**, 237–246.

Roy, K., Jablonski, D. and Valentine, J. W. (1994). Eastern Pacific molluscan provinces and latitudinal diversity gradient – no evidence for Rapoport's rule. *Proceedings of the National Academy of Sciences (United States of America)* **91**, 8871–8874.

Roy, K., Jablonski, D., Valentine, J. W. and Rosenberg, G. (1998). Marine latitudinal diversity gradients: Tests of causal hypotheses. *Proceedings of the National Academy of Sciences (United States of America)* **95**, 3699–3702.

Runnegar, B. N. (1996). Early evolution of the Mollusca: the fossil record. *In* "Origin and Evolutionary Radiation of the Mollusca" (J. Taylor, ed.), pp. 77–87. Oxford, London.

Runnegar, B. and Pojeta, J. (1974). Molluscan phylogeny: the paleontological viewpoint. *Science* **186**, 311–317.

Ruppert, E. E. and Smith, P. (1988). The functional organization of filtration nephridia. *Biological Reviews of the Cambridge Philosophical Society* **63**, 231–258.

Ruthensteiner, B., Wanninger, A., Haszprunar, G. (2001). The protonephridal system of the tusk shell, *Antalis entalis* (Mollusca, Scaphopoda). *Zoomorphology* **121**, 19–26.

Sahlman, B. (1973). Untersuchungen zur Histologie und Nahrungsbiologie der Scaphopoden. PhD dissertation, Universität Kiel, Institüt fur Meerskunde, Germany.

Salvini-Plawen, L. v. (1980). Was ist eine Trochophora? Eine Analyse der Larventypen mariner Protostomier. *Zoologische Jahrbücher. Abteilung für Anatomie und Ontogenie der Tiere* **103**, 389–423.

Salvini-Plawen, L. v. (1981). The molluscan digestive system in evolution. *Malacologia* **21**, 371–401.

Salvini-Plawen, L. v. (1985). Early evolution and the primitive groups. *In* "The Mollusca, Volume 10, Evolution" (E. R. Trueman and M. R. Clarke, eds), pp. 59–150. Academic Press, New York.

Salvini-Plawen, L. v. (1988). The structure and function of molluscan digestive systems. *In* "The Mollusca. Volume 11. Form and Function" (E. R. Trueman and M. R. Clarke, eds), pp. 301–379. Academic Press, New York.

Salvini-Plawen, L. v. and Bartolomaeus, T. (1995). Mollusca: mesenchymata with a "coelom". *In* "Body Cavities: Function and Phylogeny" (G. Lanzavecchia, R. Valvassori and M. D. Candia Carnevali, eds), pp. 75–92. Mucchi, Modena.

Salvini-Plawen, L. v. and Steiner, G. (1996). Synapomorphies and plesiomorphies in higher classification of Mollusca. *In* "Origin and Evolutionary Radiation of the Mollusca" (J. Taylor, ed.), pp. 29–51. Oxford University Press, Oxford.

Sandford, F. and Brown, C. (1997). Gastropod shell substrates of the Florida hermit-crab sponge, *Spongosorites suberitoides,* from the Gulf of Mexico. *Bulletin of Marine Science* **61**, 215–223.

Scarabino, V. (1979). Les scaphopodes bathyaux de l'Atlantique occidental (Systematique, Distribution, Adaptations). Nouvelle classification pour l'ensemble de la classe. PhD dissertation, Université d'Aix Marseille, France.

Scarabino, V. (1986a). Nuevos taxa abisales de la Class Scaphopoda (Mollusca). *Comunicaciones Zoologicas del Museo Nacional de Historia Natural de Montevideo* **11** (155), 1–19.

Scarabino, V. (1986b). Systematics of Scaphopoda (Mollusca), 1. Three new bathyal and abyssal taxa of the Order Gadilida from south and north Atlantic Ocean. *Comunicaciones Zoologicas del Museo Nacional de Historia Natural de Montevideo* **11** (161), 1–15.

Scarabino, V. (1994). Aproximacion al numero de especies de Scaphopoda actuales. *Comunicaciones De La Sociedad Malacologica Del Uruguay* **7**, 357–361.

Scarabino, V. (1995). Scaphopoda of the tropical Pacific and Indian Oceans, with a description of 3 new genera and 42 new species. *In* "Résultats des Campagnes MUSORSTOM, Volume 14" (P. Bouchet, ed.). Mémoires du Muséum National d'Histoire Naturelle, Paris, **167**, 189–379.

Schipp, R. and Hevert, F. (1981). Ultrafiltration in the branchial heart appendage of dibranchiate cephalopods: a comparative ultrastuctural and physiological study. *Journal of Experimental Biology* **92**, 23–35.

Schleip, W. (1925). "Die Furchung dispermen *Dentalium*-Eier." *Archiv für Entwicklungsmechanik der Organismen, Leipzig* **106**, 86–123.

Shevenell, A. E., Domack, E. W. and Kernan, G. M. (1996). Record of Holocene paleoclimate change along the Antarctic Peninsula: evidence from glacial marine sediments, Lallemand Fjord. *Papers and Proceedings of the Royal Society of Tasmania* **130**, 55–64.

Shimek, R. L. (1988). The functional morphology of scaphopod captacula. *The Veliger* **30**, 213–221.

Shimek, R. L. (1989). Shell morphometrics and systematics: a revision of the slender, shallow-water *Cadulus* of the Northeastern Pacific (Scaphopoda: Gadilida). *The Veliger* **32**, 233–246.

Shimek, R. L. (1990). Diet and habitat utilization in a northeastern Pacific Ocean scaphopod assemblage. *American Malacological Bulletin* **7**, 147–169.

Shimek, R. L. (1997). A new species of eastern Pacific *Fissidentalium* (Mollusca: Scaphopoda) with a symbiotic sea anemone. *The Veliger* **40**, 178–191.

Shimek, R. L. and Moreno, G. (1996). A new species of eastern Pacific *Fissidentalium* (Mollusca: Scaphopoda). *The Veliger* **39**, 71–82.

Shimek, R. L. and Steiner, G. (1997). Scaphopoda. *In* "Microscopic Anatomy of Invertebrates, Volume 6B, Mollusca II" (F. W. Harrison and A. J. Kohn, eds), pp. 719–781. Wiley-Liss, New York.

Simkiss, K. and Mason, A. Z. (1983). Metal ions: metabolic and toxic effects. *In* "The Mollusca. Volume 2. Environmental Biochemistry and Physiology" (P. W. Hochachka, ed.), pp. 101–164. Academic Press, New York.

Simroth, H. (1894a). I. Abteilung: Amphineura und Scaphopoda. *In* "Mollusca" (H. G. Bronn, ed.), pp. 356–467, plates 15–22. Winter, Leipzig.

Simroth, H. (1894b). Bemerkungen über die Morphologie der Scaphopoden. *Zeitschrift für Naturwissenschaften* **67**, 239–259.

Skelton, P. W. and Benton, M. J. (1993). Mollusca: Rostroconchia, Scaphopoda and Bivalvia. *In* "The Fossil Record." (M. J. Benton, ed.), pp. 237–263. Chapman and Hall, London.

Smith, M. H. (1967). Occurrence of haemoglobin in some molluscs. *Comparative Biochemistry and Physiology* **20**, 361–364.

Starobogatov, Y. I. (1974). Xenoconchias and their bearing on the phylogeny and systematics of some molluscan classes. *Paleontological Journal* **8**, 1–13.

Stasek, C. R. and McWilliams, W. R. (1973). The comparative morphology and evolution of the molluscan mantle edge. *The Veliger* **16**, 1–19.

Steiner, G. (1990). Beiträge zur vergleichenden Anatomie und Systematik der Scaphopoda (Mollusca). PhD dissertation, University of Vienna, Austria.

Steiner, G. (1991). Observations on the anatomy of the scaphopod mantle and the description of a new family, the Fustiariidae. *American Malacological Bulletin* **9**, 1–20.

Steiner, G. (1992a). The organisation of the pedal musculature and its connection to the dorsoventral musculature in Scaphopoda. *Journal of Molluscan Studies* **58**, 181–197.

Steiner, G. (1992b). Phylogeny and classification of Scaphopoda. *Journal of Molluscan Studies* **58**, 385–400.

Steiner, G. (1993). Spawning behavior of *Pulsellum lofotensis* (M. Sars) and *Cadulus subfusiformis* (M. Sars) (Scaphopoda, Mollusca). *Sarsia* **78**, 31–33.

Steiner, G. (1994). Variations in the number of intestinal loops in Scaphopoda (Mollusca). *Marine Ecology* **15**, 165–174.

Steiner, G. (1995). Larval and juvenile shells of four North Atlantic scaphopod species. *American Malacological Bulletin* **11**, 87–98.

Steiner, G. (1996). Suprageneric phylogeny in Scaphopoda. *In* "Origin and Evolutionary Radiation of the Mollusca" (J. D. Taylor, ed.), pp. 329–335. Oxford University Press, Oxford.

Steiner, G. (1997). Phylogeny of Scaphopoda (Mollusca) in light of new anatomical data on the Gadilinidae and some problematica, and a reply to Reynolds. *Zoologica Scripta* **29**, 73–82.

Steiner, G. (1998). The scaphopod dataset: a new character and other data revisited. *In* "Abstracts, World Congress of Malacology, Washington, D.C., 1998" (E. Bieler and P. M. Mikkelsen, eds), p. 311. Unitas Malacologica, Chicago [abstract].

Steiner, G. (1999). A new genus and species of the family Annulidentaliidae (Scaphopoda: Dentaliida) and its systematic implications. *Journal of Molluscan Studies* **65**, 151–161.

Steiner, G. and Dreyer, H. (2001). Molecular phylogeny of Scaphopoda – initial results from 18S rDNA sequences. *In* "Abstracts, World Congress of Malacology, Vienna, Austria, 2001" (L. Salvini-Plawen, J. Voltzow, H. Sattmann and G. Steiner, eds), p. 334. Unitas Malacologica, Vienna [abstract].

Steiner, G. and Kabat, A. R. (in press). Catalogue of supraspecific taxa of Scaphopoda (Mollusca). *Zoosystema* **23**.

Steiner, G. and Linse, K. (2000). Systematics and distribution of the Scaphopoda (Mollusca) in Beagle Channel (Chile). *Mitteilungen. Hamburgisches Zoologische Museum und Institüt* **97**, 13–30.

Stevens, G. C. (1989). The latitudinal gradient in geographical range: how so many species coexist in the tropics. *The American Naturalist* **133**, 240–256.

Strohl, J. (1924). Mollusken. *In* "Handbuch der Vergleichenden Physiologie" (H. Winterstein, ed.), pp. 443–607. Verlag von Gustav Fischer, Jena.

Sullivan, P. A., Robinson, W. E. and Morse, M. P. (1988). Isolation and characterization of granules from the kidney of the bivalve *Mercenaria mercenaria. Marine Biology* **99**, 359–368.

Taib, N. T. (1976). Anatomy, histology and histochemistry of the alimentary canal of *Dentalium entalis* L. with some observations on living animals. PhD dissertation, University of Glasgow.

Taib, N. T. (1980). Some observations on living animals of *Dentalium entalis* L. *Journal College Science, University Riyadh* **11**, 129–144.

Taib, N. T. (1981a). Gross anatomy of the alimentary canal of *Dentalium entalis* L. (Scaphopoda). *Journal College Science, University Riyadh* **12**, 139–145.

Taib, N. T. (1981b). Sites of absorption and food storage in the gut of *Dentalium entalis* L. *Journal College Science, University Riyadh* **12**, 147–154.

Thomas, R. D. K. (1988). Evolutionary convergence of bivalved shells: a comparative analysis of constructional constraints on their morphology. *American Zoologist* **28**, 267–276.

Thompson, D. W. (1952). "On Growth and Form, Volume 2", 2nd edition. Cambridge University Press, Cambridge.

Timmermans, L. P. M., Geilenkirchen, W. L. M. and Verdonk, N. H. (1970). Local accumulation of Feulgen-positive granules in the egg cortex of *Dentalium dentale* L. *Journal of Embryology and Experimental Morphology* **23**, 245–252.

Trueman, E. R. (1968). The burrowing process of *Dentalium* (Scaphopoda). *Journal of Zoology* **154**, 19–27.

Turchini, J. (1923). Scaphopoda. *In* "Contribution a' l'Etude de l'Histologie Comparée de la Cellule Rénale" (G. Doin, ed.), pp. 114–115. Librairie Octave Doin, Paris.

Van den Biggelaar, J. A. M. and Guerrier, P. (1983). Origin of spatial organization. *In* "The Mollusca, Volume 3, Development" (N. H. Verdonk, J. A. M. van den Biggelaar and A. S. Tompa, eds), pp. 179–213. Academic Press, New York.

Van Dongen, C. A. M. (1976a). The development of *Dentalium* with special reference to the polar lobe. 5. Differentiation of the cell pattern in lobeless embryos of *Dentalium vulgare* (Da Costa) during late larval development. *Proceedings Koninklijke Nederlandse Akademie van Wetenschappen. Series C. Biological and Medical Sciences* **79**, 245–255.

Van Dongen, C. A. M. (1976b). The development of *Dentalium* with special reference to the polar lobe. 6. Differentiation of the cell pattern in lobeless embryos of *Dentalium vulgare* (Da Costa) during late larval development. *Proceedings Koninklijke Nederlandse Akademie van Wetenschappen. Series C. Biological and Medical Sciences* **79**, 256–266.

Van Dongen, C. A. M. (1976c). The development of *Dentalium* with special reference to the polar lobe. 7. Organogenesis and histogenesis in lobeless embryos of *Dentalium vulgare* (da Costa) as compared to normal development. *Proceedings Koninklijke Nederlandse Akademie van Wetenschappen. Series C. Biological and Medical Sciences* **79**, 454–465, 4 plates.

Van Dongen, C. A. M. (1977). Mesoderm formation during normal development of *Dentalium dentale. Proceedings Koninklijke Nederlandse Akademie van Wetenschappen. Series C. Biological and Medical Sciences* **80**, 372–376.

Van Dongen, C. A. M. and Geilenkirchen, W. C. M. (1974a). The development of *Dentalium* with special reference to the polar lobe. 1. Division chronology and development of the cell pattern in *Dentalium dentale* (Scaphopoda). *Proceedings Koninklijke Nederlandse Akademie van Wetenschappen. Series* C. **77**, 57–70.

Van Dongen, C. A. M. and Geilenkirchen, W. C. M. (1974b). The development of *Dentalium* with special reference to the polar lobe. 2. Division chronology and development of the cell pattern in *Dentalium dentale* (Scaphopoda). *Proceedings Koninklijke Nederlandse Akademie van Wetenschappen. Series* C. **77**, 71–84.

Van Dongen, C. A. M. and Geilenkirchen, W. C. M. (1974c). The development of *Dentalium* with special reference to the polar lobe. 3. Division chronology and development of the cell pattern in *Dentalium dentale* (Scaphopoda). *Proceedings Koninklijke Nederlandse Akademie van Wetenschappen. Series* C. **77**, 85–100.

Van Dongen, C. A. M. and Geilenkirchen, W. C. M. (1975). The development of *Dentalium* with special reference to the polar lobe. 4. Division chronology and development of the cell pattern in *Dentalium dentale* after removal of the polar lobe at first cleavage. *Proceedings Koninklijke Nederlandse Akademie van Wetenschappen. Series* C. **78**, 358–375.

Van Loon, A. E. and Van den Biggelaar, J. A. M. (1998). Changes in cell lineage specification elucidate evolutionary relations in Spiralia. *Biological Bulletin* **195**, 367–369.

Verdonk, N. H. (1968a). The effect of removing the polar lobe in centrifuged eggs of *Dentalium. Journal of Embryology and Experimental Morphology* **19**, 33–42.

Verdonk, N. H. (1968b). The relation of the two blastomeres to the polar lobe in *Dentalium. Journal of Embryology and Experimental Morphology* **20**, 101–105.

Verdonk, N. H. and Cather, J. N. (1983). Morphogenetic determination and differentiation. *In* "The Mollusca, Volume 3, Development" (N. H. Verdonk, J. A. M. van den Biggelaar and A. S. Tompa, eds), pp. 215–252. Academic Press, New York.

Verdonk, N. H. and Van den Biggelaar, J. A. M. (1983). Early development and the formation of the germ layers. *In* "The Mollusca, Volume 3, Development" (N. H. Verdonk, J. A. M. van den Biggelaar and A. S. Tompa, eds), pp. 91–122. Academic Press, New York.

Verdonk, N. H., Geilenkirchen, W. L. M. and Timmermans, L. P. M. (1971). The localization of morphogenetic factors in uncleaved eggs of *Dentalium. Journal of Embryology and Experimental Morphology* **25**, 57–63.

Vermeij, G. J. (1987). Evolution and Escalation: an Ecological History of Life. Princeton University Press.

Vermeij, G. J. (1993). A Natural History of Shells. Princeton University Press, Princeton.

Vovelle, J. and Grasset, M. (1983). Biomineralisation des dents radulaires chez *Dentalium dentalis* L. *Haliotis* **13**, 123–130.

Wagner, P. J. (1997). Patterns of morphologic diversification among the Rostroconchia. *Paleobiology* **23**, 115–150.

Waller, T. R. (1998). Origin of the Molluscan Class Bivalvia and a phylogeny of major groups. *In* "Bivalves: An Eon of Evolution" (P. A. Johnston and J. W. Haggart, eds), pp. 1–45. University of Calgary Press, Calgary.

Wanninger, A. and Haszprunar, H. (in press). The expression of an engrailed protein during embryonic shell formation of the tusk-shell, *Antalis entalis* (Mollusca, Scaphopoda). *Evolution and Development* **3**.

Watson, R. B. (1886). Report on the *Scaphopoda* and *Gasteropoda* collected by H.M.S. *Challenger* during the Years 1873–76. *The Voyage of H.M.S. Challenger. Zoology* **15**, 1–24, 3 plates.

Watters, G. T. (1991). Utilization of a simple morphospace by polyplacophorans and its evolutionary implications. *Malacologia* **33**, 221–240.

White, K. M. (1942). The pericardial cavity and the pericardial gland of the Lamellibranchia. *Proceedings of the Malacological Society of London* **25**, 37–88.

White, T. R., Pagels, A. K. W. and Fautin, F. G. (1999). Abyssal sea anemones (Cnidaria: Actiniaria) of the northeast Pacific symbiotic with molluscs: *Anthosactis nomados,* a new species, and *Monactis vestita* (Gravier, 1918). *Proceedings of the Biological Society of Washington* **112**, 637–651.

Wilson, E. B. (1904a). Experimental studies on germinal localization. *Journal of Experimental Zoology* **1**, 1–72.

Wilson, E. B. (1904b). Experimental studies in germinal localization. II. Experiments on the Cleavage-Mosaic in *Patella* and *Dentalium. Journal of Experimental Zoology* **1**, 197–268.

Witmer, A. and Martin, A. W. (1973). The fine structure of the branchial heart appendage of the cephalopod *Octopus dofleini martini. Zeitschrift für Zellforschung und Mikroskopische Anatomie* **136,** 545–568.

Yancy, T. E. (1973). A new genus of Permian siphonodentalid scaphopods, and its bearing on the origin of the Siphonodentaliidae. *Journal of Paleontology* **47**, 1062–1064.

Yochelson, E. L. (1999). Scaphopoda. *In* "Functional Morphology of the Invertebrate Skeleton" (E. Savazzi, ed.), pp. 363–367. John Wiley and Sons, Chichester.

Yochelson, E. L., Dockery, D. and Wolf, H. (1983). Predation on sub-holocene scaphopod mollusks from southern Louisiana. *Professional Papers U.S. Geological Survey* **1282**, 1–13.

Yonge, C. M. (1937). Circulation of water in the mantle cavity of *Dentalium entalis. Proceedings of the Malacological Society of London* **22**, 333–337.

Zibrowius, H. (1998). A new type of symbiosis: *Heterocyathus japonicus* (Cnidaria: Scleractinia) living on *Fissidentalium vernedei* (Mollusca: Scaphopoda). *Zoologische Verhandelingen (Leiden)* **323**, 319–340.

Pleurotomarioidean Gastropods

M. G. Harasewych

Department of Systematic Biology, National Museum of Natural History, Smithsonian Institution, Washington, DC 20560-0118, USA
e-mail: Harasewych@nmnh.si.edu

Pleurotomarioidean gastropods are continuously present in the fossil record since the Upper Cambrian and survive into the Recent fauna, thus providing rare insights into the evolutionary history of the class Gastropoda. Pleurotomarioidea achieved greatest numerical and morphological diversity during the Paleozoic, and dominated global shallow water marine gastropod faunas during the Paleozoic and Mesozoic. Only a single family, the Pleurotomariidae, survived the end-Cretaceous Extinction, but was restricted to deep water through most of the Cenozoic. The first living

ADVANCES IN MARINE BIOLOGY VOL. 42
ISBN 0-12-026142-1

pleurotomariid was discovered during the mid-nineteenth century, along the bathyal zone of the western Atlantic. Subsequently collected specimens of these "living fossils" revealed that these animals comprise a mosaic of primitive and highly derived characters that originally inspired a new model of gastropod evolution, but more recently defy the orderly inclusion of this group within a cladistic framework of gastropod phylogeny. Molecular studies have likewise shown that the 18S rDNA gene of pleurotomariids contains novel insertions, and evolves more rapidly than in related taxa. A number of studies confirm that the Pleurotomarioidea may be included in the clade Vetigastropoda together with the Trochoidea, Fissurelloidea, Haliotoidea, and Scissurelloidea. However, neither the position of the superfamily within Vetigastropoda, nor the position of Vetigastropoda within Gastropoda is yet robustly resolved. The anatomical and molecular data are reviewed; the latter used to produce a well-resolved phylogeny of the genera within the family, and to justify the naming of the long-used informal grouping "Perotrochus Group B" as the genus Bayerotrochus. *A review of the geographic and bathymetric distributions of pleurotomariids reveals that the higher taxa segregate bathymetrically, while the species within each genus generally segregate geographically, so that most species exist in allopatry. The diet of pleurotomariids is reviewed, based on direct observations as well as analyses of gut contents, and is shown to consist primarily of sponges, but may include stalked crinoids, octocorals and, under aquarium conditions, fish and clam tissue. Despite their thin and relatively fragile shells, Pleurotomariidae survive an extraordinary number of attacks by predators, primarily crustaceans and fish. Their hypobranchial glands are capable of rapidly secreting a white fluid that probably serves as a chemical defense to repel predators. Circumstantial evidence suggests that this defensive mechanism may have originated during the Paleozoic. Despite the great antiquity of this lineage, pleurotomariids possess a substantial number of morphological, molecular and ecological novelties.*

1. INTRODUCTION

Gastropods of the superfamily Pleurotomarioidea are readily distinguished by their dextral, conispirally coiled shell with an inner nacreous layer and a characteristic emargination along the outer lip that produces a selenizone, a spiral trace of the slit that is evident throughout the adult shell (Figure 1). At least two families of pleurotomarioideans, Sinuopeidae Wenz, 1938, and Raphistomatidae Koken, 1896, occur in Upper Cambrian fossil beds, making the Pleurotomarioidea not only the oldest undisputed gastropod lineage, but also the only gastropod superfamily to be continuously represented in the fossil record from the Cambrian to the present (Woodward, 1885; Wenz, 1938; Cox, 1960b; Knight *et al.*, 1960; Tracey *et al.*, 1993).

Species and generic diversity were greatest during the Paleozoic (Figure 2A), while the number of families proliferated to 13 during the Triassic, contributing to the maximum family level diversity during this period (Knight *et al.*, 1960). The family Pleurotomariidae Swainson, 1840 (as refined by Knight *et al.*, 1960; Tracey *et al.*, 1993) has its origins in the Lower Triassic, and is the only family within the Pleurotomarioidea to survive beyond the Jurassic and into the Recent fauna. Of the more than 1500 described species and subspecies, 25 are living today (Anseeuw and Goto, 1996; Anseeuw, 1999). Pleurotomarioideans were abundant and diverse components of shallow water marine faunas throughout the Paleozoic (e.g., Thomas, 1940; Hayasaka and Hayasaka, 1953; Batten, 1958, 1966; Wolfenden, 1958; Ingles, 1963; Shikama and Nishida, 1968) and Mesozoic (e.g., Hudleston, 1895 in 1887–1896; Cox, 1960a; Kanie *et al.*, 1980; Benfrika, 1984; Ziegler, 1984). However, Cenozoic fossils are relatively rare, and tend to be limited to deep-water facies (e.g., Hirayama, 1955; Kanno, 1961; Jung, 1968, 1970; Hickman, 1976; Tomida *et al.*, 1999), while living species are usually restricted to depths ranging from 100 m to nearly 1000 m (Anseeuw and Goto, 1996; Figure 12A).

Fossil pleurotomariids were known to pre-Linnean authors (e.g., Lister, 1678), and were allied with the Trochidae in early classifications (e.g., Swainson, 1840; Adams and Adams, 1858). Although records of living pleurotomariids appear in eighteenth century Japanese literature (Kenkado, 1755), it was not until the early stages of the era of deep-sea exploration that a living pleurotomariid became known to the western world with the discovery of *Perotrochus quoyanus* (Fischer and Bernardi, 1856). This was among the earliest of the "living fossils" to be discovered in the deep sea. For over a century, each collection of additional specimens of pleurotomariids prompted intense anatomical scrutiny and a series of publications (e.g., Crosse and Fischer, 1861; Crosse, 1882; Mitsukuri, 1897; Bouvier and Fischer, 1898, 1901, 1902; Kosuge and Suzuki, 1969; Oyama, 1980). This, in turn, led to the formulation of a new paradigm of gastropod evolution, one in which anatomical bilateral symmetry was considered to be the primitive gastropod condition. Pleurotomariids have since been regarded almost universally as the most primitive living gastropods. Their symmetrical, paired, pallial cavity organs and an asymmetrically coiled shell were thought to be transitional between extinct, planispirally coiled, bilaterally symmetrical ancestors and asymmetrical modern gastropods (e.g., Thiele, 1902, 1935; Wenz, 1938; Yonge, 1947; Knight 1952; Knight *et al.*, 1960; Fretter and Graham, 1962;

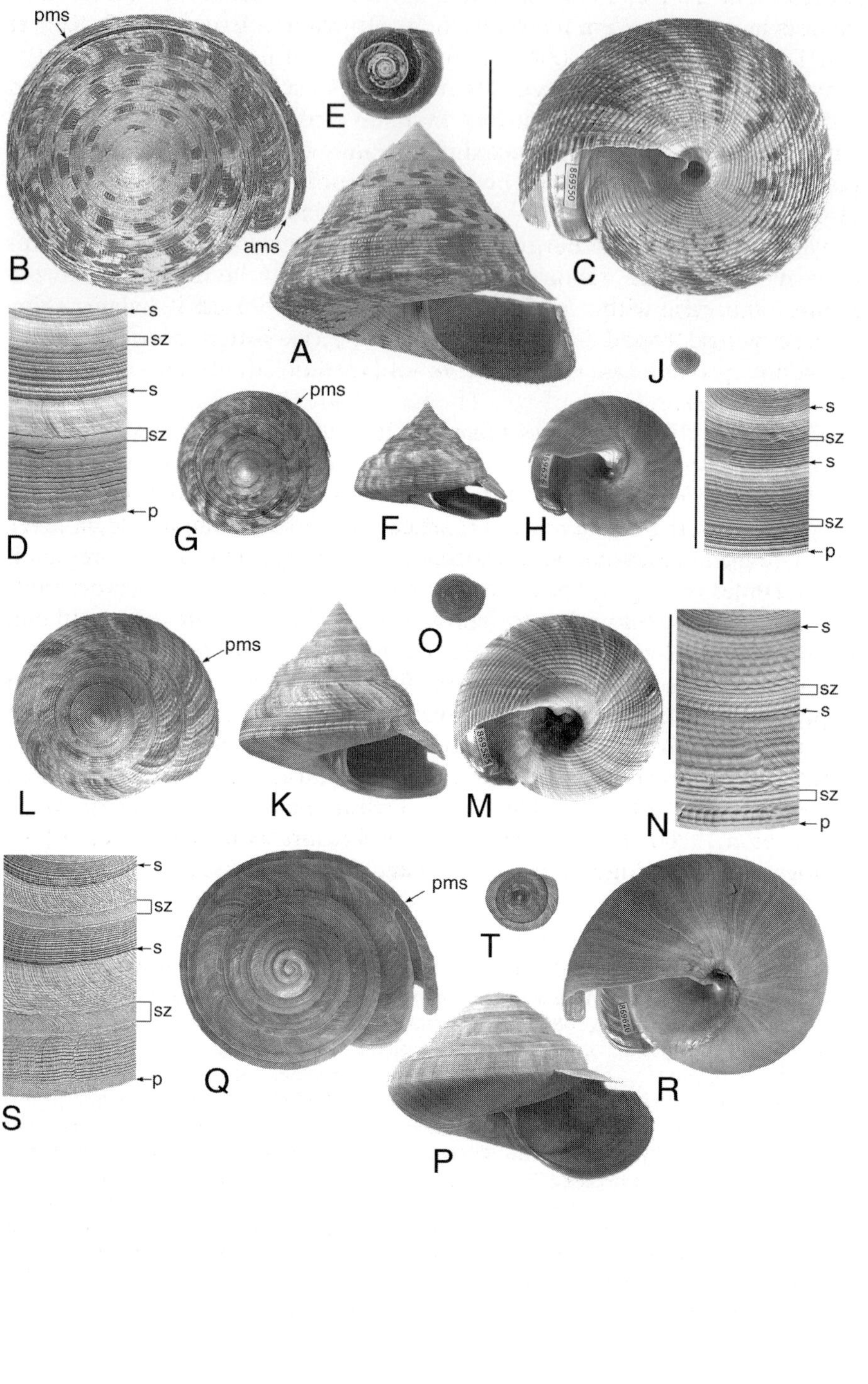
pms
ams
s
sz
p
869550
869624
869565
869620
A
B
C
D
E
F
G
H
I
J
K
L
M
N
O
P
Q
R
S
T

Morton and Yonge, 1964; Hickman, 1984a, b; Graham, 1985). This view persists in many modern textbooks (e.g., Brusca and Brusca, 1990; Rupert and Barnes, 1994), no doubt because of its heuristic value, despite the accumulation of evidence that docoglossan limpets (renamed Patellogastropoda and Eogastropoda) comprise the most basal clade within Gastropoda. More recent phylogenetic classifications generally place the Pleurotomariidae within the order Vetigastropoda (e.g., Golikov and Starobogatov, 1975; Salvini-Plawen, 1980; Haszprunar, 1988; Ponder and Lindberg, 1996, 1997; Harasewych *et al.*, 1997), frequently as a sister taxon to Trochidae. While the inclusion of the Pleurotomariidae within Vetigastropoda is well supported, its relationships to other vetigastropod taxa and the relationship of Vetigastropoda to other gastropod groups has yet to be robustly resolved, despite substantial research efforts.

The rarity of Recent pleurotomariids in museum collections (Hickman, 1984b: 228) had earlier led to suggestions of low population densities (Hickman, 1976). However, the expansion of commercial fisheries into deeper waters, the use of more sophisticated deep-sea sampling techniques for biological research, and especially the increasing use of research submersibles during the past two decades, have revealed pleurotomariids to be among the numerically dominant macroinvertebrates in rocky and steep-walled bathyal habitats, which are difficult to sample by conventional methods such as dredging and trawling. The increased sampling and opportunities for *in situ* observations, in turn, led to rapid increases in our understanding of the diversity and geographic distribution of this family in the Recent fauna (Figure 2B, e.g., Bayer, 1965; Endo, 1995; Anseeuw and Goto, 1996), its evolutionary relationships (e.g., Haszprunar, 1988; Harasewych *et al.*, 1997), and various aspects of its biology and ecology (e.g., Hickman, 1984a; Kanie *et al.*, 1986;

Figure 1 Representatives of living pleurotomariid clades. (A–E) *Entemnotrochus adansonianus* (Crosse and Fischer, 1861). Type species of *Entemnotrochus* Fischer, 1885. (A) Apertural, (B) dorsal, and (C) ventral views of shell. (D) Sculptural detail and selenizone. (E) Operculum. (F–J) *Perotrochus quoyanus* (Fischer and Bernardi, 1856). Type species of *Perotrochus* Fischer, 1885. (F) Apertural, (G) dorsal and (H) ventral views of shell. (I) Sculptural detail and selenizone. (J) Operculum. (K–O) *Mikadotrochus beyrichii* Hilgendorf, 1877. Type species of *Mikadotrochus* Lindholm, 1972. (K) Apertural, (L) dorsal and (M) ventral views of shell. (N) Sculptural detail and selenizone. (O) Operculum. (P–T) *Perotrochus midas* Bayer, 1965. Example of *Perotrochus* "Group B" (Bayer, 1965). (P) Apertural, (Q) dorsal and (R) ventral views of shell. (S) Sculptural detail and selenizone. (T) Operculum. All shells and opercula to same scale. All scale bars = 3 cm. ams, anterior margin of slit; p, periphery of whorl; pms, posterior margin of slit; s, suture; sz, selenizone.

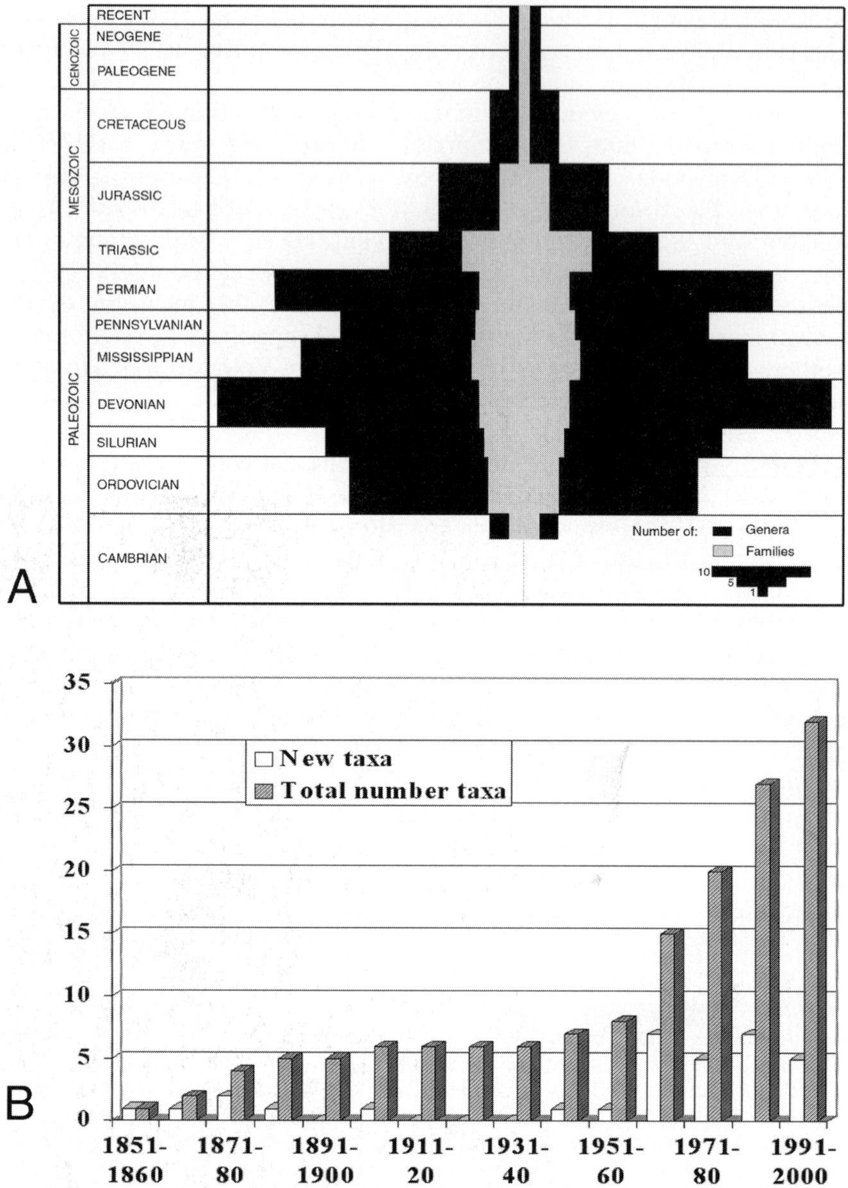

Figure 2 Diversity of Pleurotomarioidea. (A) Genus and family level diversity of the Pleurotomarioidea through Phanerozoic time. Data from Knight *et al.* (1960), *The Zoological Record* 1960–2000, and other sources. (B) Known species level diversity of Recent Pleurotomariidae (by decade) since the discovery of the first living specimen. Data from Anseeuw and Goto (1996) and Anseeuw (1999).

Plate 1 Adults of *Rhabdus rectius,* order Dentaliida, with protruding muscular foot. Note maximum diameter at anterior aperture, transparent secondary shell in upper left and lower apices, dark digestive gland in mid-region, beige kidney immediately anterior to it (center-left specimen), and protruding foot of the two center specimens. On the left the foot is being extended, on the right it is being retracted with flared epipodial lobes. Approximate shell length = 42 mm. © Patrick D. Reynolds.

Plate 2 Individual of *Gadila aberrans,* order Gadilida, family Gadilidae, subfamily Gadilinae. Note narrowing of anterior aperture and notched apex. Approximate shell length = 18 mm. ©Patrick D. Reynolds.

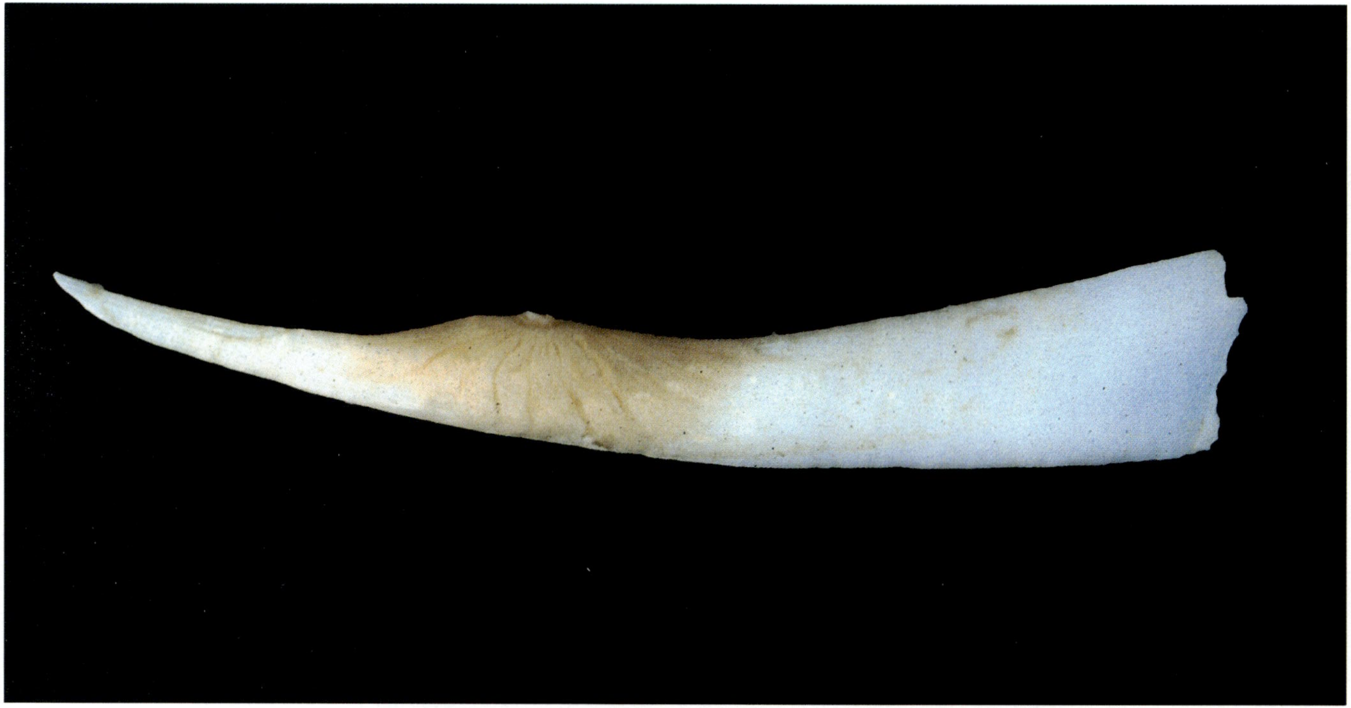

Plate 3 The scaphopod *Fissidentalium actinophorum* with attached anemone, *Anthosactis nomados*. Approximate shell length = 36 mm. ©Ronald L. Shimek.

Plate 4 The hermit crab *Orthopagurus minimus,* inhabiting the shell of *Dentalium pretiosum*. Approximate shell length = 35 mm. ©Ronald L. Shimek.

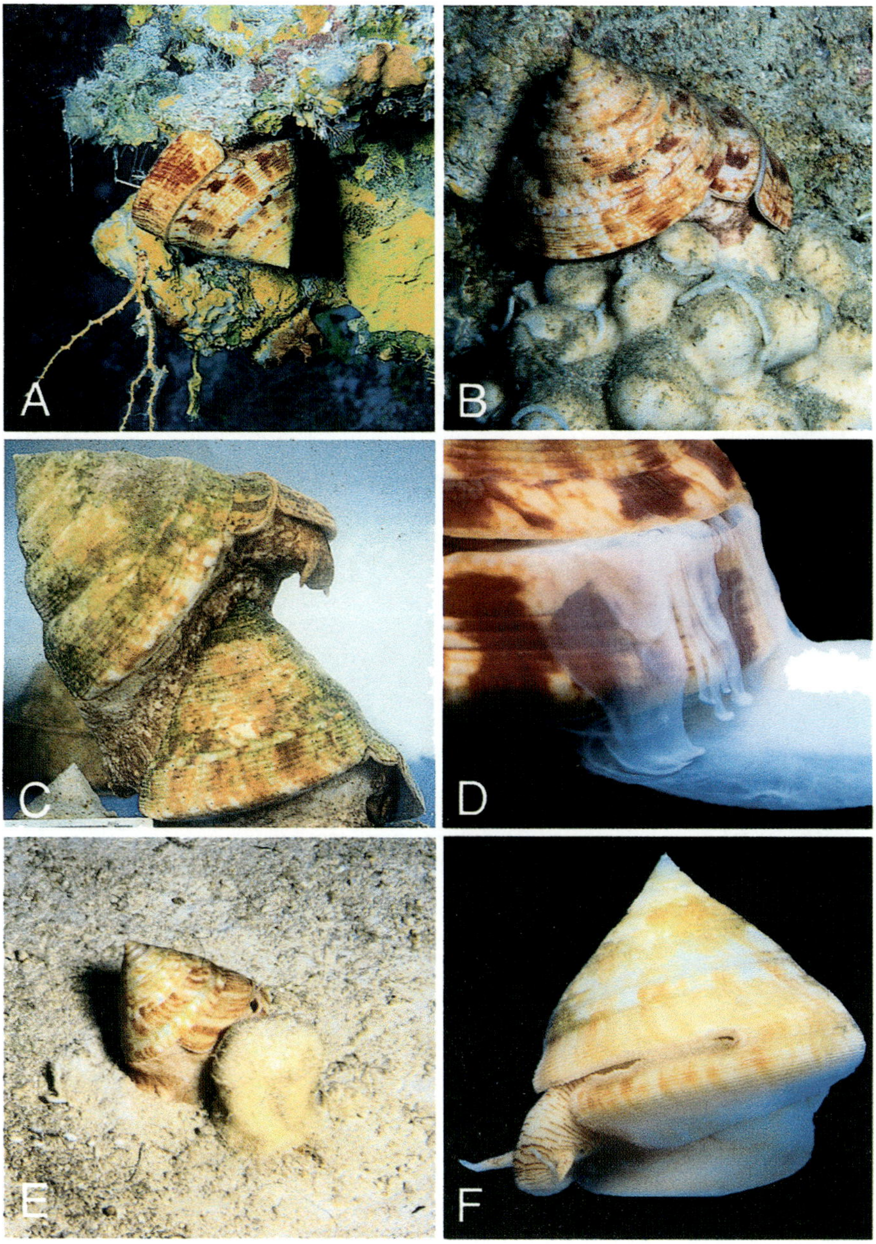
A
B
C
D
E
F

Plate 5 Photographs of living pleurotomariids. **(A–D)** *Entemnotrochus adansonianus* (Crosse and Fischer, 1861). **(A)** Off Riding Rock, San Salvador, Bahamas, 22°53′24″N, 74°54′49″W in 106 m. JOHNSON-SEA-LINK II Station JSL-II-3620, 12 September 1993. **(B)** Off Guadeloupe, Lesser Antilles, 16°16′30″N, 61°50′36″W, inn 176 m, JOHNSON-SEA-LINK I Station JSL-I-1844, 10 July 1989. **(C)** *Entemnotrochus adansonianus* in aquarium, with foot and snout exposed. Specimens collected off Navassa Island, Greater Antilles, 18°24′43″N, 75°03′00″W, in 123–141 m, JOHNSON-SEA-LINK I Station JSL-I-2322, 14 October 1988. **(D)** The secretion produced by the hypobranchial glands of *Entemnotrochus adansonianus* when the animal was disturbed in an aquarium. Specimen collected off Chubb Key, Berry Islands, Bahamas, 25°23′06″N, 77°51′36″W, in 108 m, JOHNSON-SEA-LINK I Station JSL-I-4189, 20 October 1999. **(E–F)** *Perotrochus lucaya* Bayer, 1965. **(E)** *Perotrochus lucaya* feeding on a small sponge, off Middle Point, Great Inagua Island, Bahamas, 21°01′45″N, 73°43′48″W, in 380 m, JOHNSON-SEA-LINK II Station JSL-I-2323, 15 October 1988. **(F)** *Perotrochus lucaya* crawling in an aquarium. Specimen collected off West End, Grand Bahama Island, Bahamas, 26°37′36″N, 78°58′30″W, in 399 m, JOHNSON-SEA-LINK I Station JSL-I-4191, 20 October 1999.

Plate 6 Photographs of living pleurotomariids. **(A)** *Perotrochus quoyanus* off Îlets-à-Goyaves, Guadeloupe, Lesser Antilles, 16°10′00″N 61°49′08″W, in 235 m, JOHNSON-SEA-LINK II Station JSL-II-1845, 10 July 1989. **(B–C)** *Bayerotochus pyramus*. **(B)** *Bayerotochus pyramus* feeding on octocoral, off Îlets-à-Goyaves, Guadeloupe, Lesser Antilles, 16°10′00″N, 61°49′08″W, in 306 m, JOHNSON-SEA-LINK II Station JSL-II-1845, 10 July 1989. **(C)** *Bayerotochus pyramus* feeding on encrusting sponge, off Îlets-à-Goyaves, Guadeloupe, Lesser Antilles, 16°10′12″N, 61°49′00″W, in 312 m, JOHNSON-SEA-LINK II Station JSL-II-1848, 11 July 1989. **(D–F)** *Bayerotrochus midas*. **(D)** *Bayerotrochus midas* and starfish all feeding on sponge, off Egg Island, Bahamas, 25°27′51″N, 76°54′38″W, in 752 m, JOHNSON-SEA-LINK I Station JSL-I-3622, 13 September 1993. **(E)** *Bayerotrochus midas* feeding on sponge, off SW tip of Long Island, Bahamas, 22°50′30″N, 74°53′00″W, in 645 m, JOHNSON-SEA-LINK I Station JSL-I-2324, 16 October 1988. **(F)** *Bayerotrochus midas* feeding on sponge, off Sandy Point, San Salvador, Bahamas, 23°57′04″N, 74°34′54″W, in 695 m, JOHNSON-SEA-LINK I Station JSL-I-2329, 18 October 1988.

A
B
C
D
E
F

Harasewych *et al.*, 1988; Healy, 1988; Healy and Harasewych, 1992; Kanie and Oba, 1993; Goodfriend *et al.*, 1994; Kuramochi *et al.*, 1996).

This article reviews the natural history of living pleurotomariid gastropods by integrating the substantial number of detailed, published observations on various aspects of the anatomy and ecology of many of the living taxa. This information is corroborated and/or supplemented with data garnered from the author's *in situ* observations of most western Atlantic species, dissections of well preserved specimens, and various analyses of shells and tissues. The objective is to provide a cohesive and well-corroborated synthesis of the morphology, ecology and biology of these animals that may be of use in clarifying our still imperfect understanding of the relationships of pleurotomariids to other gastropods, the evolution and functional morphology of their organs systems and genes, and the patterns of their bathymetric and geographic distribution in the world's oceans.

2. STRUCTURE AND FUNCTIONAL MORPHOLOGY

2.1. Shell morphology and ultrastructure (Figures 1, 3, 4)

The shell of pleurotomarioideans is most readily identified by it being conispirally coiled and by it having both an inner nacreous layer and an emargination or slit along its outer lip that produces a selenizone, a band that marks the position of the slit from its posterior margin along all of the whorls of the shell to within a quarter whorl of the protoconch. All pleurotomarioideans have indeterminate growth, with the exception of the Devonian family Catantostomatidae Wenz, 1938, in which the last half whorl is sharply deflected beyond the point at which the slit terminates to form a foramen (Knight *et al.*, 1960: 213, figure 128). While the majority of taxa since the Cambrian have been dextral and conispiral, the subfamily Agnesiinae Knight, 1956 [Lower Devonian to Upper Triassic] was characterized by sinistrally coiled shells. A great variety of shell forms, encompassing planispiral, discoidal, turriform and auriform lineages was evident during the Paleozoic (Knight *et al.*, 1960). Taxa with a disjunct last whorl were present during the Upper Cambrian to Lower Ordovician (Knight *et al.*, 1960), while uncoiled or open coiled taxa were reported from the Lower Devonian (Rohr and Smith, 1978). With the exception of the discoidal family Porcelliidae Broili, 1924, lineages surviving the Permian extinction had a more limited repertoire of morphology, consisting primarily of trochiform and turbiniform shells. All post-Jurassic pleurotomarioideans are members of the family

Pleurotomariidae, with 4 of 14 genera surviving into the Recent fauna. In dealing with Mesozoic taxa, paleontologists have devised a classification scheme based on the sequential discrimination of characters including: (1) the shape of the whorl section; (2) the position and width of the selenizone; (3) the form of the periphery and base; (4) the allometry of the spire; and (5) surface ornamentation and presence/features of the umbilicus (Szabó, 1980). According to these criteria, taxa with gradate and low turbiniform whorl profiles have become extinct, and only the subset of taxa with conical whorl profiles that have a broad selenizone at or near mid-whorl survive in the Recent fauna (Figure 9A, see page 266).

The pleurotomariid protoconch (Figure 3A–D) consists of $1–1\frac{1}{4}$ smooth, glassy, translucent whorls, and ranges from about 500 μm (*Perotrochus maureri*, Harasewych and Askew, 1993: figures 2–3) to 720 μm (*Mikadotrochus beyrichii*, Sasaki, 1998: figure 23) in diameter. It is slightly deflected from the coiling axis of the shell (12–19°), lacks a slit and terminates in a slightly flared lip. The transition to the teleoconch is

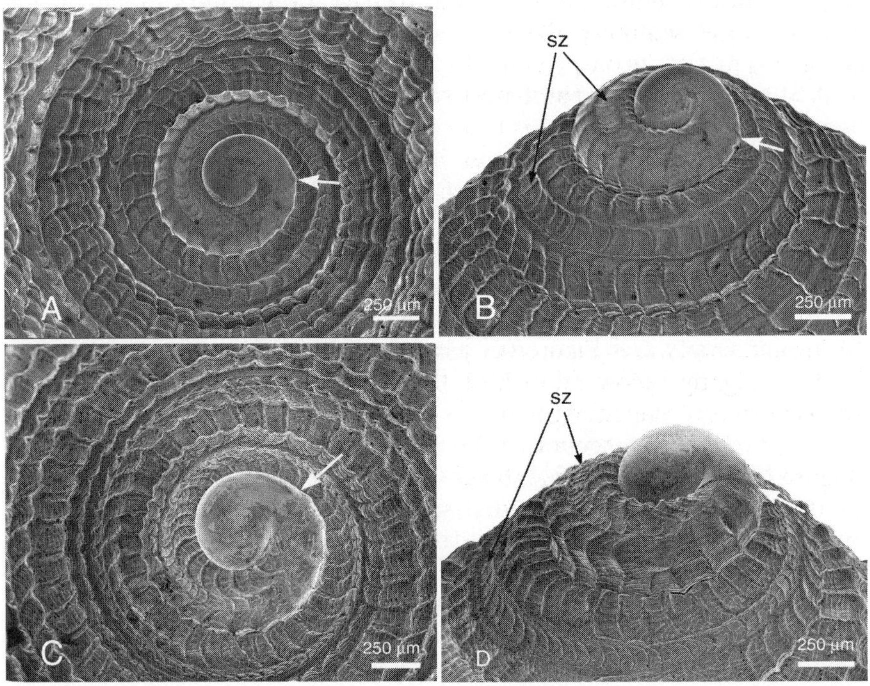

Figure 3 Morphology of the protoconch and early whorls. (A) Apical and (B) oblique views of *Entemnotrochus adansonianus* (Crosse and Fischer, 1861). (C) Apical and (D) oblique views of *Perotrochus deforgesi* Métivier, 1990. White arrows = protoconch/teleoconch boundary; sz, selenizone.

marked by an abrupt change in shell surface from glassy to chalky, and by the onset of weak spiral and pronounced axial sculpture (Figure 3A–D, arrows). Fine, irregular spiral threads extend onto strong, axial ribs (17–26 on the first whorl). The selenizone appears within the first quarter teleoconch whorl, with traces in evidence by the second axial rib. Initially, the selenizone nearly abuts the suture, but descends to a position that is near mid-whorl by the second teleoconch whorl, reaching its final relative position by the fifth or sixth teleoconch whorl, depending on the taxon.

Among the extant taxa, shells of the genus *Entemnotrochus* (Figure 1A–C) are strongly conical, generally with straight to convex profiles, have an aperture that is roughly rhomboidal, and a narrow slit that extends nearly 160° from the aperture (Figure 1B, pms) and produces a selenizone that is substantially above mid-whorl (Figure 1D, sz). The edges of the outer lip on either side of the selenizone are roughly aligned. *Entemnotrochus* have a broad, open umbilicus that extends from the base to the protoconch. Axial sculpture is prominent on the early whorls, but becomes diminished or reduced to axial growth lines in subsequent whorls. Spiral sculpture likewise is reduced to broad, weak bands or becomes entirely inconspicuous. Early whorls are bright yellow in color, while adult shells are white with irregular, dark red, axial markings. The operculum of *Entemnotrochus* (Figure 1E) is thick, dark brown, chitinous, multispiral and larger in diameter than the height of the shell aperture. Maximum shell diameter ranges from 17 to 28 cm, depending on species (Hutsell *et al.*, 1999).

The genus *Perotrochus*, as exemplified by its type species, *P. quoyanus* (Figure 1F–H) has shells that are more broadly conical, with straight to weakly concave profiles, and broad, short slits that never extend more than 60° from the aperture (Figure 1G, pms). The slit produces a selenizone that is at or slightly below mid-whorl (Figure 1I, sz). The outer lip on the abapical side of the emargination is significantly offset posteriorly (by as much as 30°) from the outer lip along the adapical side of the slit (Figure 1H). Shells lack an umbilicus, and instead have a nacreous, twisted columellar callus. Spiral sculpture remains prominent on all whorls, while axial sculpture may become reduced to weak cancellations upon the spiral cords, or become entirely absent. Shells tend to be more heavily pigmented in shades of brown to reddish brown, marked with irregular, darker, collabral banding. Opercula (Figure 1J) tend to be thinner, lighter in color and significantly smaller than half the minimum aperture dimension. Maximum shell diameters for most species are in the range of 4.2–7 cm, although some species reach 9.5 cm (Hutsell *et al.*, 1999).

The genus *Mikadotrochus* encompasses species characterized by a thick, heavy, conical shell (Figure 1K–M), with a straight to very weakly concave profile, and a very broad, short slit (Figure 1L, pms) that produces a wide selenizone below mid-whorl (Figure 1N, sz). The lip abapical to the selenizone is offset posteriorly (Figure 1M). No umbilicus is present, and the columella is thick and twisted. While spiral sculpture remains dominant on all whorls, axial ribs produce a strongly cancellated surface in many of the species. Shells tend to be strongly pigmented with browns and reds, although albino specimens are rarely encountered. Opercula (Figure 1O) are thicker and proportionally larger than in *Perotrochus*, spanning about 60% of the minimum apertural dimension. Woodward (1901: 217) reported that the operculum was absent in some specimens of *M. beyrichii*. He suggested that, due to its relatively small size, the operculum of *Mikadotrochus* (and probably also that of *Perotrochus*) is of little utility in closing the aperture for protection against predators. Rather, it is more likely used to support the shell and to protect the foot from abrasion by the shell. Maximum shell diameters for species of *Mikadotrochus* are in the range of 7–13 cm (Hutsell *et al.*, 1999).

A fourth group of pleurotomariids, nominally included in the genus *Perotrochus*, but informally segregated as "*Perotrochus* Group B" by Bayer (1965: 754), is distinguished on the basis of having large, thin, turbiniform shells (Figure 1P–R) with inflated whorls and weakly to strongly convex profiles. The slit is relatively broad and shallow (<60°; Figure 1Q, pms), and the selenizone is at or slightly below mid-whorl (Figure 1S, sz). The outer lip is strongly offset on the abapical side of the slit. The columella is thin and twisted. No umbilicus is present. Sculpture tends to be limited to fine spiral cords or very fine spiral threads that may be weakly cancellated. Pigmentation tends to be lighter than in *Perotrochus sensu stricto*, or in *Mikadotrochus* due to the thinness of the pigment-containing shell layer. Opercula (Figure 1T) are thin and amber in color, their diameters roughly three-quarters of the minimum aperture dimension. Maximum shell diameters range from 6.3 to 14 cm, depending on species (Hutsell *et al.*, 1999).

Figure 4 Shell ultrastructure of *Entemnotrochus adansonianus* (Crosse and Fischer, 1861). (A) Fracture section through ventral surface of shell, about 1 cm behind and parallel to outer lip. (B) Enlarged view of the entire outer prismatic layer and outer surface of the nacre. (C) Prismatic layer at shell edge, showing intersected crossed platy substructure. (D–E) Interfaces between intersected crossed platy substructure, regular simple prismatic crystals, and columnar nacre. (F) Cross-sectional and (G) planar views of columnar nacre. (H) Outer and (I) cross-sectional views of region containing selenizone. (J) Detail of crystal structure along the adapical side of the selenizone. (K) Detail at three-quarters of the distance between selenizone and suture, showing progressively thinning band of simple prismatic crystals between layers of columnar nacre. ncr, columnar nacre; pr, prismatic layer, spr, simple prismatic crystals, sz, selenizone.

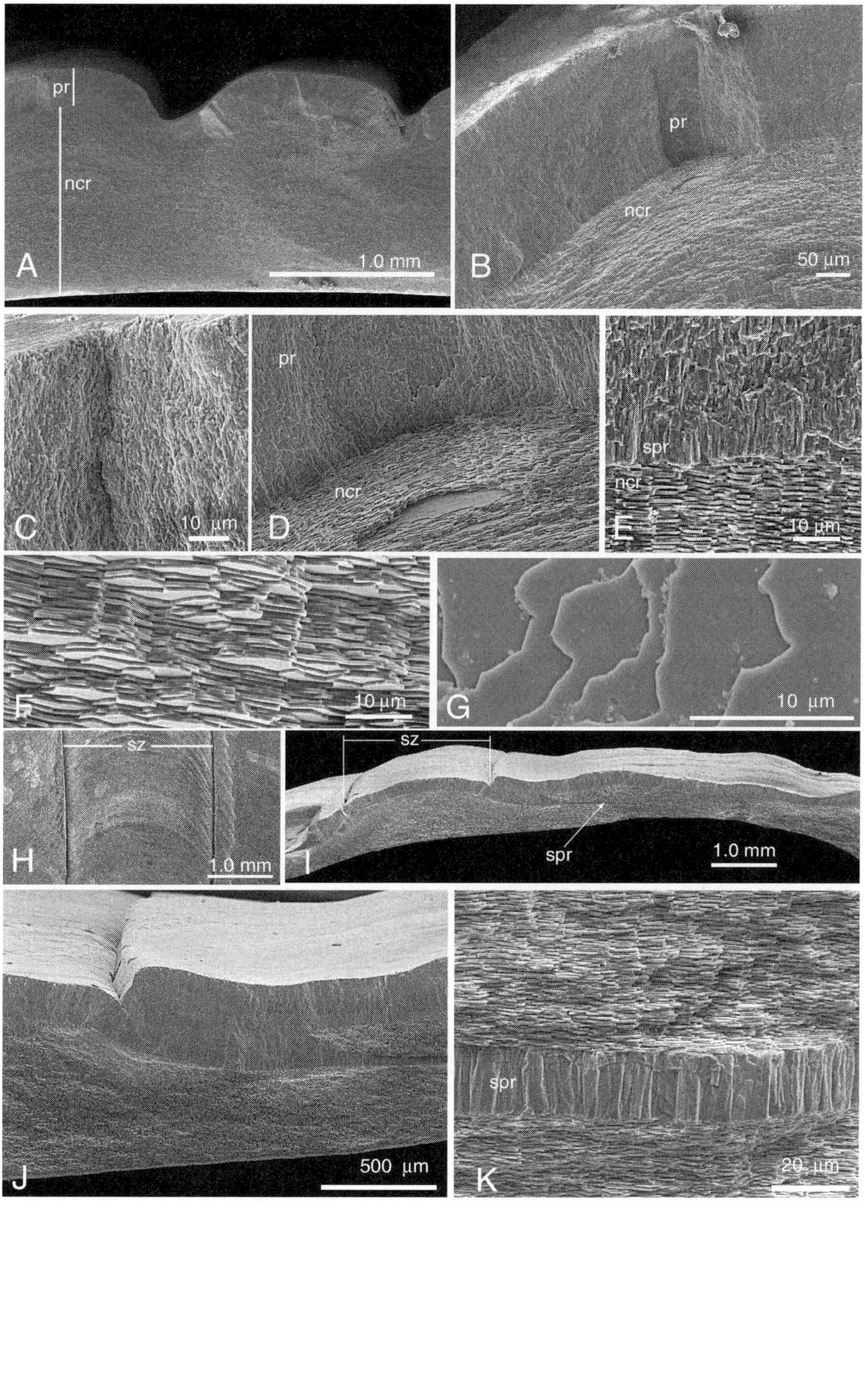
pr
ncr
1.0 mm
A
pr
ncr
50 µm
B
10 µm
C
pr
ncr
D
spr
ncr
10 µm
E
10 µm
F
10 µm
G
sz
1.0 mm
H
sz
spr
1.0 mm
I
500 µm
J
spr
20 µm
K

While the protoconch of pleurotomariids consists entirely of complex prismatic crystals (Erben and Krampitz, 1972), the teleoconch is composed of two primary layers. A thinner, outer prismatic layer (Figure 4A, B, D, pr) comprises most of the finer details of the surface sculpture and contains the shell pigments, while the thicker inner layer is nacreous (Figure 4A, B, D, ncr, F, G). Both layers are composed entirely of aragonite (Hedegaard, 1990: 144). Bøggild (1930: 299) identified these two layers in well-preserved Mesozoic taxa, while Batten (1972) confirmed their presence in five Paleozoic (Pennsylvanian) species, and reported the presence of two additional layers that he could only detect on the outer surface of a single specimen (possibly mineralized periostracum). Erben and Krampitz (1972) and Hedegaard (1990, 1997) further subdivided the outer prismatic layers of the extant genera into inner and outer regions with differing crystalline substructures. Erben and Krampitz (1972) documented ontogenetic changes in these layers in three living and one Oligocene species.

The outer prismatic shell layer of an adult specimen of *Entemnotrochus adansonianus* (Figure 4B, pr) is approximately 210 μm thick, and consists predominantly of intersected crossed platy substructure (Figure 4C), except for a thin (~15 μm) layer of regular simple prismatic crystals (Figure 4E, spr) immediately adjacent to the columnar nacre (Figure 4E, ncr). In *Mikadotrochus* and *Perotrochus* "Group B" (the latter based on *P. westralis* (as *P. teramachii*) in Hedegaard, 1990, 1997), the outermost layers consist of regular spherulitic prismatic crystals, grading into intersected crossed platy crystals. In *Mikadotrochus*, as in *Entemnotrochus*, there is a thin layer of simple prismatic crystals adjacent to the columnar nacre. In *Perotrochus* "Group B", the intersected crossed platy crystals grade into blocky prismatic crystals adjacent to the columnar nacre. The thick, homogeneous layer of columnar nacre comprises the major portion of most pleurotomariid shells. Specimens of several species in *Perotrochus* "Group B" may have an exceptionally thin outer prismatic layer, which becomes translucent, revealing the nacre beneath. It should be noted, however, that the nacreous layer does not overlie the prismatic layer along the free margin of the outer lip, so that a band consisting only of the prismatic layer is present along the margin of the slit and outer lip. This band is exceptionally broad and most evident in *Perotrochus gotoi* (see ventral views of shells in Anseeuw and Goto, 1996: 94–96).

The pattern of shell layers along the outer lip of the shell beyond the posterior margin of the slit becomes more complex, due to the superposition of the selenizone over previously deposited shell (Figure 4H–K). The material comprising the selenizone (Figure 4H–I, sz) fills the space between the slit edges on the outer surface of the shell. Along the interior surface, the prismatic layer contiguous with the selenizone extends adapically as far as the whorl suture, and abapically as far as the shell periphery, rapidly becoming thinner with increasing distance from the slit (Figure 4I–J). At a distance of a few millimeters from the slit edge, the prismatic layer is limited to only a thin (~16 μm) layer of simple prismatic crystals (Figure 4K, spr) between layers of columnar nacre. Hedegaard (1990: 70, plate 3, figure 4) described and illustrated the columnar nacre of *Perotrochus westralis* (as *P. teramachii*) as being intercalated with bands of regular, simple prismatic layers. It appears likely that the portion of the shell upon which this observation was based was located near the selenizone.

2.2. External anatomy (Plates 5 and 6, Figure 5A)

Table 1 provides an annotated list of previously published anatomical studies of pleurotomariids. The pleurotomariid body (Figure 5A) is

Table 1 Published observations on the anatomy of pleurotomariids (studies of radular morphology are listed separately in Table 2).

Taxon	Reference	Notes
Entemnotrochus adansonianus	Dall (1889)	Gross anatomy of poorly preserved specimen
Entemnotrochus rumphii	Tan (1974)	Gross anatomy
Perotrochus quoyanus	Dall (1889)	External morphology only
	Bouvier and Fischer (1898)	Radula and nervous system
Perotrochus amabilis	Fretter (1964)	Detailed anatomy on specimen lacking upper whorls, but otherwise well preserved
Perotrochus maureri	Harasewych and Askew (1993)	Anatomical features compared with *P. amabilis*
Perotrochus lucaya	Fretter (1966)	Detailed anatomy of major organ systems
Perotrochus atlanticus	Leme and Penna (1969)	Gross anatomy (as *P. notialis*)
Bayerotrochus midas	Fretter (1966)	Detailed anatomy of major organ systems
Mikadotrochus beyrichii	Woodward (1901); Bouvier and Fischer (1901, 1902); Sasaki (1998)	Each work contains independent, comprehensive descriptions of all major organ systems

conispirally coiled, consisting of $4\frac{1}{2}$ (*Mikadotrochus*, *Perotrochus* "Groups A and B") to $6\frac{1}{2}$ (*Entemnotrochus*) whorls. The mantle cavity (Figure 5B) is deep ($<\frac{1}{2}$ whorl – *Mikadotrochus*) to very deep ($>\frac{3}{4}$ whorl – *Entemnotrochus*). The slit (Figure 5A, B, ms) is shallow ($<60°$) to deep ($\sim160°$) in living animals of corresponding taxa, but may appear constricted and significantly shorter in preserved specimens. The mantle edge has long, pigmented papillae, especially along the margins of the slit, where they are capable of interdigitating along the midline. In living animals (Plates 5, 6), some portion of the slit is often closed, except for a small opening at the posterior margin that serves as the exhalant aperture of the functionally tubular mantle cavity (Plate 5B, F; see also photographs in Anseeuw and Goto, 1996: 13, 161; Okutani and Hasegawa, 2000: 38–39). Observations of western Atlantic species in aquaria revealed that the size and position of these openings may vary over time, and that multiple openings may occasionally be produced (Plate 5F).

The head has a broadly cylindrical snout (Figure 5A, B, sn) and long, bluntly rounded tentacles (Figure 5A, B, ten) with black eyes (Figure 5A, B, e) at their bases (even species that live well below the photic zone have pigmented eyes). Bifid or trifid tentacles have been reported to occur sporadically in several species (e.g., *M. beyrichii*, Woodward, 1901: 221; *P. maureri*, Harasewych and Askew, 1993: 131), and probably are the result of injuries. The ventral surface of the snout is flattened, unpigmented and covered by numerous papillae, which, in turn, are covered by a cuticle. The foot (Figure 5A, f) is long and narrow, its posterior portion trapezoidal in cross section when examined in preserved specimens, but is capable of substantial turgid inflation of the epipodial lobes in living specimens, particularly in *Perotrochus* "Group B" (Plate 6B–F). The snout and lateral surfaces of the foot of most taxa are mottled to varying degrees with dark reddish brown (which appears black at depths in excess of 25 m), while the dorsal, epipodial region, the sole of the foot and distal surface of the snout are unpigmented and usually a very light tan (Plates 5, 6). The posterior limit of the foot is not normally extended beyond the edge of the shell in living *Entemnotrochus* (Plate 5A–C), and protrudes only slightly in *Perotrochus* (Plates 5E–F, 6A) or *Mikadotrochus*. In *Perotrochus* "group B" (Plate 6B–F), however, the foot is approximately $1\frac{1}{2}$ times as long as the shell diameter and is normally engorged. The posterior portion of the epipodium enfolds dorsally until it meets the posterior margin of the shell (see photographs in Anseeuw and Goto, 1996: 9, 16). The multispiral operculum, secreted by an opercular lobe on the dorsal surface of the foot, supports the shell at approximately mid-radius. Broad epipodial folds (Figure 5A, epf) have minute papillae, but lack lappets, tentacles, or macroscopic sense organs in all species that were studied (Woodward, 1901: 220; Fretter, 1966: 605; Sasaki, 1998: 42).

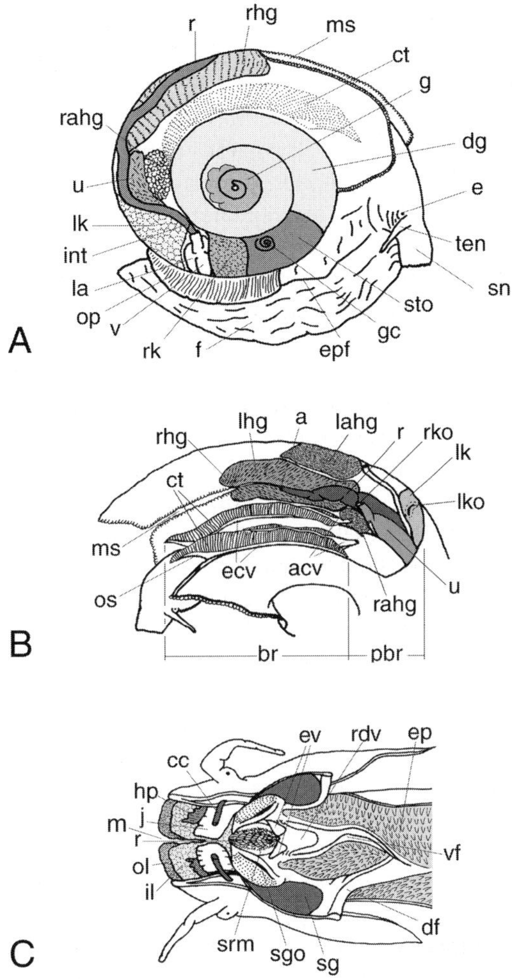

Figure 5 Gross anatomy of *Entemnotrochus adansonianus.* (A) Apical view of animal with shell removed. (B) Left lateral view of the pallial cavity, opened above left ctenidium. (C) Buccal cavity, opened along dorsal mid-line. a, anus; acv, afferent ctenidial vessel; br, branchial portion of mantle cavity; cm, columellar muscle; ct, ctenidium; df, dorsal fold; dg, digestive gland; e, eye; ecv, efferent ctenidial vessel; ep, esophageal pouch; epf, epipodial flap; ev, esophageal valve; f, foot; g, gonad; gc, coiled gastric caecum; i, intestine; la, left auricle; lahg, left accessory lobe of hypobranchial gland; lhg, left hypobranchial gland; lk, left kidney; lko, left kidney opening; ms, mantle slit; os, osphradium; pbr, post-branchial portion of mantle cavity; r, rectum; ra, right auricle; rd, radula; rdv, radular diverticulum; rahg, right accessory lobe of hypobranchial gland; rk, right kidney; rko, right kidney opening; rhg, right hypobranchial gland; sg, salivary gland; sgo, salivary gland opening; sn, snout; srm, subradular membrane; sto, stomach; ten, cephalic tentacle; v, ventricle; vf, ventral fold.

2.3. Mantle cavity organs (Figure 5A–B)

The mantle cavity of pleurotomariids (Figure 5B) is unusually deep (spanning about $\frac{1}{2}$ to $\frac{3}{4}$ whorl), and divided into the branchial (Figure 5B, br) and post-branchial (Figure 5B, pbr) regions. Unlike other gastropods, the gills of pleurotomariids do not extend to the rear of the mantle cavity. The branchial region spans the anterior $\frac{1}{2}$ (*Perotrochus* and *Mikadotrochus*) to $\frac{2}{3}$ (*Entemnotrochus*) of the mantle cavity, and contains the paired gills (ctenidia), osphradia and hypobranchial glands (each with an accessory lobe). While these organs maintain an approximately bilaterally symmetrical orientation on either side of the midline of the mantle cavity, defined by the slit and rectum, they are clearly asymmetrical in size. Organs on the left side of the body are larger and longer than their counterparts on the right. Ctenidia are bipectinate and suspended from the sides of the mantle cavity by long membranes containing efferent and afferent branchial vessels, nerves and retractor muscles. Each gill leaflet is broadly triangular in outline, and is stiffened by a skeletal support rod. Haszprunar (1987: 47) reported ctenidial bursicles in the gills of *Perotrochus caledonicus*, while Sasaki (1998: 211) noted their absence in *Mikadotrochus beyrichii*. The anterior end of each gill is not attached, but extends forward into the mantle cavity. The free ends of the ctenidia do not extend beyond the mantle edge in living animals, but may appear to do so in preserved specimens (see Fretter, 1964: figure 1; 1966: figure 1). Chemosensory osphradia are situated on the efferent membranes along the free ends of the ctenidia. The large, hypobranchial glands line the roof of the mantle cavity from just behind the posterior margin of the slit. A median furrow separates the smaller right from the larger left hypobranchial gland anteriorly, while the rectum separates them posteriorly. Each gland has an accessory lobe between its posterior margin and the base of each gill. Both the left hypobranchial gland and its accessory lobe are significantly larger than their counterparts on the right side of the body. These glands are bright white to ivory in color and are capable of rapidly secreting large volumes of a dense, opaque, white fluid that is heavier than sea water and not readily soluble in it (Plate 5D, see also Kanie *et al.*, 1986: figure 4). The post-branchial portion of the mantle cavity contains the rectum, and the openings of the left kidney and the urinogenital papilla of the right kidney.

2.4. Digestive system (Figures 5C, 6A–B, 7)

The mouth (Figures 5C, 6A, m) is situated at the center of the ventral, flattened face of the snout, and is surrounded by cuticle-covered papillae.

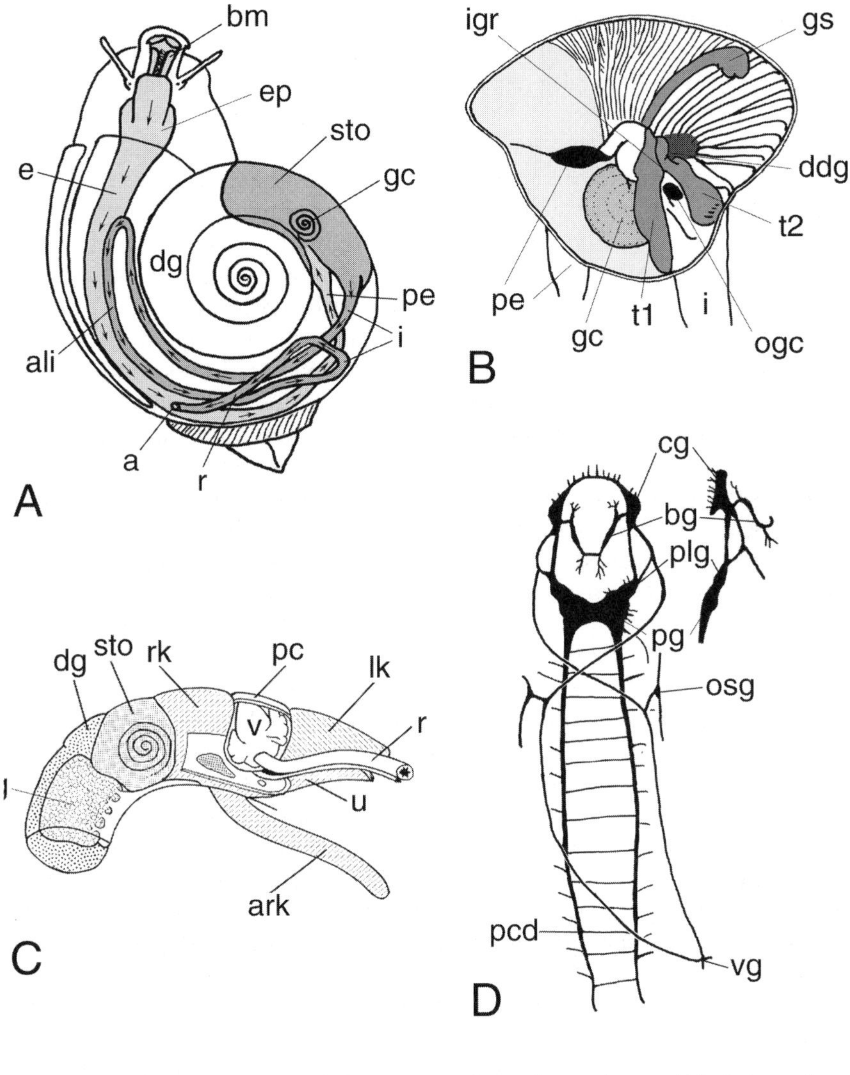

Figure 6 Gross anatomy of *Entemnotrochus adansonianus.* (A) Orientation of alimentary tract within animal, dorsal view. (B) Stomach, opened mid-dorsally. (C) Reproductive system and associated organs, viewed from right side. (D) Nervous system, viewed dorsally and laterally. a, anus; ali, anterior loop of intestine; ark, anterior lobe of right kidney; bg, buccal ganglion; bm, buccal mass; cg, cerebral ganglion; ddg, duct of digestive gland; dg, digestive gland; e, esophagus; ep, esophageal pouch; g, gonad; gc, coiled gastric caecum; i, intestine; igr, intestinal groove; lk, left kidney; m, mouth; osg, osphradial ganglion; pc, pericardium; pcd, pedal nerve cord; pe, posterior esophagus; pg, pedal ganglion; plg, pleural ganglion; r, rectum; rk, right kidney; sto, stomach; ty, typhlosole; u, urinogenital papilla; v, ventricle; vg, visceral ganglion.

It leads into a thick-walled buccal cavity (Figure 5C) within the snout that enfolds the buccal mass. The large, muscular buccal mass is surrounded by the ganglia of the nerve ring, and is attached to the body wall by numerous fine muscles. Sasaki (1998: 74, figure 25, a–d) provides detailed descriptions of the buccal musculature and cartilages of *Mikadotrochus beyrichii*. Paired jaws line the dorsal surface of the buccal cavity. Dall (1889) and Woodward (1901) believed that the jaws serve to protect the walls of the buccal cavity from abrasion by the radula and food particles, while Fretter (1966: 607) suggested that the jaws function to depress the anterior end of the odontophore as it is protruded in order to spread the radula into the feeding position. A lining of chitin protects the whole of the buccal cavity (Woodward, 1901: 232). In fact, the inner and outer lips, the jaws and the horny papillae (Fretter, 1966: figure 2) are all interconnected by chitin.

The highly specialized radula of pleurotomariids (Figure 7) has been described and illustrated by numerous authors (Table 2), and is the most complex gastropod radula, both in terms of the number of teeth per row and the number and types of highly specialized teeth. Hyman (1967: 236) coined the term "hystricoglossate" for the pleurotomariid radula because of the presence of bristle teeth that she believed to be unique to this family, but recognized that this was a variant of the rhipidoglossan radula. Hickman (1984a: 35) has since shown that *Seila*, an unrelated caenogastropod (Family Cerithiopsidae) that also feeds on sponges, has similar bristle teeth, and suggested that bristle tooth morphology is convergently derived as a specialized adaptation to feeding on sponge tissues.

The pleurotomariid radula is long (ranging from 88% of shell diameter in small *Perotrochus* to 68% of shell diameter in large specimens of *Entemnotrochus*), and consists of about 90–140 rows of teeth arranged in acutely angled, inverted V-shaped rows that produce a bifid posterior end. A transverse section through the radula (Figure 7A) will therefore intersect six or more rows of teeth. Each row of teeth is also conformationally asymmetrical (Hickman, 1981), being skewed either to the left (Figure 7D) or right (Figure 7B) side (Table 2), in order to accommodate the zipper-like alternation of the sickle teeth (Figure 7A) when the radula is retracted. Based on relatively few records, Harasewych and Askew (1993: 136) speculated that the direction of skew might prove to be constant within a species. However, additional data have shown that skew direction varies in at least some species (Table 2, *E. adansonianus* and *P. quoyanus*). The radula is contained in a straight radular sack that runs parallel to the esophagus. The radula has been divided into inner and outer (usually called marginal) tooth fields based on functional morphology (Table 2, A), and consists of six different, morphologically distinct, yet intergrading tooth types (Table 2, D). The inner field is composed of

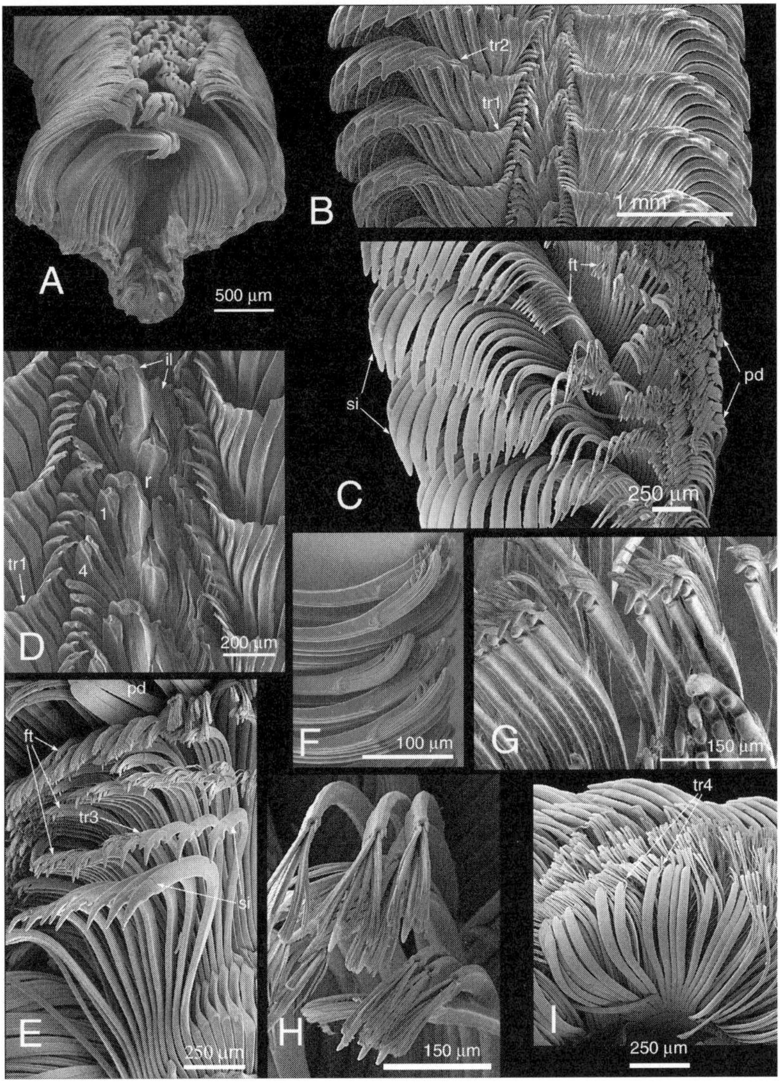

Figure 7 Radular morphology of *Entemnotrochus adansonianus*. (A) Cross-section of radular ribbon in retracted state, viewed from the posterior end. (B) Radula expanded to show rachidian, inner and outer lateral teeth, and sickle teeth. (C) Radula viewed laterally to show sickle, filament-tipped and paddle-shaped teeth. (D) Detail of rachidian, inner and outer lateral teeth. (E) Detail showing adjacent rows of, from bottom to top, outer sickle teeth, transition from sickle to filament teeth, three rows of filament-tipped teeth and paddle-shaped teeth. (F) Lateral, (G) frontal and (H) dorsal views of filament-tipped teeth. (I) Lateral view of radula showing transition from filament-tipped to paddle-shaped teeth. ft, filament-tipped teeth; pd, paddle-shaped teeth; r, rachidian tooth; si, sickle teeth; tr1, outer lateral teeth, shift of dominant cusp from inner to outer surface of tooth; tr2, transition from outer lateral to sickle teeth; tr3, transition from sickle teeth to filament-tipped teeth; tr4, transition from filament-tipped to paddle-shaped teeth. 1, 4, first and fourth outer lateral teeth.

Table 2 Comparison of radular formulae and asymmetries of pleurotomariid radulae.

Taxon	N	A:	Inner		Outer (marginal)			
		B: Rachidian	Lateral teeth			Marginal teeth		
		C: Rachidian	Central	Lamellate	Hooked	Brush	Flabelliform	
		D: Rachidian	Inner laterals	Outer laterals	Sickle	Filament-tipped	Paddle-shaped	Skew*
Entemnotrochus adansonianus[a]	7	1	1	21–23	17–19	25–33	14–16	4R/3L[a]
Entemnotrochus rumphii[b]	5	1	1	21–26	13–15	21–24	7–8	1R/1L[b]
Perotrochus quoyanus[c]	1	1	3	24	13	63	6/7	1R/1L[c]
P. lucaya[d]	1	1	3	25	13	61	6	?
P. amabilis[e]	1	1	3	24	21	63	8	R[e]
P. maureri[f]	5	1	2	24–26	21	61–63	8–9	5L[f]
P. atlanticus[g]	1	1	2	29	13	63	10	L[g]
P. caledonicus[h]	1	1	1	22	14	53	7	R[h]
Bayerotrochus midas[i]	1	1	3	26	13	63	6	L[i]
B. africanus[j]	2	1	3–4	23–25	12	50	6	?
B. diluculum[k]	2	1	3	18–26	12–14	32	?	R[k]
B. teramachii[l]	1	1	3	18	12	33	8	R[l]
Mikadotrochus beyrichii[m]	6	1	3	20–24	17–25	63–66	5–12	1R/1L[l]
M. hirasei[n]	3	1	3	20–28	24–32	58–81	6–10	L[n]
M. salmianus[o]	2	1	3	24	20–23	44–46	7–9	1R/1L[o]

? indicates that the character could not be inferred from the cited text or figure.
A = functional tooth fields. B = standard terminology for gastropod radular tooth types, e.g., Fretter and Graham (1962: 169); Brusca and Brusca (1990: 730). C = terminology of Woodward (1901). D = terminology of Hickman (1984a). *L = left-skewed asymmetry. R = right-skewed asymmetry.

[a] Both left-skewed (Plate 6D) and right-skewed (Plate 6B) radulae are illustrated herein. An additional right-skewed radula is shown in Anseeuw and Goto (1996: 179).

[b] Tan (1974) reported the radular formula (R-27-14-∞). Anseeuw and Goto (1996: 179) illustrated a right-skewed radula with a single inner lateral tooth and 23 outer lateral teeth. Kuramochi *et al.* (1996: table 2) reported counts for three specimens, and illustrated a left-skewed radula (their figures 1 and 3). Based on their figures, only a single inner lateral tooth is present in a half row of teeth.

[c] Hickman (1981: 190, 1984a: 30) illustrated a right-skewed radula, while Anseeuw and Goto (1996: 180) illustrated a left-skewed radula with seven paddle-shaped teeth. Bouvier and Fischer (1898) and Fretter (1966: 608) reported identical radular formulae.

[d] Fretter (1966) reported the radular formula.

[e] Fretter (1964) reported the radular formula. Harasewych and Askew (1993: table 2) reported that two radulae were right-skewed.

[f] Harasewych and Askew (1993) reported the radular formula and skew of five specimens.

[g] Leme and Penna (1969) reported the radular formula in their description of *P. notialis* (a synonym of *P. atlanticus* Rios & Matthews, 1968). Their figure 6 indicates a left-skewed orientation.

[h] Bouchet and Metivier (1982: 310, figure 2) reported the radular formula. Skew is based on their illustration.

[i] Fretter (1966: 608) reported the radular formula. Hickman (1981: 190, 1984a: 30) reported skew.

[j] Barnard (1963: 157) reported radular formula as approximate counts. Azuma (1964: 352) also provided approximate counts of an additional specimen.

[k] Partial radular formula and skew derived from illustrations of radula in Anseeuw and Goto (1996: 180). Kuramochi *et al.* (1996: table 2) provided partial tooth counts and SEM photos of a single, right-skewed radula (their figure 11).

[l] Kuramochi *et al.* (1996: table 2) reported the radular formula for a single specimen, determined to be right-skewed based on their figures 7 and 10.

[m] The radular formula for the three specimens examined by Woodward (1901: 252) is (R-3-21-17-63-7) as corrected by Harasewych and Askew (1993: table 2). Bouvier and Fischer (1902) reported a radular formula of (R-3-20-22-65-12), while Sasaki (1998: 47) reported the radular formula as being "roughly ca. 60-ca.40-1-ca.40-ca.60". [l] Kuramochi *et al.* (1996: table 2) provided counts for two additional specimens. Their figures 20 and 22 show a left-skewed radula.

[n] Radular formula derived from figure in Matsumoto (1929). The radula is depicted as roughly symmetrical in his drawing. [l] Kuramochi *et al.* (1996: table 2) provided counts for two additional specimens. Their figures 16 and 17 show a left-skewed radula.

[o] Kuramochi *et al.* (1996: table 2) reported tooth counts for two radulae, and illustrate one left-skewed and one right-skewed radula (their figures 13 and 14, respectively).

short, flat teeth, and includes the rachidian, and inner and outer lateral teeth, while the outer tooth field consists of the longer sickle, filament-tipped and paddle-shaped teeth (terminology of Hickman, 1984a). The inner and outer lateral teeth, as well as the sickle teeth, are homologous to the lateral teeth of other orthogastropods, while the filament-tipped and paddle-shaped teeth are homologous to marginal teeth.

The rachidian tooth (Figure 7D, r) is long and narrow, and has dorsally expanded lateral flanges, but lacks the dorsally reflected, cusp-bearing shaft present in other vetigastropods. Its main function appears to be to maintain the alignment between the lateral teeth. The rachidian tooth is generally asymmetrical and reflects the skew of the radula. The inner lateral teeth (Figure 7D, il), which range in number from 1 in *Entemnotrochus* to 3 in most *Perotrochus* and *Mikadotrochus*, are longer than the rachidian, have broad, outwardly directed, distally spatulate surfaces, and flank the rachidian on either side. Adjacent are 20–29 outer lateral teeth (Figure 7D 1,4), which begin nearest the inner lateral teeth as short, nearly flat, triangular teeth with a broad attachment area and a weak, dorsally recurved cusp at the anterior end. The outer lateral teeth become progressively longer and stouter, with the edge closest to the rachidian tooth becoming thicker with increasing distance from the rachidian. By the 15–16th tooth, the terminal cusp shifts from the anterior to the posterior edge of the tooth (Figure 7B, D, tr1). Subsequent teeth elongate and develop the sharp, recurved, multicuspate ends (Figure 7B, tr2) of the sickle teeth. The sickle teeth are the largest and most robust teeth in the radula. The more medially situated teeth are most strongly buttressed, becoming thinner, narrower, more blade-like and tricuspate closer to the outer margins of the radula (Figure 7C, E, si). The transition from sickle teeth to filament-tipped teeth is marked by the development of bristles along the convex side of the tooth opposite the most proximal cusp (Figure 7E, tr3). The bristles increase in length and number in successive teeth, while the cusps decrease in prominence (Figure 7C, E, ft, F, G, H) and are lost near the outer edges of the radula. While filament-tipped teeth are the most numerous (Table 2), they occupy a relatively small area along the margins of the radula (Figure 7C, I). The outermost teeth of each radular row are termed paddle-shaped teeth, and are broad, smooth, rounded distally, and lack cusps or filaments. The transition from filament-tipped teeth to paddle-shaped teeth (Figure 7I, tr4) spans 2–5 teeth. Based on their morphology and topological arrangement, Hickman (1984a: 35) suggested that the elements of the inner tooth field (rachidian + inner lateral teeth) function primarily to separate and orient the outer tooth field (outer lateral + sickle + filament-tipped + paddle-shaped teeth), which actively gather and prepare food. Woodward (1901: 252) inferred that the sickle teeth function in shredding and rasping tissues of the sponge prey of pleurotomariids, while filament-tipped teeth might "rasp some of the flesh from the spicules". The function of the paddle-shaped teeth is probably to provide a smoothly telescoping envelope to contain the filament-tipped teeth and to facilitate the unimpeded motion of the radula within its sheath during the rasping stroke (Harasewych and Askew, 1993: 132).

The anterior esophagus lies dorsally over the buccal mass, and is expanded to form buccal pouches that envelop the buccal mass laterally. Both the anterior and the mid-esophagus are lined with papillae. Paired salivary glands (Figure 5C, sg) flank the buccal cavity and empty into it via long, lateral openings. The mid-esophagus is elongated, while the posterior esophagus constricts to form a sphincter before joining the stomach. The stomach (Figure 6A, sto, 6B) is very large, U-shaped, with a large, spiral gastric caecum (Figures 5A, 6A, gc) composed of four whorls, and gastric shield. The large digestive glands, which comprise most of the apical 2–3 whorls of the visceral mass, open into the stomach by separate ducts. Two widely separated typhlosoles give rise to a broad intestinal groove that runs from the posterior digestive gland duct down to the intestine. The intestine (Figure 6A, int) descends into the cephalic hemocoel and runs anteriorly, parallel to the esophagus, before turning sharply on itself, and running posteriorly to the rear of the pallial cavity, where it again turns forward to pass through the pericardium and ventricle before entering the mantle cavity near its dorsal midline to form the rectum (Figures 5B, 6A, r). The rectum runs anteriorly and terminates in an anus, situated near the posterior midline of the hypobranchial glands. Arakawa *et al.* (1978) illustrated the feces of *Mikadotrochus hirasei*, reporting them to be round, 4–5 mm in diameter, and of the orthocylindric type (Arakawa, 1962).

2.5. Circulatory system (Figures 5A, 6C)

The elongated pericardium (Figures 5A, 6C, pc) is positioned transversely at the rear of the post-branchial portion of the mantle cavity, between the left and right kidneys. It contains the elongated, narrow ventricle, with the rectum passing through its long axis, two auricles, and a short, bifurcating bulbous aorta. The left auricle, fed by the efferent branchial vessel from the left ctenidium, opens into the anterior wall of the ventricle, while the right auricle, fed by the efferent branchial vessel from the right ctenidium, opens into the rear wall of the ventricle. The anterior aorta, which branches from the bulbous aorta, runs adjacent to the efferent branchial vessel before descending into the cephalic hemocoel, where it divides into the cephalic and pedal arteries below the buccal mass. The posterior aorta leads from the bulbous aorta to the visceral mass, where it branches to feed the stomach, digestive glands and gonad. The venous sinuses drain into the basibranchial sinus, situated below the posterior end of the mantle cavity. The basibranchial sinus sends blood into the afferent branchial vessels, and, via the rectal sinus, to the hypobranchial vein. After passing through the ctenidia, blood returns to the heart via the efferent branchial vessels. Fretter (1964: figure 4) provided a schematic diagram of the vascular system of *Perotrochus amabilis*.

2.6. Excretory system (Figures 5A, B, 6C)

The kidneys are large and asymmetrical, flanking the pericardium within the post-branchial portion of the mantle cavity. The left kidney (Figures 5A, B, 6C, lk) is smaller, anterior to the pericardium, and opens into the mantle cavity by a simple slit near its tapered anterior end. Its large, sac-like lumen has papillated walls that are thickest along the dorsal and posterior regions. A long, reno-pericardial canal extends from the pericardium near the right margin of the left auricle, runs along the floor of the left kidney, and opens near the left kidney opening to the mantle cavity. The right kidney (Figures 5A, 6C, rk) is situated posterior to the pericardium, and lines the anterior surface of the digestive gland and stomach. Its pallial section gives rise to a long, narrow lobe that penetrates into the cephalic hemocoel, where it runs anteriorly along the loop of the intestine to its anterior margin. The right kidney opens into the post-branchial portion of the mantle cavity via a long, cylindrical, urinogenital papilla that parallels the rectum, but opens posterior to the anus.

2.7. Reproductive system (Figures 5A, 6C)

The gonad (Figures 5A, 6C, g) of pleurotomariids lines the dorsal and columellar sides of the digestive gland. Its septate lumen empties via a simple, tubular gonadal duct that runs anteriorly to open into the right kidney (Figure 6C, ga), at the base of the urinogenital papilla (Figures 5B, 6C, u), which is glandular in mature females, but not in males. Fretter (1964: 179) observed that the urinogenital papilla of an immature female of *P. amabilis* was not glandular, and suggested that glands may develop in the wall of the papilla only when the gonads become mature. Healy and Harasewych (1992) noted that Fretter (1964) based these observations on one of the largest known specimens of *P. amabilis*, and instead hypothesized that the urinogenital papilla only becomes glandular seasonally during spawning.

Sexes of mature specimens may be discerned most readily by the color and texture of the gonad. The ovary is whitish and granular, with individual ova ranging from about 500 μm to nearly 1 mm in diameter, depending on species. Testes range from yellow to mustard brown and are much finer in texture.

Fretter (1966: 610) reported that the ova of *P. midas* are covered by a gelatinous sheath as thick as the diameter of the ovum that is produced by the ovum while still in the ovary. She hypothesized that, as in certain trochoideans, pleurotomariid eggs are embedded in a secretion from the glandular urinogenital papilla to form a spawn mass that is fertilized externally. Neither the spawn mass nor the details of oogenesis have been studied for any pleurotomariid.

The spermatozoa of pleurotomariids were first studied by Healy (1988), who noted that the sperm of *Perotrochus westralis* Whitehead, 1987 (as *P. africana* Tomlin, 1948) was primitive and unmodified, consisting of a conical acrosomal vesicle, a short, cylindrical nucleus with numerous lacunae, a midpiece of four roughly spherical mitochondria, and a long flagellum. Healy and Harasewych (1992) subsequently described in greater detail the ultrastructure of the sperm and spermatogenesis in *P. quoyanus*, and concluded that the sperm of Pleurotomariidae is more similar to that of Trochidae than to those of Haliotidae, Scissurellidae or Fissurellidae.

Although larval development has not been studied for any pleurotomariid, Bouchet and Metivier (1982: 313) reported that "*Perotrochus* species certainly have non-planktonic larval development . . .". The pleurotomariid protoconch has a large initial diameter (approximating the egg diameter), few, smooth whorls, a pronounced protoconch/teleoconch boundary, and lacks a protoconch II, as is typical of vetigastropods. These features are all indicative of the absence of a long, feeding (planktotrophic) larval stage (Shuto, 1974; Lima and Lutz, 1990). However, protoconch morphology cannot be used to distinguish between non-feeding pelagic development (as occurs in many vetigastropods), benthic development, where the larva develops within the egg mass (as occurs in some trochids, e.g., *Margarites*), or brooding (as in the scissurellid *Larochea*, and some trochids) (Sasaki, 1998: 189).

2.8. Nervous system (Figure 6D)

The nervous system has been described in considerable detail for *Perotrochus quoyanus* (Bouvier and Fischer, 1898) and *Mikadotrochus beyrichii* (Woodward, 1901: 240–247, figures 27–29; Sasaki, 1998: 48–49, figure 26d), while Fretter (1966: 610–611) provided comparative comments on the nervous systems of *Perotrochus midas* and *P. lucaya*. The nervous system of *Entemnotrochus adansonianus* is described here, and agrees in most features with that of other pleurotomariids.

The circumesophageal nerve ring is hypoathroid, with the pleural ganglia fused to the pedal ganglia, while the large cerebral ganglia are located anteriorly, at the base of the snout, near the cephalic tentacles. The labial ganglia are fused to the cerebral ganglia. The cerebro-pedal connectives are longer than, and ventral to, the cerebro-pleural connectives (Figure 6D). The cerebral ganglia innervate the tentacles and eyes, and join the buccal ganglia, situated along the dorsal surface of the buccal mass, via connectives. The buccal ganglia innervate the buccal muscles, while the cerebral commissure gives rise to numerous nerves leading to the buccal cavity and anterior face of the snout. The visceral loop is very long, and originates from the cerebro-pleural connectives, rather than from the pleural ganglia. Pleurotomariids lack distinct visceral ganglia, but the visceral loop innervates the osphradia (which have ganglia). A pair of statocysts are situated anterior to the pedal ganglia and are innervated from the cerebro-pleural connectives. The pedal cords run roughly in parallel along the length of the foot and are joined by numerous (>10) ladder-like commissures.

3. SYSTEMATIC RELATIONSHIPS

3.1. Morphology-based classifications (Figure 8)

3.1.1. *Relationships of Pleurotomarioidea to other Gastropoda*

Because of their abundance in Paleozoic and Mesozoic deposits, pleurotomarioideans have been well known to paleontologists since at least the seventeenth century (e.g., Lister, 1678). Early workers included pleurotomarioidean species in the genera *Trochus*, *Turbo* and *Cirrus* (e.g., Pulteney, 1813; Sowerby, 1816 in 1812–1846; Mantell, 1822) based on the general shape of their fossilized shells. During the 1820s, several authors began to distinguish fossils with trochiform shells that also bore a spiral band and a deep notch on the outer lip by applying permutations of the generic name *Pleurotomaria* (Sowerby, 1821 in 1812–1846; Ferussac, 1822 in 1821–1822; Blainville, 1825 in 1825–1827; Defrance, 1826). The International Commission on Zoological Nomenclature (1960: 276) subsequently ruled that the name *Pleurotomaria* should be attributed to Defrance. Early researchers at first included the genus *Pleurotomaria* in the family Trochidae, while noting that the shell morphology of some of the included species appeared more similar to the related family

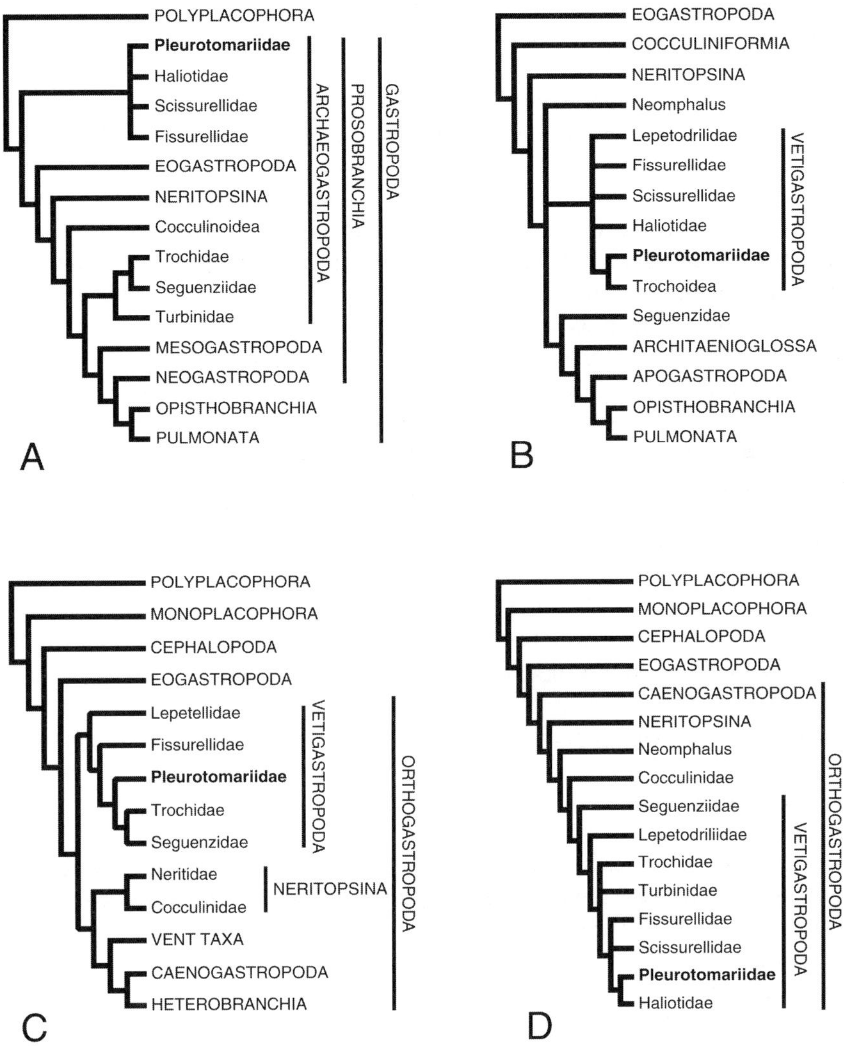

Figure 8 Relationships of Pleurotomarioidea to other Gastropoda based on morphological data. Trees constructed to represent major hypotheses of gastropod evolution, with taxon names and ranks modified to emphasize the position of the Pleurotomaridae in the respective classifications. (A) Classification of Thiele (1929) represented as a phylogenetic tree (modified from Ponder and Lindberg, 1996: figure 11.3.A). (B) Phylogenetic hypothesis of Haszprunar (1988: figure 5). (C) Phylogenetic hypothesis of Ponder and Lindberg (1996: figure 11.3.D). (D) Phylogenetic hypothesis of Sasaki (1998: figure 104).

Turbinidae (J. de C. Sowerby, 1832). In fact, Pleurotomariidae was originally proposed as a subfamily of Trochidae by Swainson (1840: 213), who proclaimed (1840: 223), "that it was impossible to arrange [fossil genera] with precision, from ignorance of the animal!"

The discovery and infrequent collection of living pleurotomariids during the second half of the nineteenth century (e.g., Fischer and Bernardi, 1856; Dall, 1889) gave rise to a series of publications documenting the novel anatomical organization and radular morphology of these animals (Tables 1 and 2). Systematists of the time varied in their views of the relationships of Pleurotomariidae, relying on the morphology of the shell or of one or a few organ systems in reaching their conclusions. Some continued to place it close to the family Trochidae (e.g., Adams and Adams, 1858), while others aligned it with the Scissurellidae, Haliotidae or Fissurellidae (e.g., Fischer, 1887: 845; Dall, 1889; Pilsbry, 1890: 69; Troschel, 1891). At the beginning of the twentieth century, a new view of gastropod evolution was advanced, based on the belief that gastropods had a bilaterally symmetrical ancestor (Thiele, 1902, 1935; Wenz, 1938, 1940). Under this hypothesis, pleurotomariids were considered to be the most primitive living gastropods. Their asymmetrically coiled shell and symmetrically paired pallial organs were regarded as transitional between extinct, planispirally coiled, bilaterally symmetrical ancestors, generally believed to be bellerophonts, and asymmetrical modern gastropods. This classification (Figure 8A) had subsequently gained almost universal acceptance (e.g., Yonge, 1947; Fretter and Graham, 1962, 1985; Taylor and Sohl, 1962; Morton and Yonge, 1964; Abbott, 1974; Boss, 1982; Hickman, 1984a, b; Vaught, 1989; Brusca and Brusca, 1990; Ruppert and Barnes, 1994; see Bieler, 1992 for a detailed review).

The rapid technological advances in deep-sea exploration during the past several decades have led to discoveries of a significant number of primitive, previously unknown gastropod groups, principally from hydrothermal vents. In addition, the unprecedented pace of development of molecular techniques and computational methods for reconstructing phylogeny has prompted a series of increasingly rigorous and quantitative reassessments of gastropod phylogeny (e.g., Golikov and Starobogatov, 1975; Salvini-Plawen, 1980; Haszprunar, 1988; Ponder and Lindberg, 1996, 1997; Sasaki, 1998). These studies all recognized that the patellogastropod limpets (in some cases together with extinct Paleozoic groups, = subclass Eogastropoda Ponder and Lindberg, 1996) were an early, monophyletic sister group to the remaining gastropods (subclass Orthogastropoda Ponder and Lindberg, 1996). So profound are the differences between Eogastropoda and the Orthogastropoda that some authors have questioned whether these limpets are gastropods (Termier and Termier, 1968; Shileyko, 1977; Tillier *et al.*, 1992).

The evolutionary relationships among the basal taxa included within Orthogastropoda, have not yet been reliably or robustly resolved, despite extensive, ongoing research. Using detailed anatomical data from many of the newly discovered deep-sea groups, Haszprunar (1988) published a hypothesis of gastropod phylogeny (Figure 8B) depicting the successive emergence of a series of higher taxa. In this classification, the Pleurotomariidae emerged as the sister taxon of Trochidae within the order Vetigastropoda (see Section 3.1.2), but far from the base of the gastropod tree. Using parsimony-based analyses and an expanded anatomical data set, Ponder and Lindberg (1996: figures 11.1.B, 11.3.D) published a phylogenetic hypothesis of gastropod evolution (Figure 8C), in which the Pleurotomariidae are also included within Vetigastropoda, which, however, emerges as the basal group within the Orthogastropoda. In order to achieve this tree topology, which avoided the "unacceptable reversals" in the number of gills, osphradia and auricles from paired to single then back to paired, the authors had to constrain several characters as irreversible. After further expanding their data set, these authors (Ponder and Lindberg, 1997: figures 2, 5) subsequently published a revised hypothesis, in which the Neritopsina (now including Cocculinidae) emerges at the base of the Orthogastropoda, and a series of hot vent families are placed at the base of a clade containing Vetigastropoda (including Pleurotomariidae).

Most recently, Sasaki (1998) re-analyzed the relationships among the lower gastropods, using newly generated anatomical data. The resulting consensus tree (Figure 8D) agreed with prior studies in separating the Eogastropoda and Orthogastropoda, and in segregating orthogastropods into the groups Vetigastropoda, Neritopsina and Caenogastropoda. However, resolution of the evolutionary relationships among these major groups, or of the families included within Vetigastropoda is not strongly supported, nor does it concur with prior trees. As in previous phylogenetic analyses, Sasaki's tree topology predicts the evolutionary loss of paired gills, osphradia and auricles, followed by their re-appearance (possibly by atavistic re-expression) within some Vetigastropoda.

3.1.2. *Composition of superfamily Pleurotomarioidea*

The ongoing studies of gastropod phylogeny have resulted in substantial revisions and speculations with regard to the taxa composing the superfamily Pleurotomarioidea. The most comprehensive work to date (Knight *et al.,* 1960) defined Pleurotomarioidea as containing 20 extinct families as well as the living families Pleurotomariidae, Scissurellidae and Haliotidae. This arrangement is followed in the majority of contemporary classifications (e.g., Hyman, 1967; Hickman, 1984b, 1998; Vaught, 1989). Based on the shared presence of several apomorphic anatomical features (including epipodial sense organs, ctenidial bursicles, an esophagus that is transversely sickle-shaped, with papillated lateral pouches, and an osphradium located along the efferent ctenidial membrane) the Pleurotomarioidea (defined to include the families Pleurotomariidae + Scissurellidae + Haliotidae) were united with the Fissurelloidea and Trochoidea, and questionably the Neomphalidae, in the order Vetigastropoda (Salvini-Plawen, 1980; Salvini-Plawen and Haszprunar, 1987). Boss (1982: 1092) added Neomphalidae to Pleurotomarioidea, but considered Fissurelloidea to be a separate superfamily. The inclusion of the Haliotidae and Scissurellidae within Pleurotomarioidea has been based primarily on the presence of a slit or series of tremata, and vestiges of bilateral symmetry in the mantle cavity. Both these groups appear in the fossil record in a recognizable form during the late Mesozoic (Hickman, 1984b: figure 2). Salvini-Plawen and Haszprunar (1987) recognized that no synapomorphies could be found to unite the living pleurotomarioidean families, and suspected that Pleurotomarioidea was polyphyletic. Haszprunar (1989) elevated each of the living families to superfamilies, and restricted Pleurotomarioidea to contain only Pleurotomariidae. Noting that the highly derived morphology of the pleurotomariid radula is not shared by Scissurellidae or Haliotidae, Haszprunar (1989: 15) hypothesized that the anatomy of the Paleozoic and Mesozoic families usually included in Pleurotomarioidea might have been more similar to that of living Scissurellidae than to Pleurotomariidae, and suggested that the fossil families previously included in Pleurotomarioidea might more appropriately be assigned to Scissurelloidea.

The most recent phylogenetic studies consider Pleurotomariidae to be the sister taxon of Trochidae + Seguenziidae (Ponder and Lindberg, 1996: figure 11.3.D), Trochidae (Haszprunar, 1988: figure 5), Haliotidae (Sasaki, 1998: figure 104) or Trochidae + Haliotidae (Ponder and Lindberg, 1997: figures 2, 5). However, many of the characters considered to be synapomorphic in the construction of these trees (e.g., fecal groove, epipodial sense organs, sensory papillae) are absent in Pleurotomariidae, and must be hypothesized to be secondarily lost within this family.

3.1.3. *Relationships of taxa within Pleurotomariidae*

The earliest records for the family Pleurotomariidae are from Upper Middle Triassic (Ladinian) strata (Tracey *et al.*, 1993). Of the 14 genera assigned to the family, more than half originated during the Jurassic and most were extinct by the end of the Cretaceous. All of the genera were defined on the basis of relatively few, easily recognizable shell characters. Classification systems for the diverse Mesozoic species tend to be based on keys, using shell characters in a particular sequence, and are phenetic. Most modern authors concede that generic distinctions are not always clear and that many taxa are not readily assigned to available genera (Cox, 1960a; Hickman, 1976; Szabó, 1980). Studies of fossil pleurotomariids tend to be of a regional, faunistic nature, with few attempts to assess relationships among the genera. Notable exceptions include the work of Szabó (1980), who produced a phenogram based on a dichotomous key of shell characters (Figure 9A), and Benfrika's (1984) dendrogram of hypothesized relationships (Figure 9B) in an unpublished thesis. Kanie *et al.* (1995: figure 4) published an evolutionary tree of the Cenozoic genera occurring in Japan.

The genus *Pleurotomaria* Defrance, 1826, was proposed based on a Jurassic type species (*Trochus anglicus* Sowerby, 1818 in 1812–1846) and is generally considered to have lived during the Lower Jurassic to the Lower Cretaceous (Knight *et al.*, 1960). Living species serve as the basis for the generic names *Perotrochus* Fischer, 1885, *Entemnotrochus* Fischer, 1885 and *Mikadotrochus* Lindholm, 1927. A number of authors have treated the living supraspecific taxa as, at best, subgenera of *Pleurotomaria* (e.g., Dall, 1889; Hickman, 1976, 1984a, b, 1998; Abbott and Dance, 1982), while others (e.g., Knight *et al.*, 1960; Bayer, 1965; Benfrika, 1984; Vaught, 1989; Koizumi, 1993; Endo, 1995; Kanie *et al.*, 1995; Anseeuw and Goto, 1996; Okutani and Hasegawa, 2000) believe that at least some of them merit generic rank.

In a review of the Recent species, Bayer (1965) not only recognized *Entemnotrochus*, *Mikadotrochus* and *Perotrochus* as valid genera, but went on to subdivide *Perotrochus* into three separate groups. He considered *Perotrochus* "Group A" to be restricted to species with small, conical shells, including *P. quoyanus*, the type species of *Perotrochus*. Species of *Perotrochus* with large, turbiniform shells and inflated whorls were segregated as *Perotrochus* "Group B", while *Perotrochus* "Group C" contained only a single species, *Perotrochus hirasei*. A synopsis of available information on living Pleurotomariidae has recently been published (Anseeuw and Goto, 1996) in which the genus *Perotrochus* is considered to include all living species other than the two assigned to *Entemnotrochus*. Okutani and Hasegawa (2000) divided the Japanese species among the genera *Entemnotrochus*, *Perotrochus* and *Mikadotrochus*, but followed Bayer (1965) by including *P. hirasei* in the genus *Perotrochus*.

There have yet to be studies of phylogenetic relationships among the Recent species based on cladistic analyses of morphological/anatomical data.

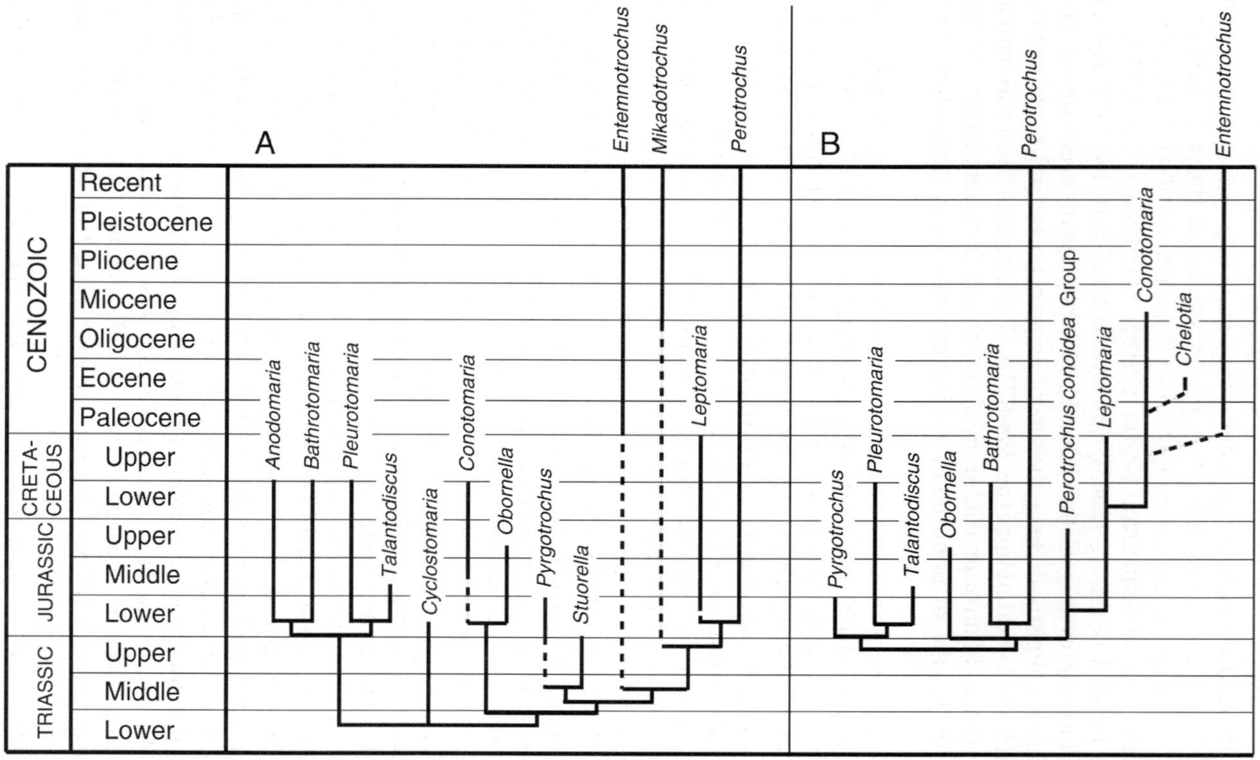

Figure 9 Relationships among living and fossil genera of Pleurotomariidae based on shell morphology and reflecting their representation in the fossil record. (A) Classification of Szabó (1980) represented as a tree, with taxon ages derived from Knight *et al.* (1960) and Benfrika (1984). (B) Phylogenetic hypothesis and taxon ages from Benfrika (1984: figure 21).

3.2. Classifications based on molecular data (Figure 10)

To date, there have been but a small number of phylogenetic studies to incorporate sequence data derived from any species of Pleurotomariidae, although some publications report on relationships of the Pleurotomarioidea based on sequences obtained from Haliotidae (e.g., Tillier *et al.*, 1992, 1994). Rosenberg *et al.* (1994) included *Perotrochus maureri* in a preliminary survey of gastropod and bivalve phylogeny based on a small and irregular sampling of taxa that were sequenced for a 150-nucleotide region from the D6 domain of the 28S RNA. Their results were anomalous in grouping the Pleurotomariidae in a clade with Neogastropoda, which together formed the sister group of the remaining caenogastropods in their study.

A more comprehensive investigation of the phylogenetic relationships of living pleurotomariids was conducted by Harasewych *et al.* (1997), who used data from both nuclear (~450 bp from the 18S rDNA gene) and mitochondrial (579 bp from the cytochrome *c* oxidase I (CO I) gene) genomes of seven species of Pleurotomariidae to determine the position of this family within the Gastropoda, as well as to delineate relationships among its genera and species. Their phylogeny of the Gastropoda based on partial 18S rDNA sequences (compare Figure 10A with Figure 8) most closely approximates the morphology-based hypothesis proposed by Haszprunar (1988). While the family Pleurotomariidae emerges within Vetigastropoda, it appears as the sister group of all remaining vetigastropod taxa, rather than of Trochidae or Haliotidae. All pleurotomariids sequenced were characterized by the presence of large insertions, especially in the regions coding for helices 10, E-10-1, and 11 of the 18S RNA, as well as by accelerated rates of sequence evolution in the 18S rDNA gene, when compared to other Vetigastropoda and Caenogastropoda in their study. These insertions were not only unique to, and diagnostic of Pleurotomariidae, but contained length variations that could themselves be used to differentiate between and infer relationships among pleurotomariid genera and species groups (Harasewych *et al.*, 1997: figures 7, 8).

Phylogenies based on partial 18S rDNA sequences (Harasewych *et al.*, 1997: figure 6), insert polymorphisms (Harasewych *et al.*, 1997: figure 7), and partial cytochrome *c* oxidase I sequences (Harasewych *et al.*, 1997: figure 10A), as well as on "total molecular evidence" (a combination of these three data sets), all reliably segregate *Perotrochus* "Group A" from *Perotrochus* "Group B", which together form a clade that comprises the sister taxon of *Entemnotrochus* (Figure 10B).

Since the publication of Harasewych *et al.* (1997), sequences for a comparable portion of the 18S rDNA were determined for one specimen of *Mikadotrochus beyrichii* (GenBank AF 417116) the type species of

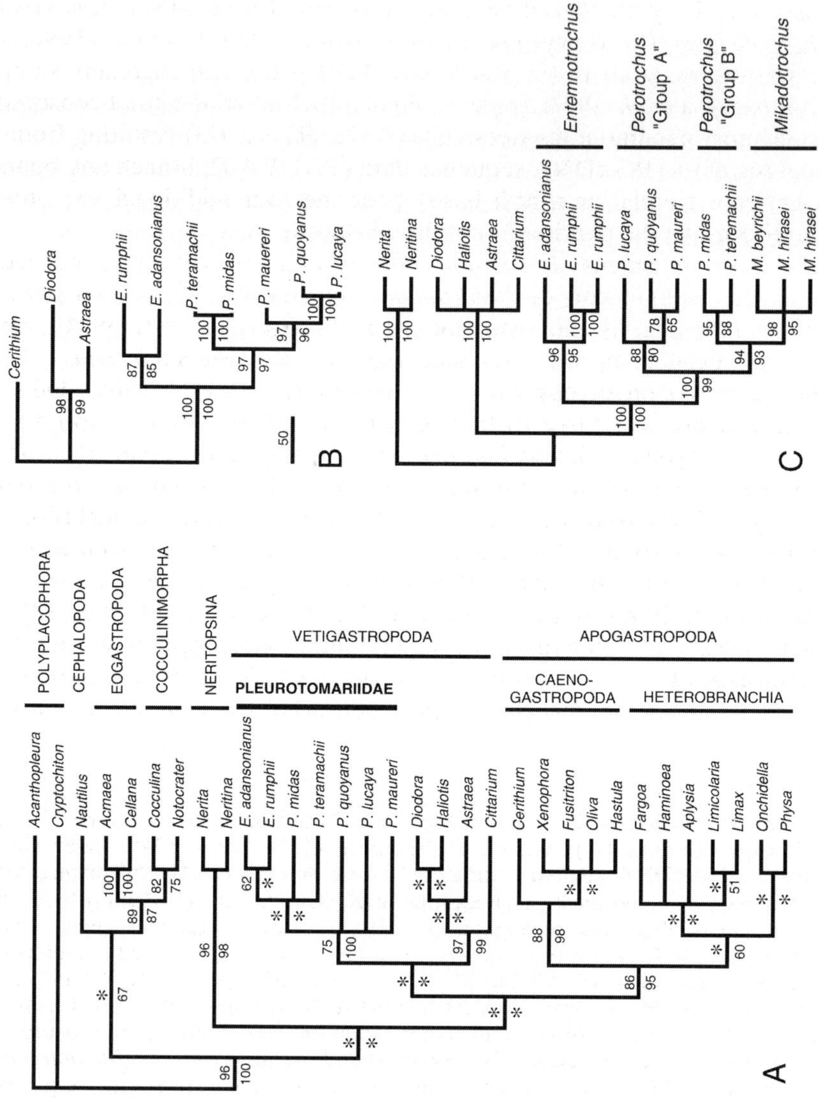
A
B
C
POLYPLACOPHORA
CEPHALOPODA
EOGASTROPODA
COCCULINIMORPHA
NERITOPSINA
VETIGASTROPODA
PLEUROTOMARIIDAE
APOGASTROPODA
CAENO-GASTROPODA
HETEROBRANCHIA
Acanthopleura
Cryptochiton
Nautilus
Acmaea
Cellana
Cocculina
Notocrater
Nerita
Neritina
E. adansonianus
E. rumphii
P. midas
P. teramachii
P. quoyanus
P. lucaya
P. maureri
Diodora
Haliotis
Astraea
Cittarium
Cerithium
Xenophora
Fusitriton
Oliva
Hastula
Fargoa
Haminoea
Aplysia
Limicolaria
Limax
Onchidella
Physa
P. mauereri
50
Entemnotrochus
Perotrochus "Group A"
Perotrochus "Group B"
Mikadotrochus
M. beyrichii
M. hirasei

Mikadotrochus, and two specimens of *Mikadotrochus hirasei* (GenBank AF 417117 and GenBank AF 417118), as well as a second specimen of *Entemnotrochus rumphii* (GenBank AF 417119). The three specimens of *Mikadotrochus* contained the large insertions characteristic of Pleurotomariidae. They all shared the pattern of insert segments characteristic of *P. midas* (see Harasewych *et al.*, 1997: figure 8), but differed in having a gap in insert segment 10 (at positions 318–320, using the alignment shown in Harasewych *et al.*, 1997: figure 3). Figure 10C shows the strict consensus of four most parsimonious trees (CI = 0.872, RI = 0.943) resulting from a reanalysis of the 18S rDNA sequence data (PAUP 4.02, branch and bound search, gaps treated as a fifth base) after the four additional sequences were aligned to the previously published data.

The three specimens of *Mikadotrochus* form a strongly supported clade that is the sister group of *Perotrochus* "Group B". These two groups form a clade that is the sister taxon of *Perotrochus* "Group A", and these in turn comprise the sister group of *Entemnotrochus*. This sequence-based phylogeny strongly corroborates Bayer's (1965) division of *Perotrochus* into three distinct groups. His *Perotrochus* "Group A" contains the type species of *Perotrochus*, and therefore retains the name *Perotrochus*. *Perotrochus* "Group C" is shown here to contain the type species of *Mikadotrochus*, and retains that generic name. As inclusion of *Perotrochus* "Group B" within *Perotrochus* would make *Perotrochus* a paraphyletic taxon, the name *Bayerotrochus* (type species: *Perotrochus midas* Bayer, 1965) is proposed here for *Perotrochus* "Group B" of Bayer (1965) in recognition of Dr Bayer's pioneering work on the Pleurotomariidae of the western Atlantic and his early recognition of this most widespread and diverse pleurotomariid clade. *Bayerotrochus* is

Figure 10 Phylogenetic relationships of Pleurotomariidae, inferred from DNA sequence data. (A) Position of Pleurotomariidae within Gastropoda based on partial 18S rDNA sequence data (after Harasewych *et al.*, 1997: figure 5B). Strict consensus of 60 most parsimonious trees (CI = 0.779, RI = 0.845) based on a data matrix that was reweighted by the rescaled consistency index (base weight = 1). (B) Relationships among living species based on "total molecular evidence", consisting of 559 bp of 18S rDNA + 579 bp of CO I + 19 indel characters (after Harasewych *et al.*, 1997: figure 10B). Single most parsimonious tree (CI = 0.729; RI = 0.685). (C) Relationships among living species based on partial 18S rDNA sequences. Newly produced sequences for *Mikadotrochus beyrichii* and *M. hirasei* aligned against previously published sequences (Harasewych *et al.*, 1997) using Clustal W. PAUP 4.02 [Branch and bound search; gap = 5th base] was used to produce this strict consensus of 4 most parsimonious trees (CI = 0.894, RI = 0.944). Bootstrap proportions are shown above the nodes, jackknife proportions below the nodes. An asterisk denotes a value below 50.

defined as comprising all species sharing a more recent common ancestor with the type species, *B. midas*, than with *Mikadotrochus beyrichii*, the type species of its sister group. Morphologically, it may be identified by its large, thin, turbiniform shell with inflated, convex whorls, its broad, shallow slit, forming a selenizone slightly below mid-whorl, its lack of an umbilicus, its thin prismatic layer with weak sculpture generally limited to spiral threads, and its relatively large operculum. *Bayerotrochus* has a distinctive, large, tumescent foot that extends beyond the shell and enfolds its posterior edge. In *Bayerotrochus*, the gene coding for the 18S RNA contains a three-base insert (aligned positions 318–320, Harasewych *et al.*, 1997: figure 3) within helix E-10-1 that is not present in any other pleurotomariid. In addition to the type species, *Bayerotrochus* contains *B. africanus* (Tomlin, 1948), *B. charlestonensis* (Askew, 1988), *B. diluculum* (Okutani, 1979), *B. metivieri* (Anseeuw, 1999), *B. pyramus* (Bayer, 1967), *B. tangaroana* (Bouchet and Métivier, 1982), *B. teramachii* (Kuroda, 1955) and *B. westralis* (Whitehead, 1987).

4. ECOLOGY

4.1. Geographic distribution (Figure 11)

Although pleurotomariids had a global distribution during the Paleozoic and Mesozoic, ranging from Alaska (Rohr and Blodgett, 1985) and the islands of Arctic Canada (Rohr and Smith, 1978) to Antarctica (Oleinik, personal communication), Recent species are confined to tropical and temperate latitudes (about 35°N to 30°S). All living species are limited to continental margins and edges of tectonic plates along the western shores of the Atlantic and Pacific Oceans, and all but the southern margin of the Indian Ocean, including northwestern Australia, southeastern Asia and the intervening islands and archipelagos (Figure 11; see also Anseeuw and Goto, 1996 for review).

The greatest species diversity occurs in the Philippines and southern Japan, while the Caribbean Sea has the greatest number of species among the western Atlantic faunas (Figure 11). These areas also have the only known post-Miocene fossil records for pleurotomariids (Hickman, 1976: figure 2), although this may be an artifact of sampling. Zezina (1997: 401) noted that the bathyal faunas along the western margins of oceans are far more diverse than those along the corresponding eastern margins (by factors as high as 25 times in the Pacific), due in part to the deleterious effects of upwelling upon benthic taxa, especially over long intervals of geologic time. The distribution of living pleurotomariids is

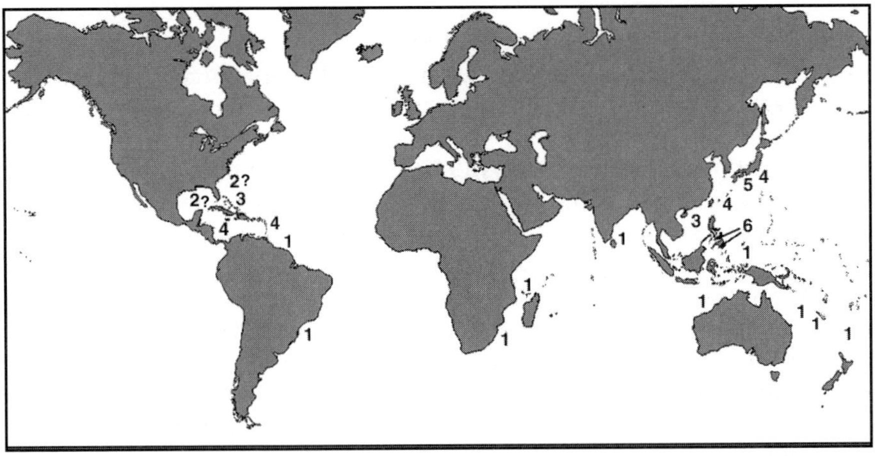

Figure 11 Map showing geographic distribution of living Pleurotomariidae. The numbers indicate the number of species that have thus far been reported from the respective regions.

highly congruent with those bathyal regions that have been colonized by warm water species (Zezina, 1997: figure 8).

Examination of the distribution of individual pleurotomariid genera reveals finer scale patterns. The genus *Entemnotrochus* is represented in the Recent fauna by two wide-ranging species, one in the western Atlantic, and the other in the western Pacific. *Entemnotrochus adansonianus* occurs in the warmer waters of the western Atlantic, from Bermuda southward, throughout the Gulf of Mexico and the Caribbean, to near the mouth of the Amazon River, while *E. rumphii* is known from the southern islands of Japan to the central Philippines and Indonesia. Species of *Perotrochus sensu stricto* (i.e., *Perotrochus* "Group A"), however, have much smaller geographic ranges. Three species occur allopatrically within the range of *E. adansonianus* (*P. amabilis* – Gulf of Mexico and the Florida Keys, *P. lucaya* – Bahamas, *P. quoyanus* – Caribbean Sea and Bermuda), while two species live beyond the range of *E. adansonianus* (*P. maureri* – off South Carolina, *P. atlanticus* – off Southern Brazil). Of the three Pacific species included in *Perotrochus*, only *P. gotoi* lives within the range of *E. rumphii*, while *P. caledonicus* and *P. deforgesi* live beyond the southern limit of its known range. More intriguing are the distributions of species of *Mikadotrochus*. Three species (*M. beyrichii, M. hirasei, M. salmianus*) appear to have broad geographic distributions ranging from central Japan to the Philippines, while two (*M. anseeuwi, M. vicdani*) are restricted to the Philippines. Okutani (1963: figure 7) documented a pattern of longitudinal replacement and allopatry for the three species occurring along the insular shelves and submarine banks off southern Japan, a pattern later confirmed by Kanie *et al.* (1995). More detailed studies over larger areas, particularly in the Philippines, may reveal fine scale allopatry among the species of *Mikadotrochus*. The genus *Bayerotrochus* (= *Perotrochus* "Group B") has the broadest geographic distribution both on a global scale and within the Indo-Pacific region. It is the only genus to be represented along the margins of the Indian Ocean. Within the western Pacific, it represents the northernmost and southernmost records for the family. Within the western Atlantic, the range of *B. charlestonensis* approximates the northern limits of *E. adansonianus*, but the southernmost records for *Bayerotrochus* are from the Caribbean Tectonic Plate, far short of the southern limits of *Entemnotrochus* or *Perotrochus*. The Caribbean species *B. midas* and *B. pyramus* are known to co-occur throughout the Antilles, and both species have been collected on the same submarine dives. Most Indian and Pacific Ocean species of *Bayerotrochus* have clearly allopatric geographic distributions, although both *B. teramachii* and *B. metivieri* have been reported from the South China Sea. Records for these species are too sparse and imprecise to determine the degree to which their ranges overlap.

The geographic distribution of pleurotomariids, once thought to be patchy and limited to islands of hard substrata, steep walls and tops of submarine banks, was attributed to the hypothesized incompatibility of the primitive mantle cavity and bipectinate gills with significant sediment load (Hickman, 1984b). However, both Thomas (1940) and Batten (1958) reported substantial dwarfing in Paleozoic pleurotomariids from depositional environments indicative of fine, suspended sediments, confirming that these animals could at least survive significant turbidity.

4.2. Bathymetric distribution and zonation (Figure 12)

Figure 12A summarizes the bathymetric ranges of the living species of Pleurotomariidae based on records from the literature and museum collections. Geographically co-occurring species often appear to have overlapping bathymetric ranges, suggesting that two or more species, usually from different genera, might be expected to live in sympatry. However, detailed observations from multiple submersible dives transecting the depth ranges of all pleurotomariid species at localities throughout the Bahamas and off Guadeloupe (Table 3, Figure 12B) clearly show that western Atlantic pleurotomariids inhabit sharply demarcated, non-overlapping bathymetric zones, each corresponding to a distinct bottom topology and habitat. *Entemnotrochus adansonianus* inhabits a relatively narrow zone of substratum that is encrusted with diverse large

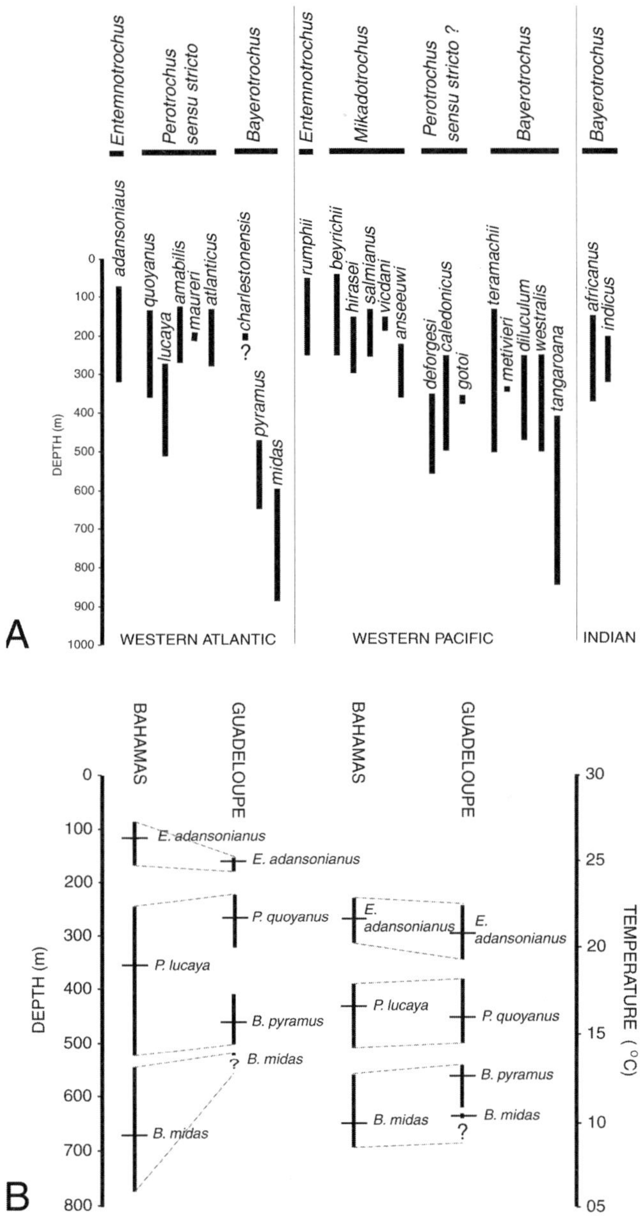

Figure 12 Distribution of living Pleurotomariidae by depth and temperature. (A) Depth ranges for living species of Pleurotomariidae based on museum and literature records. (B) Zonation of living Pleurotomariidae by depth and temperature based on *in situ* observations from the submersibles JOHNSON-SEA-LINK I and II, at two locations in the western Atlantic.

Table 3 Depth and temperature distribution of western Atlantic Pleurotomariidae (based on direct observation from submersibles).

Taxon/locality	Number of observations	Depth (meters) Mean	SD	range	Temperature (°C) Mean	SD	range
Entemnotrochus adansonianus							
Bahamas	35	114	21	87–169	21.8	1.7	20.7–23.0
Navassa	8	156	23	123–179	23.0	1.3	21.5–24.9
Guadeloupe	4	170	9	156–178	20.7	1.4	19.2–22.3
Perotrochus maureri							
Charleston Lumps	56	202	9	198–221	9.8	0.1	9.6–9.9
Perotrochus lucaya							
Bahamas	14	354	115	244–520	16.5	1.6	14.0–17.9
Perotrochus quoyanus							
Guadeloupe	11	263	30	223–322	16.0	1.4	14.1–18.0
Bayerotrochus pyramus							
Guadeloupe	5	361	31	306–399	12.5	0.9	11.4–13.7
Bayerotrochus midas							
Bahamas	26	664	83	542–774	9.9	1.4	8.5–12.9
Navassa	2	655	–	638–672	9.0	–	8.6–9.3
Guadeloupe	1	413	–	–	10.7	–	–

sponges and runs along the middle portion of a vertical to somewhat undercut wall (Plate 5A–B). This wall extends from the reef break downward to a talus slope that further descends at a pitch ranging from 15 to 45°, and is covered by rocks, rubble and fine sand. Dead shells and an occasional living *E. adansonianus* are rarely encountered at the base of the wall, while species of *Perotrochus* achieve their greatest density along the middle region of the talus slope, among small sponges and stalked crinoids (Plates 5E, 6A). At the base of the talus slope, the substrate again descends more steeply (45–85°) and consists of sharply faceted outcrops covered by a thin veneer of sediment, with infrequent outcrops of large sponges and stalked crinoids. *Bayerotrochus* inhabit the lowest levels of the talus slope (Plate 6B–D) and the upper regions of the sharply descending wall (Plate 6E–F). The depths at which the inflection points between these topographic regions occur vary widely throughout the western Atlantic and Caribbean, and no doubt account for a substantial portion of the variation in the reported bathymetric range for each species.

Transects conducted along the tropical western Atlantic might be expected to encounter the wall-dwelling *Entemnotrochus adansonianus*, a single species of *Perotrochus* along the talus slope, and one species of *Bayerotrochus* on the lower wall, the species of *Perotrochus* and *Bayerotrochus* being determined by the location of the transect. Only in the Antilles (Guadeloupe and Navassa) have four species of pleurotomariids been encountered along a single transect. The species

Bayerotrochus pyramus occurs along the lower portion of the talus slope, a habitat usually populated by *Perotrochus*. However, the temperature of the water in which *B. pyramus* was collected is more typical of the habitat of *B. midas* than of *Perotrochus* (see Figure 12B), possibly allowing for a finer partitioning of the habitat.

Despite more than a decade of observations in submersibles spanning dozens of transects, two species of western Atlantic pleurotomariids have never been found sympatrically.

Fewer data are available regarding the habitats of western Pacific species. Kanie *et al.* (1986) observed *Mikadotrochus beyrichii* on a sandy bottom around rocks, in an area of rapid current populated by stalked crinoids. Kanie *et al.* (1995) deduced that the genus *Mikadotrochus* adapted to rocky substrata after diverging from *Bayerotrochus* some time during the Miocene. Okutani and Hasegawa (2000) reported that species of *Mikadotrochus* inhabit rocky bottoms, while *Entemnotrochus rumphii* and species of *Bayerotrochus* occur on coarse, sandy bottoms. Given the greater density of species in close geographic proximity and their narrower bathymetric ranges, it is possible that Pacific pleurotomariids may partition their habitat more finely, based on factors other than temperature or depth. Rex (1977: 525) hypothesized that biological interactions rather than physical factors are responsible for different rates of zonation among groups occupying different trophic levels in the deep sea. He noted that, where interspecific competition is absent, species ranges may overlap along an environmental gradient. However, in areas where species compete for food, they will occupy mutually exclusive zones along an environmental gradient.

Despite the lack of detailed data, there are general similarities between the bathymetric patterns of the western Pacific and the western Atlantic. In both faunas, *Entemnotrochus* is the genus with the shallowest bathymetric range, *Bayerotrochus* inhabits the deepest waters, while *Perotrochus* occupies intermediate depths. *Mikadotrochus*, however, lives at the same depths as *Entemnotrochus*, but inhabits rocky rather than sandy substrata.

4.3. Biogeography

Attempts to infer the biogeographic history of the Pleurotomariidae in the context of the phylogenetic hypothesis shown in Figure 10C reveal that the most basal divergences segregate along bathymetric rather than geographic boundaries. The boundary between *Entemnotrochus* and the remaining genera corresponds to the shelf-slope transition, with *Entemnotrochus* inhabiting outer shelf depths, while the remaining genera segregate into zones along the upper continental slope.

Within the bathyal zone, *Bayerotrochus* has the broadest geographical range, spanning the Indo-Pacific, Caribbean and North Atlantic regions from which pleurotomariids have been reported. Its absence from the Atlantic coast of South America may either be an artifact of insufficient sampling at appropriate depths and on suitable substrata, or may reflect a faunal boundary between the Caribbean and South American tectonic plates. The lack of records from Bermuda and the Gulf of Mexico may likewise be a sampling artifact.

In the western Atlantic, the boundary between *Bayerotrochus* and *Perotrochus* is entirely bathymetric, as the two genera co-occur throughout the known range of *Bayerotrochus*. In this context, the single record of *B. charlestonensis* from 200 m must be considered anomalous, as this species has not been collected during two subsequent submersible cruises to the type locality. Based on available data, patterns of bathymetric zonation are less clear cut in the western Pacific, where the taxa assigned to *Perotrochus* (*P. caledonicus*, *P. deforgesi* and *P. gotoi*)[1] separate most clearly from *Mikadotrochus*, but appear to overlap in depth range with *Bayerotrochus* in the case of the *P. gotoi*, the only species presently known to overlap in geographic range with *Mikadotrochus* or *Bayerotrochus*. Similarly, the bathymetric ranges of all species of *Mikadotrochus* overlap to a large extent with both *Entemnotrochus rumphii* and *Bayerotrochus teramachii*, with which they share geographic ranges. It is unclear whether these species partition their habitats by substratum type, or whether they are truly sympatric.

Based on the fossil record, the divergence between *Entemnotrochus* and *Perotrochus* of authors (*Perotrochus* + *Mikadotrochus* + *Bayerotrochus*) dates to at least the Jurassic, when these taxa or their ancestors were members of shallow-water faunas. The bathymetric segregation of these lineages clearly predates the isolation of the Atlantic from the Pacific faunas. The time of divergence between *Bayerotrochus* and *Perotrochus*, and their bathymetric zonation along the upper continental slope is less clear. If the clade containing *Perotrochus caledonicus*, *P. deforgesi* and *P. gotoi* proves to be the sister taxon of the western Atlantic species of *Perotrochus*, this would indicate that the divergence of *Perotrochus* and *Bayerotrochus* predates the separation of the Atlantic and Pacific faunas, but that the bathymetric zonation of these genera occurred only in the

[1] These three species have been included in the genus *Perotrochus sensu lato* based on overall conchological similarity and their small size. However, they share some characteristics (e.g., shell and body pigmentation patterns) that are not present in *Perotrochus sensu stricto*. The relationship of this likely monophyletic group to *Perotrochus* and *Mikadotrochus* is not yet clear.

western Atlantic. The radiation of *Perotrochus* to form geographically allopatric western Atlantic species occurred after the bathymetric limits of the lineage had been confined. Should the clade containing *Perotrochus caledonicus*, *P. deforgesi* and *P. gotoi* prove to be more closely related to *Mikadotrochus*, it would suggest an independent and possibly more recent bathymetric zonation along the western Pacific. Although additional, more precise data will be required to reconstruct bathymetric patterns of pleurotomariid evolution in the western Pacific, it appears clear that, within genera, species tend to be geographically isolated.

4.4. Diet

With a single exception (Barnard, 1963), all published reports indicate that the gut contents and/or feces of pleurotomariids consist primarily of sponge spicules, but commonly contain foraminiferal tests and diatoms (Table 4). These findings have been variously interpreted as being

Table 4 Reports of gut contents/feces of pleurotomariid gastropods.

Species	Food	Reference
Entemnotrochus adansonianus	Sponge spicules	Herein
Entemnotrochus rumphii	Sponge spicules	Tan (1974)
Mikadotrochus beyrichii	Sponge spicules, order Poecilosclerida	Woodward (1901: 252)
Mikadotrochus hirasei	Sponge spicules, orders Haplosclerida, Poecilosclerida and Hadromerida	Arakawa *et al.* (1978: table 1)
Perotrochus quoyanus	Sponge spicules	Herein
Perotrochus amabilis	Sponge spicules, foraminiferans, diatoms and algal fragments	Fretter (1964: 182)
Perotrochus maureri[a]	Sponge spicules, foraminiferans and diatoms	Harasewych *et al.* (1988: 97)
Bayerotrochus pyramus	Sponge spicules, octocorals	Anseeuw and Goto (1996: 13)
Bayerotrochus africanus	Amorphous mass with a few tiny foraminiferans	Barnard (1963: 156)
Bayerotrochus midas	Sponge spicules, foraminiferans and diatoms	Harasewych *et al.* (1988: 97)
	Stalked crinoid	Anseeuw and Goto (1996: 123)

[a] Originally reported as *Perotrochus amabilis*.

indicative of a diet consisting of a single species of sponge (Woodward, 1901: 252), a larger diversity of sponge taxa (Arakawa *et al.,* 1978: 118), a diet consisting principally of sponges (Thiele, 1935: 1129; Hyman, 1967: 360; Yonge and Thompson, 1976: 52), a diet of encrusting invertebrates, primarily sponges (Hickman, 1984a: 29), or detrital feeding (Fretter and Graham, 1976: 1). When kept in aquaria, pleurotomariids have been observed to consume a wide variety of organisms and tissue. Both *M. beyrichii* (Matsumoto *et al.*, 1972) and *P. teramachii* (Sekido *et al.*, 1976) were reported to feed on slices of raw fish (*Trachurus japonicus*), while *M. hirasei* fed on the meat of the bivalve mollusk *Tapes philippinarum* and the starfish *Asterina* sp. (Arakawa *et al.*, 1978).

Based on *in situ* observations of feeding corroborated by gut content analyses, Harasewych *et al.* (1988) documented that *Bayerotrochus midas* and *Perotrochus maureri* (as *P. amabilis*) feed predominantly and selectively on sponge tissue, and suggested that the diatoms and foraminiferal tests present in the guts of pleurotomariids are derived from sediments that coat the surfaces of sponges. Subsequent observations, both *in situ* and in shipboard aquaria, confirmed that *E. adansonianus*, *P. quoyanus*, *P. lucaya*, *B. pyramus* and *B. midas* are all primarily spongivores. However, *B. midas* has been photographed eating a stalked crinoid (Anseeuw and Goto, 1996: 123) and *B. pyramus* was photographed feeding on an octocoral (Plate 6B). Both specimens were collected and esophageal contents confirmed that prey tissues were ingested.

Woodward (1901: 252) was the first to suggest that the pleurotomariid radula, with its multiple, distinctive tooth types, was specialized for feeding on sponges. The presence of morphologically similar, filament-tipped teeth in *Seila*, an unrelated ptenoglossan caenogastropod that also feeds on sponges, prompted Hickman (1984a: 35) to infer that this brush-like tooth morphology is a convergently evolved, functional adaptation to sponge predation. The voluminous, chitin-lined buccal cavity of pleurotomariids also appears to be an adaptation for ingesting large volumes of abrasive food.

Several authors (e.g., Yonge, 1973; Hickman, 1984a) have proposed that pleurotomariids were originally herbivorous. Yonge (1973) further suggested that the change to a carnivorous diet was associated with the ecological shift of this family from shallow water habitats, which it occupied during the Paleozoic and Mesozoic, to deeper waters by the end of the Eocene. However, the radulae of all living pleurotomariids have filament-tipped teeth, which are specialized for feeding on sponges. Thus, spongivory in pleurotomariids either predates the divergence of *Entemnotrochus* from the *Perotrochus* group during the Lower Jurassic (Benfrika, 1984: figure 21), or this highly derived tooth morphology evolved convergently in these two pleurotomariid lineages.

4.5. Predators and chemical defenses (Figures 13–15)

According to the "onshore-innovation, offshore-archaic" model for the evolution of Phanerozoic communities (Jablonski *et al.*, 1983), the disappearance of the diverse, shallow-water pleurotomariid communities of the Cretaceous, and the subsequent restriction of the family to outer slope and bathyal depths during the Cenozoic, is presumed to be a result of differential extinction of onshore as compared to offshore clades. The extreme diversification of predators, including fish and crustaceans, that was part of the major reorganization of benthic marine faunas termed the "Mesozoic Marine Revolution" by Vermeij (1977), probably contributed to the extinction of pleurotomariids in tropical, shallow water, hard bottom communities, where crushing predators are most diversified (Vermeij, 1982). While some related gastropods (e.g., Trochidae, Turbinidae) modified their shells to a more spherical shape, with tighter coils, a smaller aperture reinforced by a thicker outer lip and reinforcing sculptural elements, in order to resist predation, pleurotomariids did not incorporate such changes and, for the most part, remained thin-shelled and, in some cases, umbilicate.

The incidence of unsuccessful predation (measured as frequency of shell repair) is usually considered to be a measure of both predation pressure and the prey's resistance to crushing predation. Thus, species with a high incidence of shell repair have a high exposure to crushing predators, as well as the ability to survive their attacks. Vermeij (1989) reported that among shallow water gastropods, the incidence of unsuccessful shell breakage is greatest in the Indo-West Pacific and lowest in the Atlantic. Rarely did his reported frequency of shell repair exceed 0.5 breaks per individual, and never averaged more than 1 break per individual, even in the most prey-resistant taxa. In a survey to determine the importance of predation in structuring deep-sea communities, Vale and Rex (1989) analyzed patterns of repaired shell damage in bathyal rissoid snails, and concluded that levels averaging 0.27 breaks per individual constituted a comparatively high level of potentially lethal predation.

Therefore, it was surprising that even a cursory examination of pleurotomariid shells reveals a multitude of repaired breaks (Figure 13). In fact, a specimen without multiple repaired breaks is a rarity, and 6–9 repaired breaks are not uncommon in individual specimens of *Bayerotrochus midas*. These initial observations prompted a survey of repaired shell breakage, western Atlantic specimens collected alive, and museum specimens with opercula (to ensure that the specimens were collected living, and that breaks were not lethal) for the western Pacific species (Figure 14A–F). The results indicated an extraordinarily high level of unsuccessful predation for all species examined (all species averaged

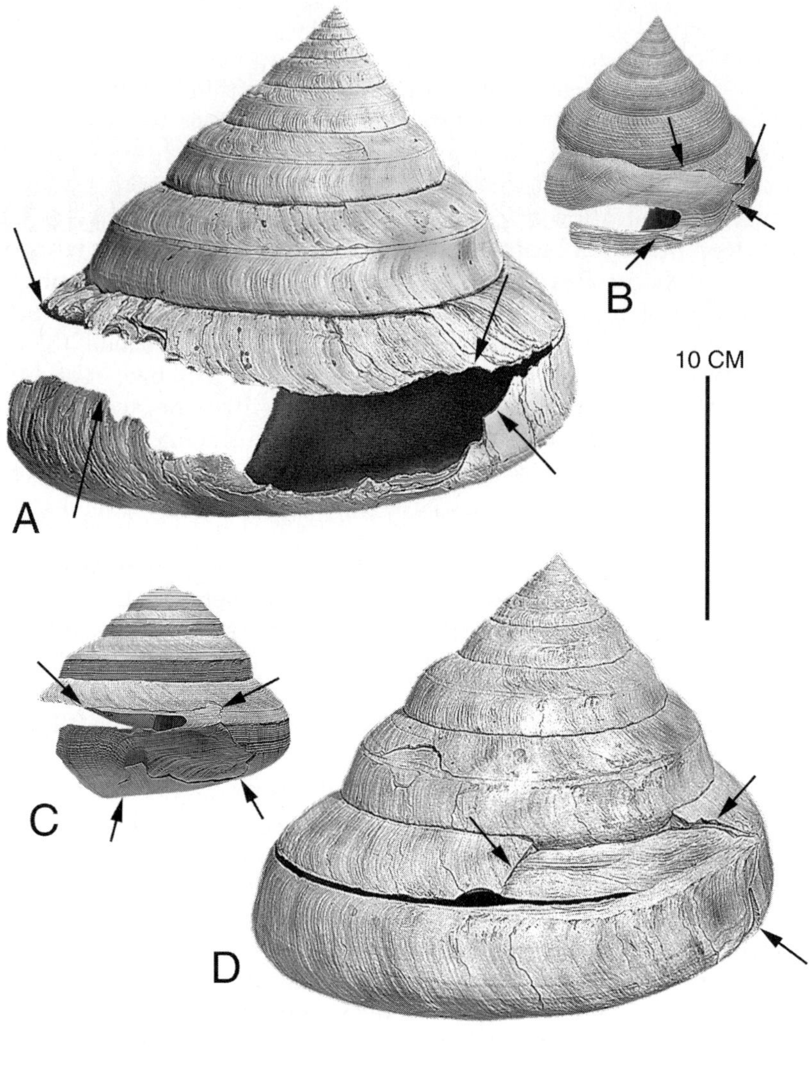

Figure 13 Repaired breaks in the shells of pleurotomariids, indicating unsuccessful attacks by predators. Peeling of the shell along the slit, indicative of attack by a crustacean, in (A) *Entemnotrochus rumphii*, (B) *Mikadotrochus hirasei* and (C) *Perotrochus midas*. Breakage due to biting attack by a fish in (D) *Entemnotrochus rumphii*. Arrows indicate limits of break from a single attack. All specimens were coated with NH_4Cl.

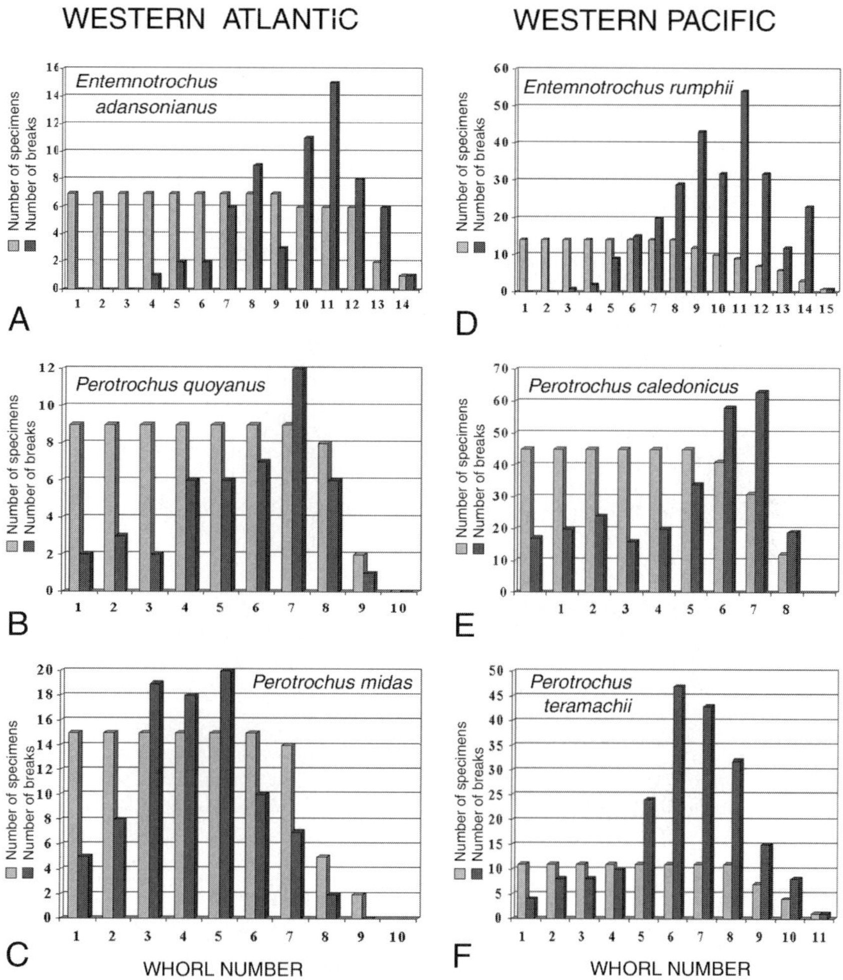

Figure 14 Frequency and distribution of repaired breaks in western Atlantic and western Pacific pleurotomariids. (A) *Entemnotrochus adansonianus*, (B) *Perotrochus quoyanus*, (C) *Perotrochus midas*, (D) *Entemnotrochus rumphii*, (E) *Perotrochus caledonicus* and (F) *Perotrochus teramachii*.

more than 1 repaired break per whorl at some stage of growth). Levels of predation for the western Pacific species were substantially higher, reaching 4–5 repaired breaks per whorl (Figure 14D, F). Nor was size a refuge from predation, as *Entemnotrochus rumphii*, which reaches 28 cm in diameter, had one of the highest frequencies of repaired breaks, especially in the later whorls. In some cases (e.g., *Perotrochus caledonicus*, Figures 14E, 15H), the frequency distribution of breaks

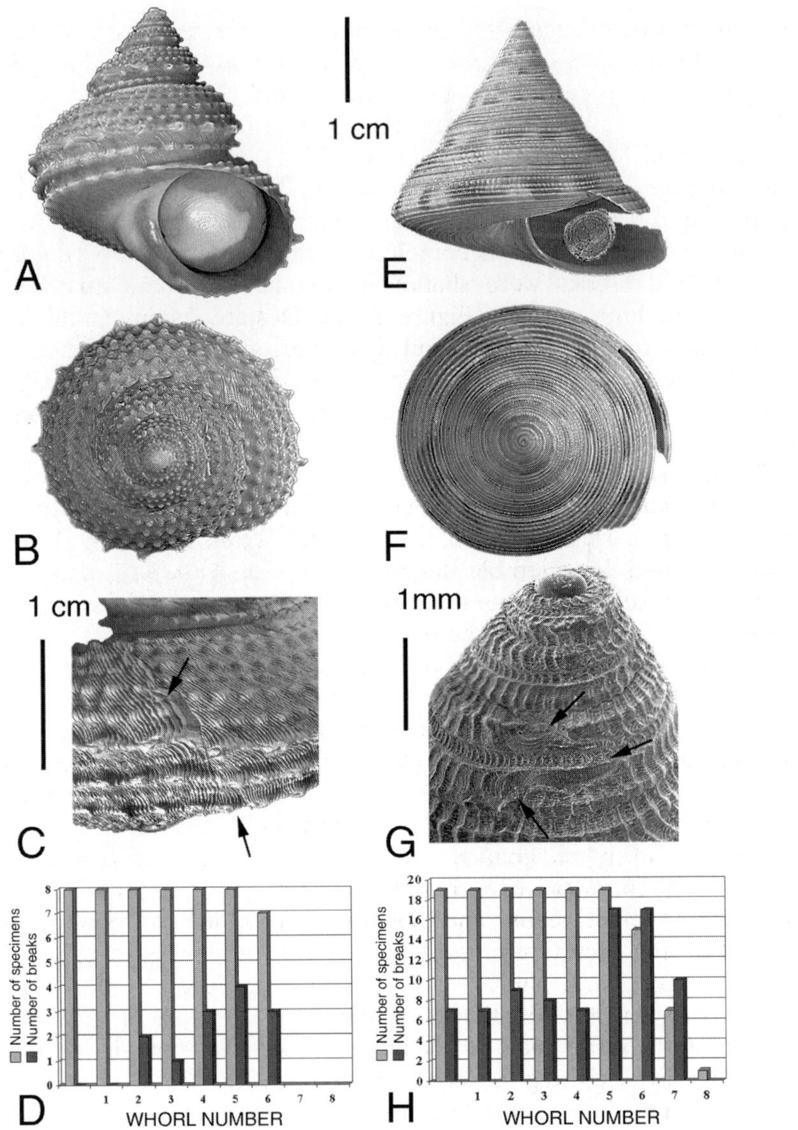

Figure 15 Repaired breaks in sympatrically occurring (A–B) *Bolma guttata* (A. Adams, 1863) and (E–F) *Perotrochus caledonicus* Bouchet and Metivier, 1982, both from N. O. "Vauban" MUSORSTOM 4, Station DW222, 30 September 1985. Off Southern New Caledonia, 22°58′S 167°33′E, in 410–440 m. (C) Example of shallow penetration of typical repaired break (arrows) in shell of *B. guttata* at $6\frac{2}{3}$ whorls. (G) Example of deep penetration of typical repaired break (arrows) in shell of *P. caledonicus* at $3\frac{1}{2}$ whorls. Relative frequency and distribution of repaired breaks in (D) *B. guttata* and (H) *P. caledonicus.*

appears bimodal, suggesting there may be two size classes of predators. Figure 15 illustrates the shell morphologies as well as the types and frequencies of repaired breaks in the trochid *Bolma guttata* A. Adams, 1863 (Figure 15A–D) and the pleurotomariid *Perotrochus caledonicus* Bouchet and Metivier, 1982 (Figure 15E–H) that were collected in the same dredge haul off southern New Caledonia. Although *Bolma* had a larger and thicker shell with stronger spiral and axial sculpture, as well as a calcified operculum that completely sealed the aperture (Figure 15A, B), repaired breaks were shallower (Figure 15C), less numerous, and confined to later whorls (Figure 15D). Despite having a thinner, less structurally reinforced shell, and a thinner, smaller, uncalcified operculum, the co-occurring *Perotrochus caledonicus* was able to survive more deeply penetrating shell breakage (Figure 15G) more frequently, and from a smaller size (Figure 15H).

Preston *et al.* (1993) observed the attacks of three species of crabs (*Carcinus maenus*, *C. pagurus* and *Liocarcinus puber*) on *Calliostoma ziziphynum*, a shallow-water trochid, and found that only 13–50% of attacks caused considerable damage to the shell aperture and, of these, only half resulted in the *Calliostoma* being eaten. The frequency of unsuccessful attacks by predators could therefore be substantially higher than indicated by the number of repaired breaks.

Observations of living animals, both in their habitat and in aquaria, revealed that when the animal is disturbed and especially when the shell is damaged, the hypobranchial glands of all species of Pleurotomariidae produce voluminous amounts (15–30 ml within 2 min by a specimen of *E. adansonianus* 10 cm in diameter) of a bluish white fluid that emanates from the rear of the slit and coats the shell (Plate 5D; see also Kanie *et al.*, 1986: figure 4). Preliminary experiments in aquaria suggest that this fluid is irritating to crustaceans, and possibly fish, while analyses of purified extracts of this secretion reveal the principal component to be structurally related to the alarm pheromone Navanone, produced by the nudibranch *Navanax inermis* (A. White, pers. comm.).

It would appear that all pleurotomariids are capable of deterring predators by secreting a repellent that causes them to release their prey. Given the steep-walled environment in which many of these animals live, the act of dropping a pleurotomariid will often cause it to roll 5–20 m down slope (distances based on observations of pleurotomariids dislodged and subsequently recovered after coming to rest), clearly out of the reach of the predator. As all pleurotomariids have two pairs of large hypobranchial glands that are capable of secreting large quantities of fluid in response to shell damage, this mechanism must have evolved prior to the divergence of *Entemnotrochus* and the remaining genera prior to the late Jurassic.

In a study of crushing predation among Pennsylvanian gastropods from north Central Texas, Schindel *et al.* (1982) observed that the frequency of shell repair was highest for trochiform shells, and concluded that this shell form was most resistant to crushing predation.

A perusal of the list of studied taxa (Schindel *et al.*, 1982: table 2) revealed that all specimens with trochiform shells were pleurotomarioideans spanning the families Raphistomatidae, Eotomariidae Wenz, 1938, Phymatopleuridae Batten, 1956, and Lophospiridae Wenz, 1938. While there was a strong correlation between spire height and frequency of repaired breaks among the pleurotomarioidean taxa, it is possible that this mode of chemical defense against predators might date from the Paleozoic and may have been more widespread among the pleurotomarioidean families.

ACKNOWLEDGEMENTS

Much of this work would not have been possible without the support and participation of Jack Lightbourne, Walter Paine, Yoshihiro Goto, Marilee McNielus and the late Connie Boone, in a number of the research cruises to study living pleurotomariids. I especially thank Craig Young and Shirley Pomponi for their gracious hospitality during a number of their research cruises. I am also grateful to Robert Costello, Steven Pappas, Glen Goodfriend, Paul Calomon, Patrick Anseeuw, Stephen Cairns, John Wise, Andrew McArthur and Guido Pastorino for their assistance during the research cruises, and for many insightful discussions over the past decade.

Special thanks are due to Tim Askew, Director of Marine Operations and to the Captains and crews of the *R/V Edwin Link*, the *R/V Seward Johnson*, and *R/V Sea Diver*, and to the pilots and crews of the research submersibles *Johnson-Sea-Link I*, *II*, *Clelia* and *Nekton Delta* for their assistance, and especially for sharing their years of personal observations on pleurotomariids. The assistance of Yolanda Villacampa, Susann Braden and Julie Whitis with many aspects of data collection, including literature searches, scanning electron microscopy, measurements and counts of repaired breaks, is recognized with thanks.

I would also like to acknowledge my gratitude to Dr Frederick M. Bayer for his pioneering works on the pleurotomariid fauna of the tropical western Atlantic and for generously sharing his notes and observations.

This is Smithsonian Marine Station at Fort Pierce Contribution Number 251.

REFERENCES

Abbott, R. T. (1974). "American Seashells, Second Edition". Van Nostrand Reinhold Co., New York.

Abbott, R. T. and Dance, S. P. (1982). "Compendium of Seashells". American Malacologists, Melbourne.

Adams, H. and Adams, A. (1858). "The Genera of Recent Mollusca; Arranged According to their Organization, Volume 2". John Van Voorst, London.

Anseeuw, P. (1999). *Perotrochus indicus*: A new species of pleurotomariid from India. *Gloria Maris* **37**, 88–95.

Anseeuw, P. and Goto, Y. (1996). "The Living Pleurotomariidae". Elle Scientific Publications, Osaka, Japan.

Arakawa, K. Y. (1962). A coprological study on the molluscan faeces (a preliminary note). *Venus* **22**, 151–172, plates 9–11.

Arakawa, Y., Nakano, D., Tsukuda, O. and Hoshino, T. (1978). On faecal pellets and food habits of emperor's slit shell, *Mikadotrochus hirasei* (Pilsbry). *Venus* **37**, 116–120.

Azuma, M. (1964). Notes on the radula of *Perotrochus africanus* (Tomlin, 1948). *Venus* **22**, 350–355.

Barnard, K. H. (1963). Notes on the animals of *Gyrina gigantea* (Lam.) and *Pleurotomaria africana* Tomlin. *Proceedings of the Malacological Society of London* **35**, 155–158.

Batten, R. L. (1958). Permian Gastropoda of the southwestern United States. 2. Pleurotomariacea: Portlockiellidae, Phymatopleuridae and Eotomariidae. *American Museum of Natural History Bulletin* **114**, 153–246.

Batten, R. L. (1966). "The Lower Carboniferous Gastropod Fauna from the Hotwells Limestone of Compton Martin, Somerset". Palaeontographical Society, Monograph, London.

Batten, R. L. (1972). The ultrastructure of five common Pennsylvanian pleurotomarian gastropod species of eastern United States. *American Museum Novitates* **2501**, 1–34.

Bayer, F. M. (1965). New pleurotomariid gastropods from the western Atlantic, with a summary of Recent species. *Bulletin of Marine Science* **15**, 737–796.

Benfrika el Mostafa (1984). "Systematique, Biogéographie et Écologie des Pleurotomariidés Post-paléozoiques". Thèse, Université de Paris, Centre d'Orsay.

Bieler, R. (1992). Gastropod phylogeny and systematics. *Annual Review of Ecology and Systematics* **23**, 311–338.

Blainville, H. M. de (1825–1827). "Manuel de Malacologie et de Conchyliologie, etc., Volumes 1 and 2". Paris and Strasbourg.

Bøggild, O. B. (1930). The shell structure of the mollusks. *Det Kongelige Danske Videnskabernes Selskabs Skrifter, Naturvidenskabelige og Mathematiske Afdeling*, 9. Række, **2**, 231–326, plates 1–15.

Boss, K. J. (1982). Mollusca. *In* "Synopsis and Classification of Living Organisms, Volume 2" (S. P. Parker, ed.), pp. 1092–1096. McGraw-Hill, New York.

Bouchet, P. and Metivier, B. (1982). Living Pleurotomariidae (Mollusca: Gastropoda) from the South Pacific. *New Zealand Journal of Zoology* **9**, 309–318.

Bouvier, E. L. and Fischer, H. (1898). Étude Monographique des Pleurotomaires Actuels. *Archives de Zoologie Expérimentale et Génerale* Series 3, **6**, 115–180, plates 10–13; reprinted in (1899) *Bulletin of the Museum of Comparative Zoology* *at Harvard College* **32**, 193–246, plates 2–4, and (1899) *Journal de Conchyliologie* **47**, 77–151, plates 4–7.

Bouvier, E. L. and Fischer, H. (1901). Sur l'organisation interne du *Pleurotomaria beyrichii* Hilg. *Comptes Rendus Academie des Sciences (Paris)* **132**, 845–847.

Bouvier, E. L. and Fischer, H. (1902). L'organisation et les affinités des Gastéropodes primitifs d'après l'Étude anatomique du *Pleurotomaria beyrichii. Journal de Conchyliologie* **50**, 117–272, plates 2–6.

Brusca, R. C. and Brusca, G. J. (1990). "Invertebrates". Sinauer Associates, Inc., Sunderland, Massachusetts.

Cox, L. R. (1960a). The British Cretaceous Pleurotomariidae. *British Museum (Natural History) Geological Bulletin* **4**, 385–423.

Cox, L. R. (1960b). Thoughts on the classification of the Gastropoda. *Proceedings of the Malacological Society of London* **33**, 239–261.

Crosse, H. (1882). Les Pleurotomaires de l'Epoque actuelle. *Journal de Conchyliologie* **30**, 5–22.

Crosse, H. and Fischer, P. (1861). Observations sur le genre Pleurotomaire, et description d'une deuxième espèce vivante appartenant au même genre. *Journal de Conchyliologie* **9**, 155–167, plate 5, figs 1, 2.

Dall, W. H. (1889). Reports on the results of dredging, under the supervision of Alexander Agassiz, in the Gulf of Mexico (1877–1878) and in the Caribbean Sea (1879–1880), by the U.S. Coast Survey Steamer "Blake", Lieut.-Commander C. D. Sigsbee, U.S.N., and Commander J. R. Bartlett, U.S.N., Commanding 29 – Report on the Mollusca. Part 2 – Gastropoda and Scaphopoda. *Bulletin of the Museum of Comparative Zoology at Harvard College* **18**, 1–492, plates 10–50.

Defrance, M. J. L. (1826). Genus *Pleurotomaria. In* "Dictionnaire des Sciences Naturelles, dans Lequel on Traite Méthodiquement des Differens Êtres de la Nature, Considérés Soit en Eux-mêmes d'Après l'État Actuel de nos Connoissances, soit Relativement à l'Utilité qu'en Peuvent Retirer la Medicine, l'Agriculture, le Commerce et les Artes: Suivi d'une Bbiographie des plus Célèbres Naturalists, par Plusieurs Professeurs du Muséum National d'Histoire Naturelle et des Autres Principales de Écoles de Paris, Volume 41", (F. Cuvier, ed.) p. 381. Paris and Strasbourg.

Endo, H. (1995). "Slit Shells: Pleurotomariidae". Endo Shell Museum, Manazuru.

Erben, H. K. and Krampitz, G. (1972). Ultrastruktur und Aminosäuren-Verhältnisse in den Schalen der rezenten Pleurotomariidae (Gastropoda). *Biomineralisation* **6**, 12–31.

Férussac, A. É. J. P. F. d'A. de (1821–1822). "Tableaux Systématiques des Animaux Mollusques Classés en Familles Naturelles: Dans Lesquels on a Établi la Concordance de tous les Systèmes: Suivis d'un Prodrome Général pour tous les Mollusques Terrestres ou Fluviatiles Vivants ou Fossiles". A. Bertrand, Paris.

Fischer, P. (1887). "Manuel de Conchyliologie et de Paléontologie Conchyliologique ou Histoire Naturelle des Mollusques Vivants et Fossiles". Savy, Paris.

Fischer, P. and Bernardi, A. C. (1856). Description d'un Pleurotomaire vivante. *Journal de Conchyliologie* **5**, 160–166, plate 5.

Fretter, V. (1964). Observations on the anatomy of *Mikadotrochus amabilis* Bayer. *Bulletin of Marine Science* **14**, 172–184.

Fretter, V. (1966). Biological investigations of the deep sea. 16. Observations on the anatomy of *Perotrochus. Bulletin of Marine Science* **16**, 603–614.

Fretter, V. and Graham, A. (1962). "British Prosobranch Molluscs, their Functional Anatomy and Ecology". The Ray Society, London.

Fretter, V. and Graham, A. (1976). The prosobranch molluscs of Britain and Denmark. Part 1 – Pleurotomariacea, Fissurellacea and Patellacea. *Journal of Molluscan Studies*, Supplement **1**, iv + 37.

Golikov, A. N. and Starobogatov, Y. I. (1975). Systematics of Prosobranch Gastropods. *Malacologia* **15**, 185–232.

Goodfriend, G., Kashgarian, M. and Harasewych, M. G. (1994). Use of aspartic acid racemisation and post-bomb ^{14}C to reconstruct growth rate and longevity of the deep-water slit shell *Entemnotrochus adansonianus*. *Geochimica et Cosmochimica Acta* **59**, 1125–1129.

Graham, A. (1985). Evolution within the Gastropoda: Prosobranchia. *In* "The Mollusca. Volume 10 – Evolution" (E. R. Trueman and M. R. Clarke, eds), pp. 151–186. Academic Press, Orlando.

Harasewych, M. G. and Askew, T. M. (1993). *Perotrochus maureri*, a new species of pleurotomariid from the western Atlantic (Gastropoda: Pleurotomariidae). *Nautilus* **106**, 130–136.

Harasewych, M. G., Pomponi, S. A. and Askew, T. M. (1988). Spongivory in pleurotomariid gastropods. *Nautilus* **102**, 92–98.

Harasewych, M. G., Adamkewicz, S. L., Blake, J. A., Saudek, D., Spriggs, T. and Bult, C. J. (1997). Phylogeny and relationships of pleurotomariid gastropods (Mollusca: Gastropoda): an assessment based on partial 18S rDNA and cytochrome *c* oxidase I sequences. *Molecular Marine Biology and Biotechnology* **6**, 1–20.

Haszprunar, G. (1987). The fine structure of the ctenidial sense organs (bursicles) of Vetigastropoda (Zeugobranchia, Trochoidea) and their phylogenetic significance. *Journal of Molluscan Studies* **53**, 46–51.

Haszprunar, G. (1988). On the origin and evolution of major gastropod groups, with special reference to the Streptoneura. *Journal of Molluscan Studies* **54**, 367–441.

Haszprunar, G. (1989). New slit limpets (Scissurellacea and Fissurellacea) from hydrothermal vents. Part 2. Anatomy and relationships. *Los Angeles County Museum, Contributions to Science* **408**, 1–17.

Hayasaka, I. and Hayasaka, S. (1953). Fossil assemblages of molluscs and brachiopods of unusually large sizes from the Permian of Japan. *Transactions and Proceedings of the Palaeontological Society of Japan, New Series*, **10**, 34–44.

Healy, J. M. (1988). Ultrastructural observations on the spermatozoa of *Pleurotomaria africana* Tomlin (Gastropoda). *Journal of Molluscan Studies* **54**, 309–316.

Healy, J. M. and Harasewych, M. G. (1992). Spermatogenesis in *Perotrochus quoyanus* ((Fischer and Bernardi) Gastropoda, Pleurotomariidae). *Nautilus* **106**, 1–14.

Hedegaard, C. (1990). "Shell Structures of the Recent Archaeogastropoda, Volumes 1 and 2". Thesis for the Cand. Scient. Degree at the Department of Ecology and Genetics, University of Aarhus, Denmark.

Hedegaard, C. (1997). Shell Structures of the Recent Vetigastropoda. *Journal of Molluscan Studies* **63**, 369–377.

Hickman, C. S. (1976). *Pleurotomaria* (Archeogastropoda) in the Eocene of the northeastern Pacific: a review of Cenozoic biogeography and ecology of the genus. *Journal of Paleontology* **50**, 1090–1102.

Hickman, C. S. (1981). Evolution and function of asymmetry in the archeogastropod radula. *The Veliger* **23**, 189–194.

Hickman, C. S. (1984a). Form and function of the radulae of pleurotomariid gastropods. *The Veliger* **27**, 29–36.

Hickman, C. S. (1984b). *Pleurotomaria*: Pedigreed Perseverance? *In* "Living Fossils" (N. Eldredge and S. M. Stanley, eds), pp. 225–231. Springer-Verlag, New York.

Hickman, C. S. (1998). Superfamily Pleurotomarioidea. *In* "Mollusca: The Southern Synthesis. Fauna of Australia. Volume 5. Part B" (P. L. Beesley, G. J. B. Ross and A. Wells, eds), pp. 664–669. CSIRO Publishing, Melbourne.

Hirayama, K. (1955). Discovery of *Perotrochus* in the Miocene deposits of Tochigi Prefecture, Japan. *Venus* **18**, 266–272.

Hudleston, W. H. (1887–1896). "The British Jurassic Gasteropoda, Part 1. A Monograph of the Inferior Oolite Gasteropoda". Palæontographical Society, Monograph, London.

Hutsell, K. C., Hutsell, L. L. and Pisor, D. L. (1999). "Registry of World Record Size Shells". Snail's Pace Publications, San Diego.

Hyman, L. H. (1967). "Mollusca I. Aplacophora, Polyplacophora, Monoplacophora, Gastropoda, The Coelomate Bilateria. The Invertebrates, Volume 6". McGraw-Hill Book Company, New York.

Ingles, J. C. (1963). Geometry, paleontology and petrography of the Thornton reef complex, Silurian of northeastern Illinois. *American Association of Petroleum Geologists Bulletin* **47**, 405–440.

International Commission on Zoological Nomenclature. (1960). Opinion 582. Validation of the generic name *Pleurotomaria* as from Defrance 1826 (Class Gastropoda). *Bulletin of Zoological Nomenclature* **17**, 276–280.

Jablonski, D., Sepkowski, J. J., Botjer, D. J. and Sheehan, P. M. (1983). Onshore–offshore patterns in the evolution of Phanerozoic shelf communities. *Science* **222**, 1123–1125.

Jung, P. (1968). Fossil *Pleurotomaria* and *Haliotis* from Barbados and Carriacou, West Indies. *Ecologae Geologicae Helvetiae* **61**, 593–605.

Jung, P. (1970). Fossil *Pleurotomaria* from Jamaica. *Caribbean Journal of Science* **10**, 83–86.

Kanie, Y. and Oba, T. (1993). Oxygen and carbon isotopic measurements on the shell of *Mikadotrochus beyrichii* (Pleurotomariidaqe: Mollusca) from Sagami Bay, central Japan. *Proceedings of the Jamstec Symposium on Deep Sea Research* **9**, 233–236.

Kanie, Y., Takahashi, T. and Mizuno, Y. (1980). Color patterns of Cretaceous pleurotomariid gastropods from Hokkaido. *Science Reports, Yokosuka City Museum* **27**, 37–42.

Kanie, Y., Sugiura, A. and Egawa, K. (1986). Submarine observations on *Mikadotrochus beyrichii* (Archeogastropoda) in Sagami-Nada, Central Honshu, Japan. *Venus* **45**, 203–209.

Kanie, Y., Hattori, M., Kuramochi, T. and Ikeda, H. (1995). Ecology and phylogenetic classification of the Pleurotomariidae (Gastropoda: Mollusca). *Jamstec Journal of Deep Sea Research* **11**, 279–284.

Kanno, S. (1961). Miocene "*Pleurotomaria*" and its associated fauna from Tochigi Prefecture, Japan. *Japanese Journal of Geology and Geography* **32**, 111–118, plate 6.

Kenkado, K. (1755). "Kikai Zufu". Rice paper scroll.

Knight, J. B. (1952). Primitive fossil gastropods and their bearing on gastropod classification. *Smithsonian Miscellaneous Collections* **117**, 1–56.

Knight, J. B., Cox, L. R., Keen, A. M., Batten, R. L., Yochelson, E. L. and Robertson, R. (1960). Gastropoda, Systematic Descriptions. *In* "Treatise of

Invertebrate Paleontology. I – Mollusca 1:I169–I331" (R. C. Moore, ed.). University of Kansas Press, Lawrence.

Koizumi, H. (1993). Generic comparison with *Perotrochus* and *Mikadotrochus* (Pleurotomariidae). *Chiribotan* **24**, 9–11.

Kosuge, S. and Suzuki, M. (1969). *Entemnotrochus rumphii* (Schepman, 1879), newly collected from the South China Sea. *Venus* **28**, 155–158, plate 12.

Kuramochi, T., Kanie, Y., Hattori, M. and Uematsu, K. (1996). Radula morphologies of the Pleurotomariidae Gastropoda from Japanese waters. *JAMSTEC Journal of Deep Sea Research* **12**, 197–215.

Leme, J. L. M. and Penna, L. (1969). Ocorrência de *Mikadotrochus* no Brasil, con descrição de uma nova espécie (Gastropoda, Pleurotomariidae). *Papéis Avulsos de Zoologia, Sao Paulo* **22**, 225–230.

Lima, G. M. and Lutz, R. A. (1990). The relationship of larval shell morphology to mode of development in marine prosobranch snails. *Journal of the Marine Biological Association of the United Kingdom* **70**, 611–639.

Lister, M. (1678). "Historiæ animalium Angliæ tres tractus. Unus de Araneis. Alter de Cochleis tum terrestribus tum fluviatilibus. Tertius de cochlies marinis". Royal Society, London.

Mantell, G. (1822). "The fossils of the South Downs, or Illustrations of the Geology of Sussex". H. G. Bohn, London.

Matsumoto, Y. (1929). Radula description of *Pleurotomaria hirasei*. *Venus* **1**, 138–139.

Matsumoto, Y., Kataoka, T. and Sekido, M. (1972). On the behavior of Beyrich's slit shell, *Mikadotrochus beyrichii* (Hilgendorf) kept in the aquarium. *Venus* **30**, 147–152.

Mitsukuri, K. (1897). A living specimen of *Pleurotomaria beyrichii*. *Annotationes Zoologicae Japonenses* **1**, 67–68.

Morton, J. E. and Yonge, C. M. (1964). Classification and Structure of the Mollusca. *In* "Physiology of Mollusca, Volume 1" (K. M. Wilbur and C. M. Yonge, eds), pp. 1–58. Academic Press, New York.

Okutani, T. (1963). Preliminary notes on molluscan assemblages of the submarine banks around the Izu Islands. *Pacific Science* **17**, 73–89.

Okutani, T. and Hasegawa, K. (2000). Superfamily Pleurotomarioidea. *In* "Marine Mollusks in Japan" (T. Okutani, ed.), pp. 37–44. Tokai University Press, Tokyo.

Oyama, K. (1980). A collection of the third *Entemnotrochus rumphii* Schepman from Taiwan. *Bulletin of Malacology, Republic of China* **7**, 131–133.

Pilsbry, H. A. (1890). Family Pleurotomariidae. *Manual of Conchology* **12**, 69–72, plates 56–57.

Ponder, W. F. and Lindberg, D. R. (1996). Gastropod phylogeny – challenges for the 90s. *In* "Origin and Evolutionary Radiation of the Mollusca" (J. Taylor, ed.), pp. 135–154. Oxford University Press, Oxford.

Ponder, W. F. and Lindberg, D. R. (1997). Towards a phylogeny of gastropod molluscs: an analysis using morphological characters. *Zoological Journal of the Linnean Society* **119**, 83–265.

Preston, S. J., Roberts, D. and Montgomery, W. I. (1993). Shell scarring in *Calliostoma ziziphynum* (Prosobranchia: Trochidae) from Stangford Lough, Northern Ireland. *Journal of Molluscan Studies* **59**, 211–222.

Pulteney, R. (1813). Catalogues of the birds, shells and some of the more rare plants of Dorsetshire. *In* "History and Antiquities of the County of Dorset, Volume 3" (J. Hutchins, ed.), pp. 1–110, plates 1–24. Nichols, Son, and Bentley, London.

Rex, M. A. (1977). Zonation in deep-sea gastropods: the importance of biological interactions to rates of zonation. *In* "Biology of Benthic Organisms" (B. F. Keegan, P. O. Ceidigh and P. J. S. Boaden, eds), pp. 521–530. Pergamon Press, New York.

Rohr, D. M. and Blodgett, R. B. (1985). Upper Ordovician Gastropoda from west-central Alaska. *Journal of Paleontology* **59**, 667–673.

Rohr, D. M. and Smith, R. E. (1978). Lower Devonian Gastropoda from the Canadian Arctic Islands. *Canadian Journal of Earth Sciences* **15**, 1228–1241.

Rosenberg, G., Kuncio, R. S., Davis, G. M. and Harasewych, M. G. (1994). Preliminary ribosomal RNA phylogeny of gastropod and unionoidean bivalve mollusks. *Nautilus, Supplement* **2**, 111–121.

Rupert, E. E. and Barnes, R. D. (1994). "Invertebrate Zoology, 6th edition". Saunders College Publishing, New York.

Sasaki, T. (1998). Comparative anatomy and phylogeny of the Recent Archaeogastropoda (Mollusca: Gastropoda). *Bulletin of the University Museum, University of Tokyo* **38**, 1–224.

Salvini-Plawen, L. v. (1980). A reconsideration of systematics in the Mollusca (phylogeny and higher classification). *Malacologia* **19**, 249–278.

Salvini-Plawen, L. v. and Haszprunar, G. (1987). The Vetigastropoda and the systematics of streptoneurous Gastropoda (Mollusca). *Journal of Zoology, London* **211**, 747–770.

Schindel, D. E., Vermeij, G. J. and Zipser, E. (1982). Frequencies of repaired shell fractures among the Pennsylvanian gastropods of north-central Texas. *Journal of Paleontology* **56**, 729–740.

Sekido, M., Tsukada, O., Matsumoto, Y. and Kataoka, T. (1976). On the keeping of Teramachi's Slit Shell, *Perotrochus teramachii*. *Journal of the Japanese Association of Zoological Gardens and Aquariums* **18**, 49–52.

Shikama, T. and Nishida, T. (1968). On some species of Carboniferous pleurotomariaceans from Akiohsi (Molluscan paleontology of the Akiyoshi limestone group, III). *Transactions and Proceedings of the Paleontological Society of Japan* (New Series) **69**, 211–217.

Shileyko, A. A. (1977). The symmetry of the Docoglossa and the problem of the origin of the order. *Bulletin of the Moscow Naturalists Society, Biology Series* **81**, 60–65.

Shuto, T. (1974). Larval ecology of prosobranch gastropods and its bearing on biogeography and paleontology. *Lethaia* **7**, 239–256.

Sowerby, J. (1812–1846). "The Mineral Conchology of Great Britain". Sowerby, London.

Sowerby, J. de C. (1832). *Pleurotomaria*. *In* Sowerby, G. B. (1822–1840). "The Genera of Recent and Fossil Shells, for the Use of Students of Conchology and Geology" (G. B. Sowerby, ed.). Sowerby, London.

Swainson, W. (1840). "A Treatise on Malacology: or the Natural Classification of Shells and Shell Fish". John Taylor, London.

Szabó, J. (1980). Lower and Middle Jurassic gastropods from the Bakony Mountains (Hungary). Part 2. Pleurotomariacea and Fissurellacea. *Annales Historico-Naturales Musei Nationalis Hungarici* **72**, 49–71.

Tan, T. H. (1974). A preliminary study on the anatomy of *Pleurotomaria (Entemnotrochus) rumphii* Schepman. *Bulletin of the Malacological Society of China* **1**, 15–20.

Taylor, D. W. and Sohl, N. F. (1962). An outline of gastropod classification. *Malacologia* **1**, 7–32.

Termier, H. and Termier, G. (1968). "Évolution et Biocinése. Les Invértebrés dans l'histoire du Monde vivant". Masson and Cie, Paris.

Thiele, J. (1902). Die systematische Stellung der Solenogastren und die Phylogenie der Mollusken. *Zeitschrift für wissenschaftliche Zoologie* **72**, 249–266, plates 18–27.

Thiele, J. (1929–1935). "Handbuch der Systematischen Weichtierkunde". Gustav Fischer, Jena.

Thomas, E. G. (1940). Revision of the Scottish Carboniferous Pleurotomariidae. *Transactions of the Geological Society of Glasgow* **20**, 30–72.

Tillier, S., Masselot, M., Philippe, H. and Tillier, A. (1992). Phylogénie moléculaire des Gastropoda (Mollusca) fondée sur le séquençage partiel de l'ARN ribosomique 28 S. *Comptes Rendus de l'Academie des Sciences (Paris)* **314** (Ser. 3), 79–85.

Tillier, S., Masselot, M., Guerdoux, J. and Tillier, A. (1994). Monophyly of major gastropod taxa tested from partial 28S rRNA sequences, with emphasis on Euthyneura and hot-vent limpets Peltospiroidea. *Nautilus, Supplement* **2**, 122–140.

Tomida, S., Teraoka, A., Okumura, Y. and Karazawa, H. (1999). A fossil pleurotomariid gastropod from the Miocene Bihoku Group in southwestern Japan and its paleontological significance. *Bulletin of the Mizanumi Fossil Museum* **26**, 115–119.

Tracey, S., Todd, J. A. and Erwin, D. H. (1993). Mollusca: Gastropoda. *In* "The Fossil Record 2". (M. J. Benton, ed.), pp. 137–167. Chapman and Hall, London.

Troschel, F. H. (1891). "Das Gebiss der Schnecken zur Begründung einer Natürlichen Classification, Volume 2, Part 7", pp. 251–334, plates 25–28. Nicolaische Verlags, Berlin.

Vale, F. K. and Rex, M. A. (1989). Repaired shell damage in a complex of rissoid gastropods from the upper continental slope south of New England. *Nautilus* **103**, 105–108.

Vaught, K. C. (1989). "A Classification of Living Mollusca". American Malacologists, Melbourne.

Vermeij, G. J. (1977). The Mesozoic marine revolution: evidence from snails, predation and grazers. *Paleobiology* **3**, 245–258.

Vermeij, G. J. (1982). Gastropod shell form, breakage, and repair in relation to predation by the crab *Calappa*. *Malacologia* **23**, 1–12.

Vermeij, G. J. (1989). Interoceanic differences in adaptation: effects of history and productivity. *Marine Ecology Progress Series* **57**, 293–305.

Wenz, W. (1938). Gastropoda. Allgemeiner Teil und Prosobranchia. *In* "Handbuch der Paläozoologie" (O. H. Schindewolf, ed.), pp. 1–948. Gebrüder Borntraeger, Berlin.

Wenz, W. (1940). Ursprung und frühe Stamm Geschichte der Gastropoden. *Archiv für Molluskenkunde* **72**, 1–10.

Wolfenden, E. B. (1958). Paleoecology of the Carboniferous reef complex and shelf limestones in northwestern Derbyshire, England. *Bulletin of the Geological Society of America* **69**, 871–898.

Woodward, H. (1885). On Recent and fossil Pleurotomariae. *Geological Magazine*, Decade 3 **2**, 433–439, plate 11.

Woodward, M. F. (1901). The anatomy of *Pleurotomaria beyrichii*, Hilg. *The Quarterly Journal of the Microscopical Society* **44**, 215–268, plates 13–16.

Yonge, C. M. (1947). The pallial organs in the aspidobranch Gastropoda and their evolution throughout the Mollusca. *Philosophical Transactions of the Royal Society, London. B* **232**, 443–518.

Yonge, C. M. (1973). Observation of the pleurotomariid *Entemnotrochus adansoniana* in its natural habitat. *Nature* **241**, 66–68.
Yonge, C. M. and Thompson, T. E. (1976). "Living Marine Mollusks". William Collins Sons and Co., Ltd, London.
Zezina, O. N. (1997). Biogeography of the Bathyal Zone. *Advances in Marine Biology* **32**, 389–420.
Ziegler, V. (1984). Family Pleurotomariidae Swainson 1840 (Gastropoda, Archeogastropoda) from the Bohemian Cretaceous Basin. *Sbornik Narodniho Muzea v Praze Rada b Prirodni Vedy* **39**, 255–291.

Taxonomic Index

Note: page numbers in *italics* indicate tables and diagrams

Subject Index

Note page numbers in *italics* indicate diagrams and tables.